Tucholsky Wagner Zola Scott Sydow Freud Schlegel
Turgenev Wallace Fonatne Twain Walther von der Vogelweide Fouqué Friedrich II. von Preußen
Weber Freiligrath Frey
Fechner Fichte Weiße Rose von Fallersleben Kant Ernst Frommel
Hölderlin Richthofen
Engels Fielding Eichendorff Tacitus Dumas
Fehrs Faber Flaubert
Feuerbach Maximilian I. von Habsburg Fock Eliasberg Zweig Ebner Eschenbach
Ewald Eliot Vergil
Goethe Elisabeth von Österreich London
Mendelssohn Balzac Shakespeare Dostojewski Ganghofer
Trackl Stevenson Lichtenberg Rathenau Doyle Gjellerup
Tolstoi Hambruch
Mommsen Thoma Lenz Hanrieder Droste-Hülshoff
Dach Verne von Arnim Hägele Hauff Humboldt
Karrillon Reuter Rousseau Hagen Hauptmann Gautier
Garschin Defoe Baudelaire
Damaschke Descartes Hebbel
Hegel Kussmaul Herder
Wolfram von Eschenbach Dickens Schopenhauer Rilke George
Bronner Darwin Melville Grimm Jerome
Campe Horváth Aristoteles Bebel Proust
Bismarck Vigny Barlach Voltaire Federer Herodot
Gengenbach Heine
Storm Casanova Tersteegen Grillparzer Georgy
Chamberlain Lessing Langbein Cilm Gryphius
Brentano Lafontaine
Strachwitz Claudius Schiller Kralik Iffland Sokrates
Katharina II. von Rußland Bellamy Schilling
Gerstäcker Raabe Gibbon Tschechow
Löns Hesse Hoffmann Gogol Wilde Vulpius
Luther Heym Hofmannsthal Gleim
Roth Klee Hölty Morgenstern Goedicke
Luxemburg Heyse Klopstock Kleist
Machiavelli La Roche Puschkin Homer Mörike Musil
Horaz
Navarra Aurel Musset Kierkegaard Kraft Kraus
Nestroy Marie de France Lamprecht Kind Kirchhoff Hugo Moltke
Nietzsche Nansen Laotse Ipsen Liebknecht
Marx Ringelnatz
von Ossietzky Lassalle Gorki Klett Leibniz
May vom Stein Lawrence Irving
Petalozzi Knigge
Platon Pückler Michelangelo Kock Kafka
Sachs Poe Liebermann Korolenko
de Sade Praetorius Mistral Zetkin

The publishing house tredition has created the series **TREDITION CLASSICS**. It contains classical literature works from over two thousand years. Most of these titles have been out of print and off the bookstore shelves for decades.

The book series is intended to preserve the cultural legacy and to promote the timeless works of classical literature. As a reader of a **TREDITION CLASSICS** book, the reader supports the mission to save many of the amazing works of world literature from oblivion.

The symbol of **TREDITION CLASSICS** is Johannes Gutenberg (1400 – 1468), the inventor of movable type printing.

With the series, tredition intends to make thousands of international literature classics available in printed format again – worldwide.

All books are available at book retailers worldwide in paperback and in hardcover. For more information please visit: www.tredition.com

tredition was established in 2006 by Sandra Latusseck and Soenke Schulz. Based in Hamburg, Germany, tredition offers publishing solutions to authors and publishing houses, combined with worldwide distribution of printed and digital book content. tredition is uniquely positioned to enable authors and publishing houses to create books on their own terms and without conventional manufacturing risks.

For more information please visit: www.tredition.com

An Elementary Study of Chemistry

William McPherson

Imprint

This book is part of the TREDITION CLASSICS series.

Author: William McPherson
Cover design: toepferschumann, Berlin (Germany)

Publisher: tredition GmbH, Hamburg (Germany)
ISBN: 978-3-8491-7420-0

www.tredition.com
www.tredition.de

ANTOINE LAURENT LAVOISIER

Famous for his care in quantitative experiments, for demonstrating the true nature of combustion, for introducing system into the naming and grouping of chemical substances. Executed (1794) during the French Revolution because of his connection with the government

This picture is taken from a French engraving of 1799. The panel represents Lavoisier as he is being arrested in his laboratory by the Revolutionary Committee

PREFACE

In offering this book to teachers of elementary chemistry the authors lay no claim to any great originality. It has been their aim to prepare a text-book constructed along lines which have become recognized as best suited to an elementary treatment of the subject. At the same time they have made a consistent effort to make the text clear in outline, simple in style and language, conservatively modern in point of view, and thoroughly teachable.

The question as to what shall be included in an elementary text on chemistry is perhaps the most perplexing one which an author must answer. While an enthusiastic chemist with a broad understanding of the science is very apt to go beyond the capacity of the elementary student, the authors of this text, after an experience of many years, cannot help believing that the tendency has been rather in the other direction. In many texts no mention at all is made of fundamental laws of chemical action because their complete presentation is quite beyond the comprehension of the student, whereas in many cases it is possible to present the essential features of these laws in a way that will be of real assistance in the understanding of the science. For example, it is a difficult matter to deduce the law of mass action in any very simple way; yet the elementary student can readily comprehend that reactions are reversible, and that the point of equilibrium depends upon, rather simple conditions. The authors believe that it is worth while to [Pg iv] present such principles in even an elementary and partial manner because they are of great assistance to the general student, and because they make a foundation upon which the student who continues his studies to more advanced courses can securely build.

The authors have no apologies to make for the extent to which they have made use of the theory of electrolytic dissociation. It is inevitable that in any rapidly developing science there will be differences of opinion in regard to the value of certain theories. There can be no question, however, that the outline of the theory of dissociation here presented is in accord with the views of the very great majority of the chemists of the present time. Moreover, its introduction to the extent to which the authors have presented it simplifies rather than increases the difficulties with which the development of the principles of the science is attended.

The oxygen standard for atomic weights has been adopted throughout the text. The International Committee, to which is assigned the duty of yearly reporting a revised list of the atomic weights of the elements, has adopted this standard for their report, and there is no longer any authority for the older hydrogen standard. The authors do not believe that the adoption of the oxygen standard introduces any real difficulties in making perfectly clear the methods by which atomic weights are calculated.

The problems appended to the various chapters have been chosen with a view not only of fixing the principles developed in the text in the mind of the student, but also of enabling him to answer such questions as arise in his laboratory work. They are, therefore, more or less practical in character. It is not necessary that all of them should [Pg v] be solved, though with few exceptions the lists are not long. The answers to the questions are not directly given in the text as a rule, but can be inferred from the statements made. They therefore require independent thought on the part of the student.

With very few exceptions only such experiments are included in the text as cannot be easily carried out by the student. It is expected that these will be performed by the teacher at the lecture table. Directions for laboratory work by the student are published in a separate volume.

While the authors believe that the most important function of the elementary text is to develop the principles of the science, they recognize the importance of some discussion of the practical application of these principles to our everyday life. Considerable space is therefore devoted to this phase of chemistry. The teacher should supplement this discussion whenever possible by having the class visit different factories where chemical processes are employed.

Although this text is now for the first time offered to teachers of elementary chemistry, it has nevertheless been used by a number of teachers during the past three years. The present edition has been largely rewritten in the light of the criticisms offered, and we desire to express our thanks to the many teachers who have helped us in this respect, especially to Dr. William Lloyd Evans of this laboratory, a teacher of wide experience, for his continued interest and help-fulness. We also very cordially solicit correspondence with teachers who may find difficulties or inaccuracies in the text.

The authors wish to make acknowledgments for the photographs and engravings of eminent chemists from which [Pg vi] the cuts included in the text were taken; to Messrs. Elliott and Fry, London, England, for that of Ramsay; to The Macmillan Company for those of Davy and Dalton, taken from the Century Science Series; to the L. E. Knott Apparatus Company, Boston, for that of Bunsen.

THE AUTHORS

OHIO STATE UNIVERSITY

COLUMBUS, OHIO

CONTENTS

LIST OF FULL-PAGE ILLUSTRATIONS

AN ELEMENTARY STUDY OF CHEMISTRY

CHAPTER I

INTRODUCTION

The natural sciences. Before we advance very far in the study of nature, it becomes evident that the one large study must be divided into a number of more limited ones for the convenience of the investigator as well as of the student. These more limited studies are called the *natural sciences.*

Since the study of nature is divided in this way for mere convenience, and not because there is any division in nature itself, it often happens that the different sciences are very intimately related, and a thorough knowledge of any one of them involves a considerable acquaintance with several others. Thus the botanist must know something about animals as well as about plants; the student of human physiology must know something about physics as well as about the parts of the body.

Intimate relation of chemistry and physics. Physics and chemistry are two sciences related in this close way, and it is not easy to make a precise distinction between them. In a general way it may be said that they are both concerned with inanimate matter rather than with living, and more particularly with the changes which such matter [Pg 2] may be made to undergo. These changes must be considered more closely before a definition of the two sciences can be given.

Physical changes. One class of changes is not accompanied by an alteration in the composition of matter. When a lump of coal is broken the pieces do not differ from the original lump save in size. A rod of iron may be broken into pieces; it may be magnetized; it may be heated until it glows; it may be melted. In none of these changes has the composition of the iron been affected. The pieces of iron, the magnetized iron, the glowing iron, the melted iron, are just as truly iron as was the original rod. Sugar may be dissolved in water, but neither the sugar nor the water is changed in composition. The resulting liquid has the sweet taste of sugar; moreover the water may be evaporated by heating and the sugar recovered unchanged. Such changes are called *physical changes*.

DEFINITION: *Physical changes are those which do not involve a change in the composition of the matter.*

Chemical changes. Matter may undergo other changes in which its composition is altered. When a lump of coal is burned ashes and invisible gases are formed which are entirely different in composition and properties from the original coal. A rod of iron when exposed to moist air is gradually changed into rust, which is entirely different from the original iron. When sugar is heated a black substance is formed which is neither sweet nor soluble in water. Such changes are evidently quite different from the physical changes just described, for in them new substances are formed in place of the ones undergoing change. Changes of this kind are called *chemical changes*. [Pg 3]

DEFINITION: *Chemical changes are those which involve a change in the composition of the matter.*

How to distinguish between physical and chemical changes. It is not always easy to tell to which class a given change belongs, and many cases will require careful thought on the part of the student. The test question in all cases is, Has the composition of the substance been changed? Usually this can be answered by a study of the properties of the substance before and after the change, since a change in composition is attended by a change in properties. In some cases, however, only a trained observer can decide the question.

Changes in physical state. One class of physical changes should be noted with especial care, since it is likely to prove misleading. It is a familiar fact that ice is changed into water, and water into steam, by heating. Here we have three different substances,—the solid ice, the liquid water, and the gaseous steam,—the properties of which differ widely. The chemist can readily show, however, that these three bodies have exactly the same composition, being composed of the same substances in the same proportion. Hence the change from one of these substances into another is a physical change. Many other substances may, under suitable conditions, be changed from solids into liquids, or from liquids into gases, without change in composition. Thus butter and wax will melt when heated; alcohol and gasoline will evaporate when exposed to the air. *The three states — solid, liquid, and gas — are called the three physical states of matter.*

Physical and chemical properties. Many properties of a substance can be noted without causing the substance to undergo chemical change, and are therefore called its *physical properties.* Among these are its physical state, color, odor, taste, size, shape, weight. Other properties are only [Pg 4] discovered when the substance undergoes chemical change. These are called its *chemical properties.* Thus we find that coal burns in air, gunpowder explodes when ignited, milk sours when exposed to air.

Definition of physics and chemistry. It is now possible to make a general distinction between physics and chemistry.

DEFINITION: *Physics is the science which deals with those changes in matter which do not involve a change in composition.*

DEFINITION: *Chemistry is the science which deals with those changes in matter which do involve a change in composition.*

Two factors in all changes. In all the changes which matter can undergo, whether physical or chemical, two factors must be taken into account, namely, *energy* and *matter.*

Energy. It is a familiar fact that certain bodies have the power to do work. Thus water falling from a height upon a water wheel turns the wheel and in this way does the work of the mills. Magnetized iron attracts iron to itself and the motion of the iron as it moves towards the magnet can be made to do work. When coal is burned it causes the engine to move and transports the loaded cars from place to place. When a body has this power to do work it is said to possess energy.

Law of conservation of energy. Careful experiments have shown that when one body parts with its energy the energy is not destroyed but is transferred to another body or system of bodies. Just as energy cannot be destroyed, neither can it be created. If one body gains a certain amount of energy, some other body has lost an equivalent amount. [Pg 5] These facts are summed up in the law of conservation of energy which may be stated thus: *While energy can be changed from one form into another, it cannot be created or destroyed.*

Transformations of energy. Although energy can neither be created nor destroyed, it is evident that it may assume many different forms. Thus the falling water may turn the electric generator and produce a current of electricity. The energy lost by the falling water is thus transformed into the energy of the electric current. This in turn may be changed into the energy of motion, as when the current is used for propelling the cars, or into the energy of heat and light, as when it is used for heating and lighting the cars. Again, the energy of coal may be converted into energy of heat and subsequently of motion, as when it is used as a fuel in steam engines.

Since the energy possessed by coal only becomes available when the coal is made to undergo a chemical change, it is sometimes called *chemical energy*. It is this form of energy in which we are especially interested in the study of chemistry.

Matter. Matter may be defined as that which occupies space and possesses weight. Like energy, matter may be changed oftentimes from one form into another; and since in these transformations all the other physical properties of a substance save weight are likely to change, the inquiry arises, Does the weight also change? Much careful experimenting has shown that it does not. The weight of the products formed in any change in matter always equals the weight of the substances undergoing change.

Law of conservation of matter. The important truth just stated is frequently referred to as the law of conservation [Pg 6] of matter, and this law may be briefly stated thus: *Matter can neither be created nor destroyed, though it can be changed from one form into another.*

Classification of matter. At first sight there appears to be no limit to the varieties of matter of which the world is made. For convenience in study we may classify all these varieties under three heads, namely, *mechanical mixtures, chemical compounds,* and *elements.*

Fig. 1

Mechanical mixtures. If equal bulks of common salt and iron filings are thoroughly mixed together, a product is obtained which, judging by its appearance, is a new substance. If it is examined more closely, however, it will be seen to be merely a mixture of the salt and iron, each of which substances retains its own peculiar properties. The mixture tastes just like salt; the iron particles can be seen and their gritty character detected. A magnet rubbed in the mixture draws out the iron just as if the salt were not there. On the other hand, the salt can be separated from the iron quite easily. Thus, if several grams of the mixture are placed in a test tube, and the tube half filled with water and thoroughly shaken, the salt dissolves in the water. The iron particles can then be filtered from the liquid by pouring the entire mixture upon a piece of filter paper folded so as to fit into the interior of a funnel (Fig. 1). The paper retains the solid but allows the clear liquid, known as the *filtrate*, to drain through. The iron particles left upon the filter paper will be found to be identical with [Pg 7] the original iron. The salt can be recovered from the filtrate by evaporation of the water. To accomplish this the filtrate is poured into a small evaporating dish and gently heated (Fig. 2) until the water has disappeared, or *evaporated*. The solid left in the dish is identical in every way with the original salt. Both the iron and the salt have thus been recovered in their original condition. It is evident that no new substance has been formed by rubbing the salt and iron together. The product is called a *mechanical mixture*. Such mixtures are very common in nature, almost all minerals, sands, and soils being examples of this class of substances. It is at once apparent that there is no law regulating the composition of a mechanical mixture, and no two mixtures are likely to have exactly the same composition. The ingredients of a mechanical mixture can usually be separated by mechanical means, such as sifting, sorting, magnetic attraction, or by dissolving one constituent and leaving the other unchanged.

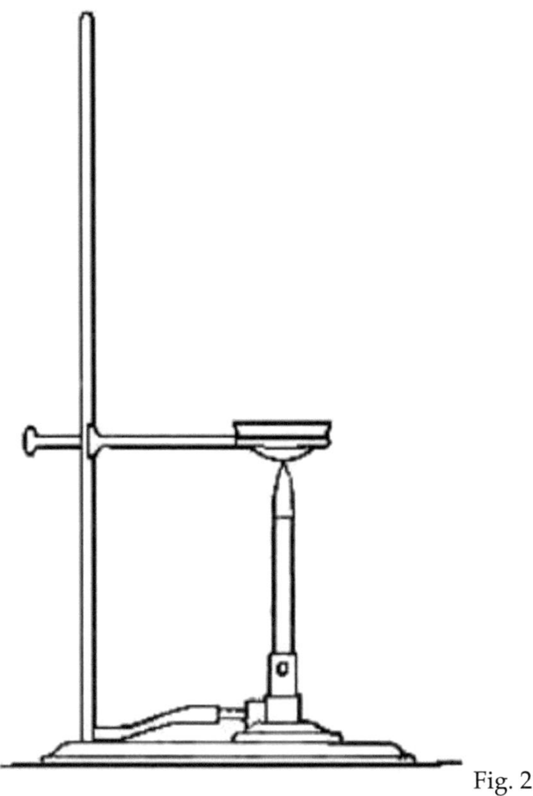

Fig. 2

DEFINITION: *A mechanical mixture is one in which the constituents retain their original properties, no chemical action having taken place when they were brought together.*

Chemical compounds. If iron filings and powdered sulphur are thoroughly ground together in a mortar, a yellowish-green substance results. It might easily be taken to be a new body; but as in the case of the iron and salt, the ingredients can readily be separated. A magnet draws out the iron. Water does not dissolve the sulphur, but other liquids do, as, for example, the liquid called carbon disulphide. [Pg 8] When the mixture is treated with carbon disulphide the iron is left unchanged, and the sulphur can be obtained again, after filtering off the iron, by evaporating the liquid. The substance is, therefore, a mechanical mixture.

If now a new portion of the mixture is placed in a dry test tube and carefully heated in the flame of a Bunsen burner, as shown in Fig. 3, a striking change takes place. The mixture begins to glow at some point, the glow rapidly extending throughout the whole mass. If the test tube is now broken and the product examined, it will be found to be a hard, black, brittle substance, in no way recalling the iron or the sulphur. The magnet no longer attracts it; carbon disulphide will not dissolve sulphur from it. It is a new substance with new properties, resulting from the chemical union of iron and sulphur, and is called iron sulphide. Such substances are called *chemical compounds*, and differ from mechanical mixtures in that the substances producing them lose their own characteristic properties. We shall see later that the two also differ in that the composition of a chemical compound never varies.

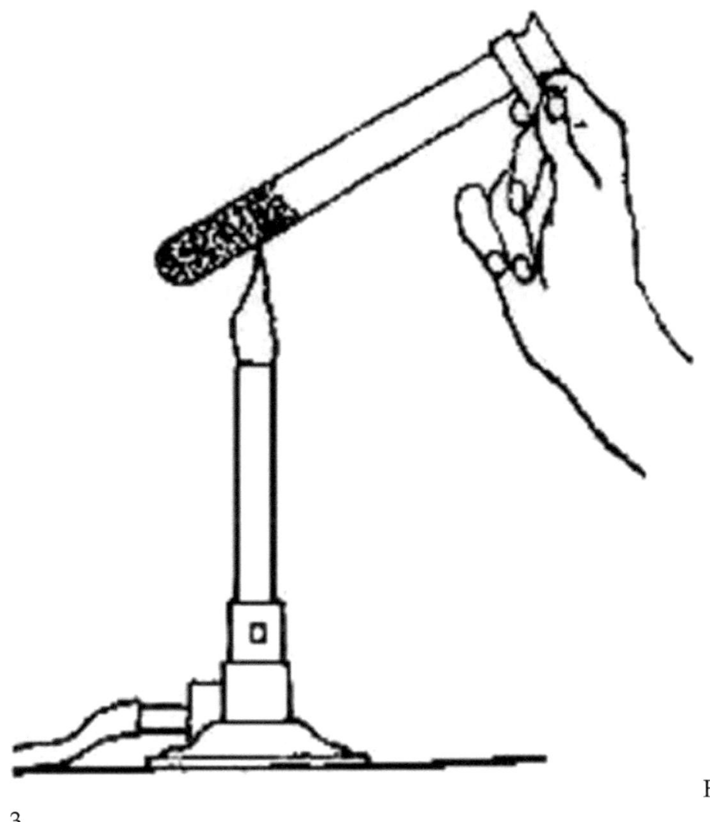

Fig. 3

DEFINITION: *A chemical compound is a substance the constituents of which have lost their own characteristic properties, and which cannot be separated save by a chemical change.*

Elements. It has been seen that iron sulphide is composed of two entirely different substances,—iron and sulphur. The question arises, Do these substances in turn contain other substances, that is, are they also chemical compounds? [Pg 9] Chemists have tried in a great many ways to decompose them, but all their efforts have failed. Substances which have resisted all efforts to decompose them into other substances are called *elements*. It is not always easy to prove that a given substance is really an element. Some way as yet untried may be successful in decomposing it into other simpler forms of matter, and the supposed element will then prove to be a compound. Water, lime, and many other familiar compounds were at one time thought to be elements.

DEFINITION: *An element is a substance which cannot be separated into simpler substances by any known means.*

Kinds of matter. While matter has been grouped in three classes for the purpose of study, it will be apparent that there are really but two distinct kinds of matter, namely, compounds and elements. A mechanical mixture is not a third distinct kind of matter, but is made up of varying quantities of either compounds or elements or both.

Alchemy. In olden times it was thought that some way could be found to change one element into another, and a great many efforts were made to accomplish this transformation. Most of these efforts were directed toward changing the commoner metals into gold, and many fanciful ways for doing this were described. The chemists of that time were called *alchemists*, and the art which they practiced was called *alchemy*. The alchemists gradually became convinced that the only way common metals could be changed into gold was by the wonderful power of a magic substance which they called the *philosopher's stone*, which would accomplish this transformation by its mere touch and would in addition give perpetual youth to its fortunate possessor. No one has ever found such a stone, and no one has succeeded in changing one metal into another.

Number of elements. The number of substances now considered to be elements is not large—about eighty in all. Many of these are rare, and very few of them make any [Pg 10] large fraction of the materials in the earth's crust. Clarke gives the following estimate of the composition of the earth's crust:

Oxygen	47.0%	Calcium	3.5%
Silicon	27.9	Magnesium	2.5
Aluminium	8.1	Sodium	2.7
Iron	4.7	Potassium	2.4

Other elements   1.2%

A complete list of the elements is given in the Appendix. In this list the more common of the elements are marked with an asterisk. It is not necessary to study more than a third of the total number of elements to gain a very good knowledge of chemistry.

Physical state of the elements. About ten of the elements are gases at ordinary temperatures. Two—mercury and bromine—are liquids. The others are all solids, though their melting points vary through wide limits, from cæsium which melts at 26° to elements which do not melt save in the intense heat of the electric furnace.

Occurrence of the elements. Comparatively few of the elements occur as uncombined substances in nature, most of them being found in the form of chemical compounds. When an element does occur by itself, as is the case with gold, we say that it occurs in the *free state* or *native*; when it is combined with other substances in the form of compounds, we say that it occurs in the *combined state*, or *in combination*. In the latter case there is usually little about the compound to suggest that the element is present in it; for we have seen that elements lose their own peculiar properties when they enter into combination with other elements. It would never be suspected, for example, that the reddish, earthy-looking iron ore contains iron. [Pg 11]

Names of elements. The names given to the elements have been selected in a great many different ways. (1) Some names are very old and their original meaning is obscure. Such names are iron, gold, and copper. (2) Many names indicate some striking physical property of the element. The name bromine, for example, is derived from a Greek word meaning a stench, referring to the extremely unpleasant odor of the substance. The name iodine comes from a word meaning violet, alluding to the beautiful color of iodine vapor. (3) Some names indicate prominent chemical properties of the elements. Thus, nitrogen means the producer of niter, nitrogen being a constituent of niter or saltpeter. Hydrogen means water former, signifying its presence in water. Argon means lazy or inert, the element being so named because of its inactivity. (4) Other elements are named from countries or localities, as germanium and scandium.

Symbols. In indicating the elements found in compounds it is inconvenient to use such long names, and hence chemists have adopted a system of abbreviations. These abbreviations are known as *symbols*, each element having a distinctive symbol. (1) Sometimes the initial letter of the name will suffice to indicate the element. Thus I stands for iodine, C for carbon. (2) Usually it is necessary to add some other characteristic letter to the symbol, since several names may begin with the same letter. Thus C stands for carbon, Cl for chlorine, Cd for cadmium, Ce for cerium, Cb for columbium. (3) Sometimes the symbol is an abbreviation of the old Latin name. In this way Fe (ferrum) indicates iron, Cu (cuprum), copper, Au (aurum), gold. The symbols are included in the list of elements given in the Appendix. They will become familiar through constant use. [Pg 12]

Chemical affinity the cause of chemical combination. The agency which causes substances to combine and which holds them together when combined is called *chemical affinity*. The experiments described in this chapter, however, show that heat is often necessary to bring about chemical action. The distinction between the cause producing chemical action and the circumstances favoring it must be clearly made. Chemical affinity is always the cause of chemical union. Many agencies may make it possible for chemical affinity to act by overcoming circumstances which stand in its way. Among these agencies are heat, light, and electricity. As a rule, solution also promotes action between two substances. Sometimes these agencies may overcome chemical attraction and so occasion the decomposition of a compound.

EXERCISES

1. To what class of changes do the following belong? (*a*) The melting of ice; (*b*) the souring of milk; (*c*) the burning of a candle; (*d*) the explosion of gunpowder; (*e*) the corrosion of metals. What test question must be applied in each of the above cases?

2. Give two additional examples (*a*) of chemical changes; (*b*) of physical changes.

3. Is a chemical change always accompanied by a physical change? Is a physical change always accompanied by a chemical change?

4. Give two or more characteristics of a chemical change.

5. (*a*) When a given weight of water freezes, does it absorb or evolve heat? (*b*) When the resulting ice melts, is the total heat change the same or different from that of freezing?

6. Give three examples of each of the following: (*a*) mechanical mixtures; (*b*) chemical compounds; (*c*) elements.

7. Give the derivation of the names of the following elements: thorium, gallium, selenium, uranium. (Consult dictionary.)

8. Give examples of chemical changes which are produced through the agency of heat; of light; of electricity.

[Pg 13]

CHAPTER II

OXYGEN

History. The discovery of oxygen is generally attributed to the English chemist Priestley, who in 1774 obtained the element by heating a compound of mercury and oxygen, known as red oxide of mercury. It is probable, however, that the Swedish chemist Scheele had previously obtained it, although an account of his experiments was not published until 1777. The name oxygen signifies acid former. It was given to the element by the French chemist Lavoisier, since he believed that all acids owe their characteristic properties to the presence of oxygen. This view we now know to be incorrect.

Occurrence. Oxygen is by far the most abundant of all the elements. It occurs both in the free and in the combined state. In the free state it occurs in the air, 100 volumes of dry air containing about 21 volumes of oxygen. In the combined state it forms eight ninths of water and nearly one half of the rocks composing the earth's crust. It is also an important constituent of the compounds which compose plant and animal tissues; for example, about 66% by weight of the human body is oxygen.

Preparation. Although oxygen occurs in the free state in the atmosphere, its separation from the nitrogen and other gases with which it is mixed is such a difficult matter that in the laboratory it has been found more convenient to prepare it from its compounds. The most important of the laboratory methods are the following: [Pg 14]

1. *Preparation from water.* Water is a compound, consisting of 11.18% hydrogen and 88.82% oxygen. It is easily separated into these constituents by passing an electric current through it under suitable conditions. The process will be described in the chapter on water. While this method of preparation is a simple one, it is not economical.

2. *Preparation from mercuric oxide.* This method is of interest, since it is the one which led to the discovery of oxygen. The oxide, which consists of 7.4% oxygen and 92.6% mercury, is placed in a small, glass test tube and heated. The compound is in this way decomposed into mercury which collects on the sides of the glass tube, forming a silvery mirror, and oxygen which, being a gas, escapes from the tube. The presence of the oxygen is shown by lighting the end of a splint, extinguishing the flame and bringing the glowing coal into the mouth of the tube. The oxygen causes the glowing coal to burst into a flame.

In a similar way oxygen may be obtained from its compounds with some of the other elements. Thus manganese dioxide, a black compound of manganese and oxygen, when heated to about 700°, loses one third of its oxygen, while barium dioxide, when heated, loses one half of its oxygen.

3. *Preparation from potassium chlorate (usual laboratory method).* Potassium chlorate is a white solid which consists of 31.9% potassium, 28.9% chlorine, and 39.2% oxygen. When heated it undergoes a series of changes in which all the oxygen is finally set free, leaving a compound of potassium and chlorine called potassium chloride. The change may be represented as follows:

/ potassium \ (potassium / potassium \ (potassium { chlorine } chlorate) = { } chloride) + oxygen \ oxygen / \ chlorine /

JOSEPH PRIESTLEY (English) (1733-1804)

School-teacher, theologian, philosopher, scientist; friend of Benjamin Franklin; discoverer of oxygen; defender of the phlogiston theory; the first to use mercury in a pneumatic trough, by which means he first isolated in gaseous form hydrochloric acid, sulphur dioxide, and ammonia

[Pg 15]

The evolution of the oxygen begins at about 400°. It has been found, however, that if the potassium chlorate is mixed with about one fourth its weight of manganese dioxide, the oxygen is given off at a much lower temperature. Just how the manganese dioxide brings about this result is not definitely known. The amount of oxygen obtained from a given weight of potassium chlorate is exactly the same whether the manganese dioxide is present or not. So far as can be detected the manganese dioxide undergoes no change.

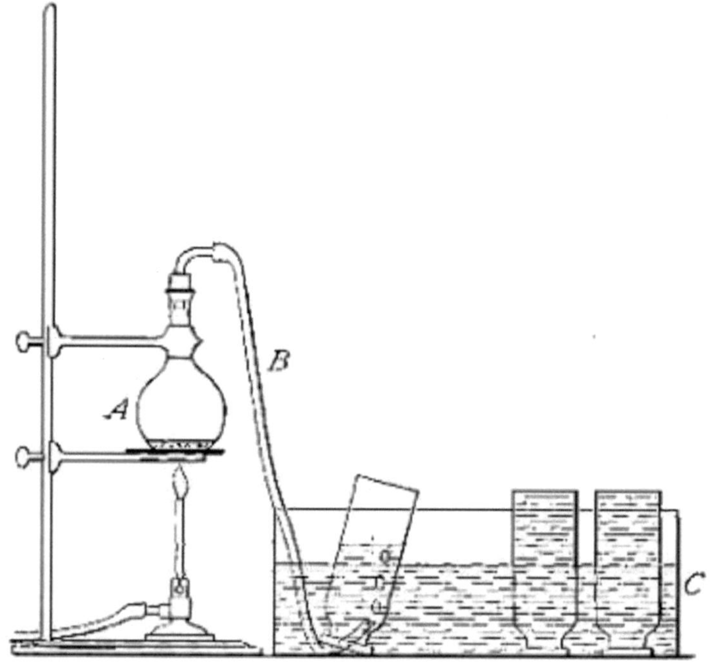

Fig. 4

Directions for preparing oxygen. The manner of preparing oxygen from potassium chlorate is illustrated in the accompanying diagram (Fig. 4). A mixture consisting of one part of manganese dioxide and four parts of potassium chlorate is placed in the flask *A* and gently heated. The oxygen is evolved and escapes through the tube *B*. It is collected by bringing over the end of the tube the mouth of a bottle completely filled with water and inverted in a vessel of water, as shown in the figure. The gas rises in the bottle and displaces the water. In the preparation of large quantities of oxygen, a copper retort (Fig. 5) is often substituted for the glass flask.

Fig. 5

In the preparation of oxygen from potassium chlorate and manganese dioxide, the materials used must be pure, otherwise a violent explosion may occur. The purity of the materials is tested by heating a small amount of the mixture in a test tube.

The collection of gases. The method used for collecting oxygen illustrates the general method used for collecting such gases as are [Pg 16] insoluble in water or nearly so. The vessel C (Fig. 4), containing the water in which the bottles are inverted, is called a *pneumatic trough.*

Commercial methods of preparation. Oxygen can now be purchased stored under great pressure in strong steel cylinders (Fig. 6). It is prepared either by heating a mixture of potassium chlorate and manganese dioxide, or by separating it from the nitrogen and other gases with which it is mixed in the atmosphere. The methods employed for effecting this separation will be described in subsequent chapters.

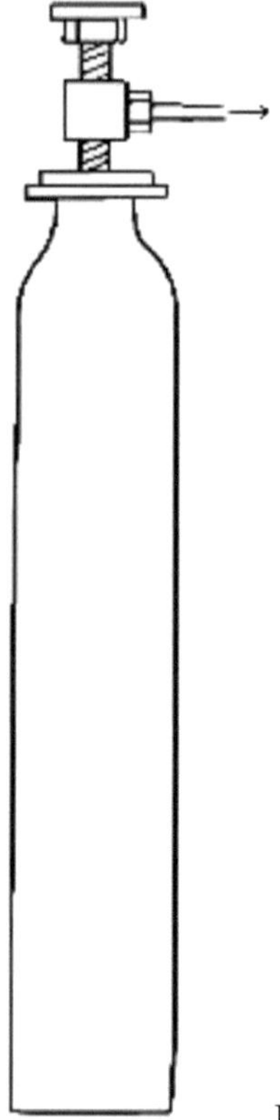

Fig. 6

Physical properties. Oxygen is a colorless, odorless, tasteless gas, slightly heavier than air. One liter of it, measured at a temperature of 0° and under a pressure of one atmosphere, weighs 1.4285 g., while under similar conditions one liter of air weighs 1.2923 g. It is but slightly soluble in water. Oxygen, like other gases, may be liquefied by applying very great pressure to the highly cooled gas. When the pressure is removed the liquid oxygen passes again into the gaseous state, since its boiling point under ordinary atmospheric pressure is -182.5°.

Chemical properties. At ordinary temperatures oxygen is not very active chemically. Most substances are either not at all affected by it, or the action is so slow as to escape notice. At higher temperatures, however, it is very active, and unites directly with most of the elements. This activity may be shown by heating various substances until just ignited and then bringing them into vessels of the gas, when they will burn with great brilliancy. Thus a glowing splint introduced into a jar of oxygen bursts into flame. Sulphur burns in the air with a very weak flame and feeble light; in oxygen, however, the flame is increased in size and [Pg 17] brightness. Substances which readily burn in air, such as phosphorus, burn in oxygen with dazzling brilliancy. Even substances which burn in air with great difficulty, such as iron, readily burn in oxygen.

The burning of a substance in oxygen is due to the rapid combination of the substance or of the elements composing it with the oxygen. Thus, when sulphur burns both the oxygen and sulphur disappear as such and there is formed a compound of the two, which is an invisible gas, having the characteristic odor of burning sulphur. Similarly, phosphorus on burning forms a white solid compound of phosphorus and oxygen, while iron forms a reddish-black compound of iron and oxygen.

Oxidation. The term *oxidation* is applied to the chemical change which takes place when a substance, or one of its constituent parts, combines with oxygen. This process may take place rapidly, as in the burning of phosphorus, or slowly, as in the oxidation (or rusting) of iron when exposed to the air. It is always accompanied by the liberation of heat. The amount of heat liberated by the oxidation of a definite weight of any given substance is always the same, being entirely independent of the rapidity of the process. If the oxidation takes place slowly, the heat is generated so slowly that it is difficult to detect it. If the oxidation takes place rapidly, however, the heat is generated in such a short interval of time that the substance may become white hot or burst into a flame.

Combustion; kindling temperature. When oxidation takes place so rapidly that the heat generated is sufficient to cause the substance to glow or burst into a flame the process is called *combustion*. In order that any substance may undergo combustion, it is necessary that it should be [Pg 18] heated to a certain temperature, known as the *kindling temperature*. This temperature varies widely for different bodies, but is always definite for the same body. Thus the kindling temperature of phosphorus is far lower than that of iron, but is definite for each. When any portion of a substance is heated until it begins to burn the combustion will continue without the further application of heat, provided the heat generated by the process is sufficient to bring other parts of the substance to the kindling temperature. On the other hand, if the heat generated is not sufficient to maintain the kindling temperature, combustion ceases.

Oxides. The compounds formed by the oxidation of any element are called *oxides*. Thus in the combustion of sulphur, phosphorus, and iron, the compounds formed are called respectively oxide of sulphur, oxide of phosphorus, and oxide of iron. In general, then, *an oxide is a compound of oxygen with another element*. A great many substances of this class are known; in fact, the oxides of all the common elements have been prepared, with the exception of those of fluorine and bromine. Some of these are familiar compounds. Water, for example, is an oxide of hydrogen, and lime an oxide of the metal calcium.

Products of combustion. The particular oxides formed by the combustion of any substance are called *products of combustion* of that substance. Thus oxide of sulphur is the product of the combustion of sulphur; oxide of iron is the product of the combustion of iron. It is evident that the products of the combustion of any substance must weigh more than the original substance, the increase in weight corresponding to the amount of oxygen taken up in the act of combustion. For example, when iron burns the oxide of iron formed weighs more than the original iron. [Pg 19]

In some cases the products of combustion are invisible gases, so that the substance undergoing combustion is apparently destroyed. Thus, when a candle burns it is consumed, and so far as the eye can judge nothing is formed during combustion. That invisible gases are formed, however, and that the weight of these is greater than the weight of the candle may be shown by the following experiment.

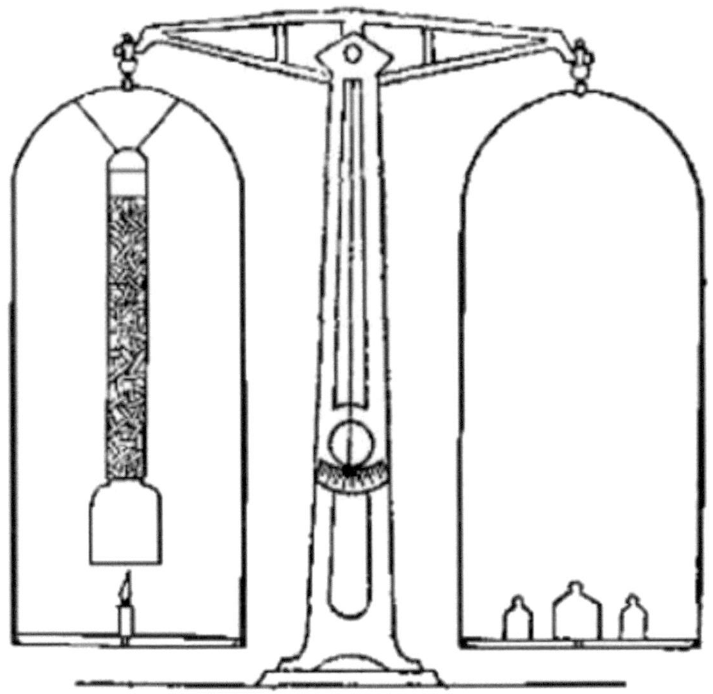

Fig. 7

A lamp chimney is filled with sticks of the compound known as sodium hydroxide (caustic soda), and suspended from the beam of the balance, as shown in Fig. 7. A piece of candle is placed on the balance pan so that the wick comes just below the chimney, and the balance is brought to a level by adding weights to the other pan. The candle is then lighted. The products formed pass up through the chimney and are absorbed by the sodium hydroxide. Although the candle burns away, the pan upon which it rests slowly sinks, showing that the combustion is attended by an increase in weight.

Combustion in air and in oxygen. Combustion in air and in oxygen differs only in rapidity, the products formed being exactly the same. That the process should take place less rapidly in the former is readily understood, for the air is only about one fifth oxygen, the remaining four fifths being inert gases. Not only is less oxygen available, but much of the heat is absorbed in raising the temperature of the inert gases surrounding the substance undergoing combustion, and the temperature reached in the combustion is therefore less.

Phlogiston theory of combustion. The French chemist Lavoisier (1743-1794), who gave to oxygen its name was the first to show that combustion is due to union with oxygen. Previous to his time combustion was supposed to be due to the presence of a substance or principle called *phlogiston*. One substance was thought to be more combustible than another because it contained more phlogiston. Coal, for example, was thought to be very rich in phlogiston. The ashes [Pg 20] left after combustion would not burn because all the phlogiston had escaped. If the phlogiston could be restored in any way, the substance would then become combustible again. Although this view seems absurd to us in the light of our present knowledge, it formerly had general acceptance. The discovery of oxygen led Lavoisier to investigate the subject, and through his experiments he arrived at the true explanation of combustion. The discovery of oxygen together with the part it plays in combustion is generally regarded as the most important discovery in the history of chemistry. It marked the dawn of a new period in the growth of the science.

Combustion in the broad sense. According to the definition given above, the presence of oxygen is necessary for combustion. The term is sometimes used, however, in a broader sense to designate any chemical change attended by the evolution of heat and light. Thus iron and sulphur, or hydrogen and chlorine under certain conditions, will combine so rapidly that light is evolved, and the action is called a combustion. Whenever combustion takes place in the air, however, the process is one of oxidation.

Spontaneous combustion. The temperature reached in a given chemical action, such as oxidation, depends upon the rate at which the reaction takes place. This rate is usually increased by raising the temperature of the substances taking part in the action.

When a slow oxidation takes place under such conditions that the heat generated is not lost by being conducted away, the temperature of the substance undergoing oxidation is raised, and this in turn hastens the rate of oxidation. The rise in temperature may continue in this way until the kindling temperature of the substance is reached, when combustion begins. Combustion occurring in this way is called *spontaneous combustion*.

Certain oils, such as the linseed oil used in paints, slowly undergo oxidation at ordinary temperatures, and not infrequently the origin of fires has been traced to the spontaneous combustion of oily rags. The spontaneous combustion of hay has been known to set barns on fire. Heaps of coal have been found to be on fire when spontaneous combustion offered the only possible explanation.

[Pg 21]

Importance of oxygen. 1. Oxygen is essential to life. Among living organisms only certain minute forms of plant life can exist without it. In the process of respiration the air is taken into the lungs where a certain amount of oxygen is absorbed by the blood. It is then carried to all parts of the body, oxidizing the worn-out tissues and changing them into substances which may readily be eliminated from the body. The heat generated by this oxidation is the source of the heat of the body. The small amount of oxygen which water dissolves from the air supports all the varied forms of aquatic animals.

2. Oxygen is also essential to decay. The process of decay is really a kind of oxidation, but it will only take place in the presence of certain minute forms of life known as bacteria. Just how these assist in the oxidation is not known. By this process the dead products of animal and vegetable life which collect on the surface of the earth are slowly oxidized and so converted into harmless substances. In this way oxygen acts as a great purifying agent.

3. Oxygen is also used in the treatment of certain diseases in which the patient is unable to inhale sufficient air to supply the necessary amount of oxygen.

OZONE

Preparation. When electric sparks are passed through oxygen or air a small percentage of the oxygen is converted into a substance called *ozone*, which differs greatly from oxygen in its properties. The

same change can also be brought about by certain chemical processes. Thus, if some pieces of phosphorus are placed in a bottle and
partially covered with water, the presence of ozone may soon be
detected in the air contained in the bottle. The conversion of oxygen
into ozone is attended by a change in volume, 3 volumes of oxygen
forming 2 volumes of ozone. If the resulting ozone is heated to
about 300°, the [Pg 22] reverse change takes place, the 2 volumes of
ozone being changed back into 3 volumes of oxygen. It is possible
that traces of ozone exist in the atmosphere, although its presence
there has not been definitely proved, the tests formerly used for its
detection having been shown to be unreliable.

Properties. As commonly prepared, ozone is mixed with a large
excess of oxygen. It is possible, however, to separate the ozone and
thus obtain it in pure form. The gas so obtained has the characteristic odor noticed about electrical machines when in operation. By
subjecting it to great pressure and a low temperature, the gas condenses to a bluish liquid, boiling at -119°. When unmixed with other
gases ozone is very explosive, changing back into oxygen with the
liberation of heat. Its chemical properties are similar to those of
oxygen except that it is far more active. Air or oxygen containing a
small amount of ozone is now used in place of oxygen in certain
manufacturing processes.

The difference between oxygen and ozone. Experiments show
that in changing oxygen into ozone no other kind of matter is either
added to the oxygen or withdrawn from it. The question arises then,
How can we account for the difference in their properties? It must
be remembered that in all changes we have to take into account
energy as well as *matter*. By changing the amount of energy in a substance we change its properties. That oxygen and ozone contain
different amounts of energy may be shown in a number of ways; for
example, by the fact that the conversion of ozone into oxygen is
attended by the liberation of heat. The passage of the electric sparks
through oxygen has in some way changed the energy content of the
element and thus it has acquired new properties. *Oxygen and ozone
must, therefore, be regarded as identical so far as the kind of matter of
which they are composed is concerned. Their different properties are due to
their different energy contents.*

Allotropic states or forms of matter. Other elements besides oxygen may exist in more than one form. These different forms of the same element are called *allotropic states* or *forms* of the element. These forms differ not only in physical properties but also in their energy contents. Elements often exist in a variety of forms which look quite different. These differences may be due to accidental causes, such as the size or shape of the particles or the way in which the element was prepared. Only such forms, however, as have different energy contents are properly called allotropic forms. [Pg 23]

MEASUREMENT OF GAS VOLUMES

Standard conditions. It is a well-known fact that the volume occupied by a definite weight of any gas can be altered by changing the temperature of the gas or the pressure to which it is subjected. In measuring the volume of gases it is therefore necessary, for the sake of accuracy, to adopt some standard conditions of temperature and pressure. The conditions agreed upon are (1) a temperature of 0°, and (2) a pressure equal to the average pressure exerted by the atmosphere at the sea level, that is, 1033.3 g. per square centimeter. These conditions of temperature and pressure are known as the *standard conditions*, and when the volume of a gas is given it is understood that the measurement was made under these conditions, unless it is expressly stated otherwise. For example, the weight of a liter of oxygen has been given as 1.4285 g. This means that one liter of oxygen, measured at a temperature of 0° and under a pressure of 1033.3 g. per square centimeter, weighs 1.4285 g.

The conditions which prevail in the laboratory are never the standard conditions. It becomes necessary, therefore, to find a way to calculate the volume which a gas will occupy under standard conditions from the volume which it occupies under any other conditions. This may be done in accordance with the following laws.

Law of Charles. This law expresses the effect which a change in the temperature of a gas has upon its volume. It may be stated as follows: *For every degree the temperature of a gas rises above zero the volume of the gas is increased by 1/273 of the volume which it occupies at zero; likewise for every degree the temperature of the gas falls below zero the volume of the gas is decreased by 1/273 of the volume which it occupies* *at zero, provided in both cases that the pressure to which the gas is subjected remains constant.*

If V represents the volume of gas at $0°$, then the volume at $1°$ will be $V + 1/273\ V$; at $2°$ it will be $V + 2/273\ V$; or, in general, the volume v, at the temperature t, will be expressed by the formula

$$(1)\ v = V + t/273\ V,$$

$$\text{or } (2)\ v = V(1 + (t/273)).$$

Since $1/273 = 0.00366$, the formula may be written

$$(3)\ v = V(1 + 0.00366t).$$
[Pg 24]

Since the value of V (volume under standard conditions) is the one usually sought, it is convenient to transpose the equation to the following form:

$$(4)\ V = v/(1 + 0.00366t).$$

The following problem will serve as an illustration of the application of this equation.

The volume of a gas at $20°$ is 750 cc.; find the volume it will occupy at $0°$, the pressure remaining constant.

In this case, $v = 750$ cc. and $t = 20$. By substituting these values, equation (4) becomes

$$V = 750/(1 + 0.00366 \times 20) = 698.9 \text{ cc.}$$

Law of Boyle. This law expresses the relation between the volume occupied by a gas and the pressure to which it is subjected. It may be stated as follows: *The volume of a gas is inversely proportional* *to the pressure under which it is measured, provided the temperature of the gas remains constant.*

If V represents the volume when subjected to a pressure P and v represents its volume when the pressure is changed to p, then, in accordance with the above law, $V : v :: p : P$, or $VP = vp$. In other words, for a given weight of a gas the product of the numbers representing its volume and the pressure to which it is subjected is a constant.

Since the pressure of the atmosphere at any point is indicated by the barometric reading, it is convenient in the solution of the problems to substitute the latter for the pressure measured in grams per square centimeter. The average reading of the barometer at the sea level is 760 mm., which corresponds to a pressure of 1033.3 g. per square centimeter. The following problem will serve as an illustration of the application of Boyle's law.

A gas occupies a volume of 500 cc. in a laboratory where the barometric reading is 740 mm. What volume would it occupy if the atmospheric pressure changed so that the reading became 750 mm.?

Substituting the values in the equation $VP = vp$, we have 500 × 740 = v × 750, or v = 493.3 cc.

Variations in the volume of a gas due to changes both in temperature and pressure. Inasmuch as corrections must be made as a rule [Pg 25] for both temperature and pressure, it is convenient to combine the equations given above for the corrections for each, so that the two corrections may be made in one operation. The following equation is thus obtained:

$$(5)\ V_s = vp/(760(1 + 0.00366t)),$$

in which V_s represents the volume of a gas under standard conditions and v, p, and t the volume, pressure, and temperature respectively at which the gas was actually measured.

The following problem will serve to illustrate the application of this equation.

A gas having a temperature of 20° occupies a volume of 500 cc. when subjected to a pressure indicated by a barometric reading of 740 mm. What volume would this gas occupy under standard conditions?

In this problem $v = 500$, $p = 740$, and $t = 20$. Substituting these values in the above equation, we get

$$V_s = (500 \times 740)/(760\,(1 + 0.00366 \times 20)) = 453.6 \text{ cc.}$$

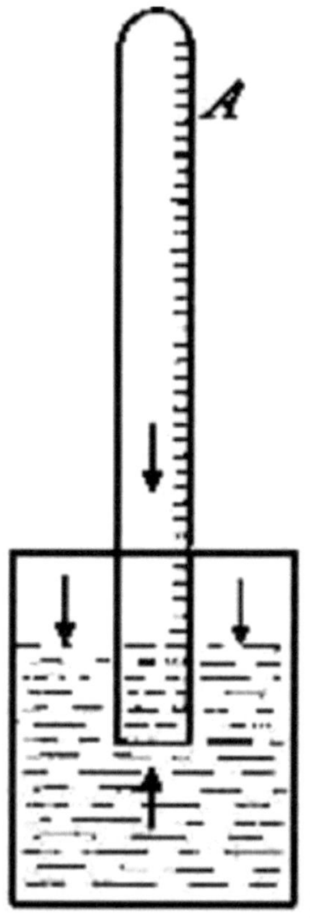

Fig. 8

Variations in the volume of a gas due to the pressure of aqueous vapor. In many cases gases are collected over water, as explained under the preparation of oxygen. In such cases there is present in the gas a certain amount of water vapor. This vapor exerts a definite pressure, which acts in opposition to the atmospheric pressure and which therefore must be subtracted from the latter in determining the effective pressure upon the gas. Thus, suppose we wish to determine the pressure to which the gas in tube A (Fig. 8) is subjected. The tube is raised or lowered until the level of the water inside and outside the tube is the same. The atmosphere presses down upon the surface of the water (as indicated by the arrows), thus forcing the water upward within the tube with a pressure equal to the atmospheric pressure. The full force of this upward pressure, however, is not spent in compressing the gas within the tube, for since it is collected over water it contains a certain amount of water vapor. This water vapor exerts a pressure (as indicated by the arrow within the tube) in opposition to [Pg 26] the upward pressure. It is plain, therefore, that the effective pressure upon the gas is equal to the atmospheric pressure less the pressure exerted by the aqueous vapor. The pressure exerted by the aqueous vapor increases with the temperature. The figures representing the extent of this pressure (often called the *tension of aqueous vapor*) are given in the Appendix. They express the pressure or tension in millimeters of mercury, just as the atmospheric pressure is expressed in millimeters of mercury. Representing the pressure of the aqueous vapor by a, formula (5) becomes

(6) $V_s = v(p - a)/(760(1 + 0.00366t))$.

The following problem will serve to illustrate the method of applying the correction for the pressure of the aqueous vapor.

The volume of a gas measured over water in a laboratory where the temperature is 20° and the barometric reading is 740 mm. is 500 cc. What volume would this occupy under standard conditions?

The pressure exerted by the aqueous vapor at 20° (see table in Appendix) is equal to the pressure exerted by a column of mercury 17.4 mm. in height. Substituting the values of v, t, p, and a in formula (6), we have

$$(6)\ V_s = 500(740 - 17.4)/(760(1 + 0.00366 \times 20)) = 442.9 \text{ cc.}$$

Adjustment of tubes before reading gas volumes. In measuring the volumes of gases collected in graduated tubes or other receivers, over a liquid as illustrated in Fig. 8, the reading should be taken after raising or lowering the tube containing the gas until the level of the liquid inside and outside the tube is the same; for it is only under these conditions that the upward pressure within the tube is the same as the atmospheric pressure.

EXERCISES

1. What is the meaning of the following words? phlogiston, ozone, phosphorus. (Consult dictionary.)

2. Can combustion take place without the emission of light?

3. Is the evolution of light always produced by combustion?

4. (*a*) What weight of oxygen can be obtained from 100 g. of water? (*b*) What volume would this occupy under standard conditions? [Pg 27]

5. (*a*) What weight of oxygen can be obtained from 500g. of mercuric oxide? (*b*) What volume would this occupy under standard conditions?

6. What weight of each of the following compounds is necessary to prepare 50 l. of oxygen? (*a*) water; (*b*) mercuric oxide; (*c*) potassium chlorate.

7. Reduce the following volumes to 0°, the pressure remaining constant: (*a*) 150 cc. at 10°; (*b*) 840 cc. at 273°.

8. A certain volume of gas is measured when the temperature is 20°. At what temperature will its volume be doubled?

9. Reduce the following volumes to standard conditions of pressure, the temperature remaining constant: (*a*) 200 cc. at 740 mm.; (*b*) 500 l. at 380 mm.

10. What is the weight of 1 l. of oxygen when the pressure is 750 mm. and the temperature 0°?

11. Reduce the following volumes to standard conditions of temperature and pressure: (*a*) 340 cc. at 12° and 753 mm; (*b*) 500 cc. at 15° and 740 mm.

12. What weight of potassium chlorate is necessary to prepare 250 l. of oxygen at 20° and 750 mm.?

13. Assuming the cost of potassium chlorate and mercuric oxide to be respectively $0.50 and $1.50 per kilogram, calculate the cost of materials necessary for the preparation of 50 l. of oxygen from each of the above compounds.

14. 100 g. of potassium chlorate and 25 g. of manganese dioxide were heated in the preparation of oxygen. What products were left in the flask, and how much of each was present?

[Pg 28]

CHAPTER III

HYDROGEN

Historical. The element hydrogen was first clearly recognized as a distinct substance by the English investigator Cavendish, who in 1766 obtained it in a pure state, and showed it to be different from the other inflammable airs or gases which had long been known. Lavoisier gave it the name hydrogen, signifying water former, since it had been found to be a constituent of water.

Occurrence. In the free state hydrogen is found in the atmosphere, but only in traces. In the combined state it is widely distributed, being a constituent of water as well as of all living organisms, and the products derived from them, such as starch and sugar. About 10% of the human body is hydrogen. Combined with carbon, it forms the substances which constitute petroleum and natural gas.

It is an interesting fact that while hydrogen in the free state occurs only in traces on the earth, it occurs in enormous quantities in the gaseous matter surrounding the sun and certain other stars.

Preparation from water. Hydrogen can be prepared from water by several methods, the most important of which are the following.

1. *By the electric current.* As has been indicated in the preparation of oxygen, water is easily separated into its constituents, hydrogen and oxygen, by passing an electric current through it under certain conditions.

2. *By the action of certain metals.* When brought into contact with certain metals under appropriate conditions, [Pg 29] water gives up a portion or the whole of its hydrogen, its place being taken by the metal. In the case of a few of the metals this change occurs at ordinary temperatures. Thus, if a bit of sodium is thrown on water, an action is seen to take place at once, sufficient heat being generated to melt the sodium, which runs about on the surface of the water. The change which takes place consists in the displacement of one half of the hydrogen of the water by the sodium, and may be represented as follows:

_ _ _ _ | hydrogen | | sodium | sodium + | hydrogen
|(water) = | hydrogen |(sodium hydroxide) + hydrogen
|_oxygen _| |_oxygen _|

The sodium hydroxide formed is a white solid which remains dissolved in the undecomposed water, and may be obtained by evaporating the solution to dryness. The hydrogen is evolved as a gas and may be collected by suitable apparatus.

Other metals, such as magnesium and iron, decompose water rapidly, but only at higher temperatures. When steam is passed over hot iron, for example, the iron combines with the oxygen of the steam, thus displacing the hydrogen. Experiments show that the change may be represented as follows:

_ _ | hydrogen | _ _ _ _ iron + | hydrogen |(water) = | iron
|(iron oxide) + | hydrogen | |_oxygen _| |_oxygen _|
|_hydrogen_|

The iron oxide formed is a reddish-black compound, identical with that obtained by the combustion of iron in oxygen.

Directions for preparing hydrogen by the action of steam on iron. The apparatus used in the preparation of hydrogen from iron and [Pg 30] steam is shown in Fig. 9. A porcelain or iron tube B, about 50 cm. in length and 2 cm. or 3 cm. in diameter, is partially filled with fine iron wire or tacks and connected as shown in the figure. The tube B is heated, slowly at first, until the iron is red-hot. Steam is then conducted through the tube by boiling the water in the flask A. The hot iron combines with the oxygen in the steam, setting free the hydrogen, which is collected over water. The gas which first passes over is mixed with the air previously contained in the flask and tube, and is allowed to escape, *since a mixture of hydrogen with oxygen or air explodes violently when brought in contact with a flame.* It is evident that the flask A must be disconnected from the tube before the heat is withdrawn.

That the gas obtained is different from air and oxygen may be shown by holding a bottle of it mouth downward and bringing a lighted splint into it. The hydrogen is ignited and burns with an almost colorless flame.

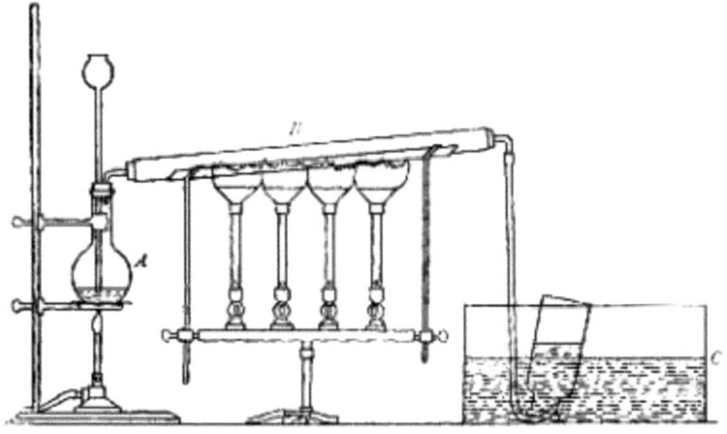

Fig. 9

Preparation from acids (*usual laboratory method*). While hydrogen can be prepared from water, either by the action of the electric current or by the action of certain metals, these methods are not economical and are therefore but little used. In the laboratory hydrogen is generally prepared from compounds known as acids, all of which contain hydrogen. When acids are brought in contact with certain metals, the metals dissolve and set free the hydrogen [Pg 31] of the acid. Although this reaction is a quite general one, it has been found most convenient in preparing hydrogen by this method to use either zinc or iron as the metal and either hydrochloric or sulphuric acid as the acid. Hydrochloric acid is a compound consisting of 2.77% hydrogen and 97.23% chlorine, while sulphuric acid consists of 2.05% hydrogen, 32.70% sulphur, and 65.25% oxygen.

The changes which take place in the preparation of hydrogen from zinc and sulphuric acid (diluted with water) may be represented as follows:

_ _ _ _ | hydrogen |(sulphuric | zinc |(zinc zinc + | sulphur | acid) = | sulphur | sulphate) + hydrogen |_oxygen _| |_oxygen _|

In other words, the zinc has taken the place of the hydrogen in sulphuric acid. The resulting compound contains zinc, sulphur, and oxygen, and is known as zinc sulphate. This remains dissolved in the water present in the acid. It may be obtained in the form of a white solid by evaporating the liquid left after the metal has passed into solution.

When zinc and hydrochloric acid are used the following changes take place:

_ _ _ _ | hydrogen |(hydrochloric | zinc |(zinc zinc + |_chlorine_| acid) = |_chlorine_| chloride) + hydrogen

When iron is used the changes which take place are exactly similar to those just given for zinc.

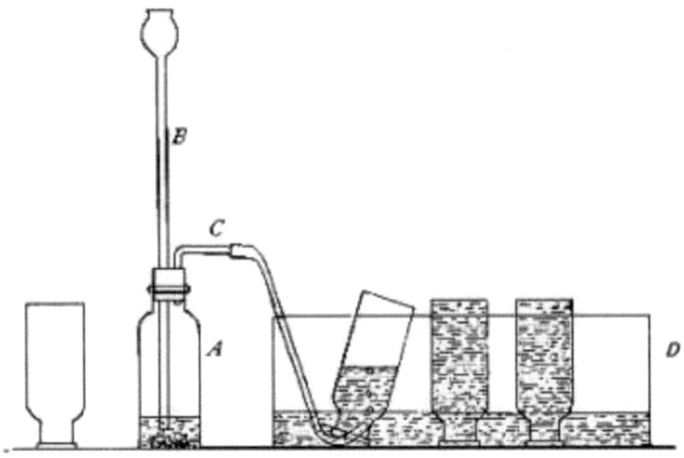

Fig. 10.

Directions for preparing hydrogen from acids. The preparation of hydrogen from acids is carried out in the laboratory as follows: The metal is placed in a flask or wide-mouthed bottle *A* (Fig. 10) and the acid is added slowly through the funnel tube *B*. The metal dissolves in the acid, while the hydrogen which is liberated escapes through the exit tube *C* and is collected over water. It is evident that the hydrogen [Pg 32] which passes over first is mixed with the air from the bottle *A*. Hence care must be taken not to bring a flame near the exit tube, since, as has been stated previously, such a mixture explodes with great violence when brought in contact with a flame.

Precautions. Both sulphuric acid and zinc, if impure, are likely to contain small amounts of arsenic. Such materials should not be used in preparing hydrogen, since the arsenic present combines with a portion of the hydrogen to form a very poisonous gas known as arsine. On the other hand, chemically pure sulphuric acid, i.e. sulphuric acid that is entirely free from impurities, will not act upon chemically pure zinc. The reaction may be started, however, by the addition of a few drops of a solution of copper sulphate or platinum tetrachloride.

Physical properties. Hydrogen is similar to oxygen in that it is a colorless, tasteless, odorless gas. It is characterized by its extreme lightness, being the lightest of all known substances. One liter of the gas weighs only 0.08984 g. On comparing this weight with that of an equal volume of oxygen, viz., 1.4285 g., the latter is found to be 15.88 times as heavy as hydrogen. Similarly, air is found to be 14.38 times as heavy as hydrogen. Soap bubbles blown with hydrogen rapidly rise in the air. On account of its lightness it is possible to pour it upward from one bottle into another. Thus, if the bottle *A* (Fig. 11) is filled with hydrogen, placed mouth downward by the side of bottle *B*, [Pg 33] filled with air, and is then gradually inverted under *B* as indicated in the figure, the hydrogen will flow upward into bottle *B*, displacing the air. Its presence in bottle *B* may then be shown by bringing a lighted splint to the mouth of the bottle, when the hydrogen will be ignited by the flame. It is evident, from this experiment, that in order to retain the gas in an open bottle the bottle must be placed mouth downward.

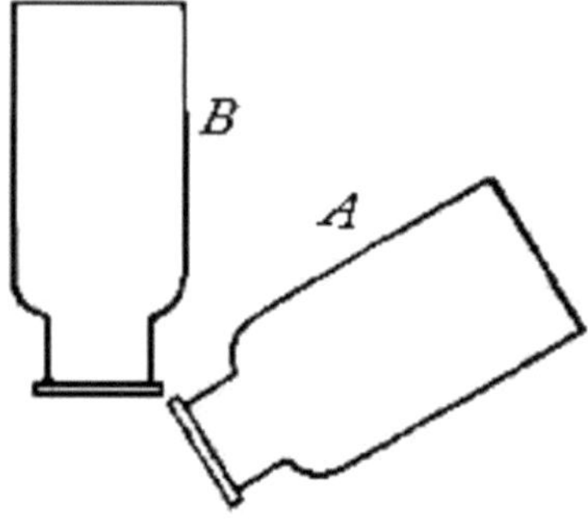

Fig. 11

Hydrogen is far more difficult to liquefy than any other gas, with the exception of helium, a rare element recently found to exist in the atmosphere. The English scientist Dewar, however, in 1898 succeeded not only in obtaining hydrogen in liquid state but also as a solid. Liquid hydrogen is colorless and has a density of only 0.07. Its boiling point under atmospheric pressure is -252°. Under diminished pressure the temperature has been reduced to -262°. The solubility of hydrogen in water is very slight, being still less than that of oxygen.

Pure hydrogen produces no injurious results when inhaled. Of course one could not live in an atmosphere of the gas, since oxygen is essential to respiration.

Chemical properties. At ordinary temperatures hydrogen is not an active element. A mixture of hydrogen and chlorine, however, will combine with explosive violence at ordinary temperature if exposed to the sunlight. The union can be brought about also by heating. The product formed in either case is hydrochloric acid. Under suitable conditions hydrogen combines with nitrogen to form ammonia, and with sulphur to form the foul-smelling gas, hydrogen sulphide. The affinity of hydrogen for oxygen is so great that [Pg 34] a mixture of hydrogen and oxygen or hydrogen and air explodes with great violence when heated to the kindling temperature (about 612°). Nevertheless under proper conditions hydrogen may be made to burn quietly in either oxygen or air. The resulting hydrogen flame is almost colorless and is very hot. The combustion of the hydrogen is, of course, due to its union with oxygen. The product of the combustion is therefore a compound of hydrogen and oxygen. That this compound is water may be shown easily by experiment.

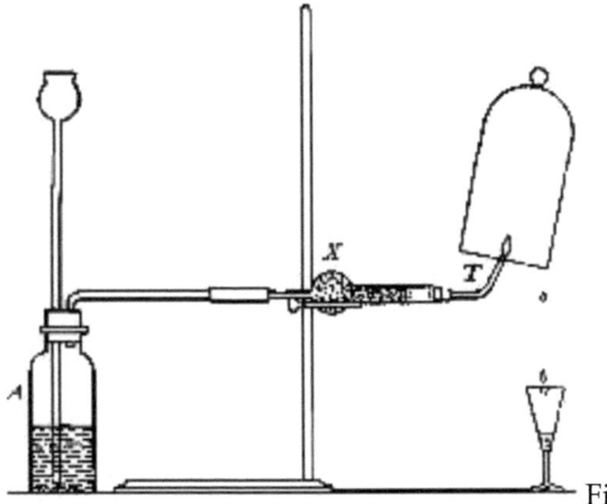

Fig. 12

Directions for burning hydrogen in air. The combustion of hydrogen in air may be carried out safely as follows: The hydrogen is generated in the bottle A (Fig. 12), is dried by conducting it through the tube X, filled with some substance (generally calcium chloride) which has a great attraction for moisture, and escapes through the tube T, the end of which is drawn out to a jet. The hydrogen first liberated mixes with the air contained in the generator. If a flame is brought near the jet before this mixture has all escaped, a violent and very dangerous explosion results, since the entire apparatus is filled with the explosive mixture. On the other hand, if the flame is not applied until all the air has been expelled, the hydrogen is ignited and burns quietly, since only the small amount of it which escapes from the jet can come in contact with the oxygen of the air at any one time. By holding a cold, dry bell jar or bottle over the flame, in the manner shown in the figure, the steam formed by the combustion of the hydrogen is condensed, the water collecting in drops on the sides of the jar.

[Pg 35]

Precautions. In order to avoid danger it is absolutely necessary to prove that the hydrogen is free from air before igniting it. This can be done by testing small amounts of the escaping gas. A convenient and safe method of doing this is to fill a test tube with the gas by inverting it over the jet. The hydrogen, on account of its lightness, collects in the tube, displacing the air. After holding it over the jet for a few moments in order that it may be filled with the gas, the tube is gently brought, mouth downward, to the flame of a burner placed not nearer than an arm's length from the jet. If the hydrogen is mixed with air a slight explosion occurs, but if pure it burns quietly in the tube. The operation is repeated until the gas burns quietly, when the tube is quickly brought back over the jet for an instant, whereby the escaping hydrogen is ignited by the flame in the tube.

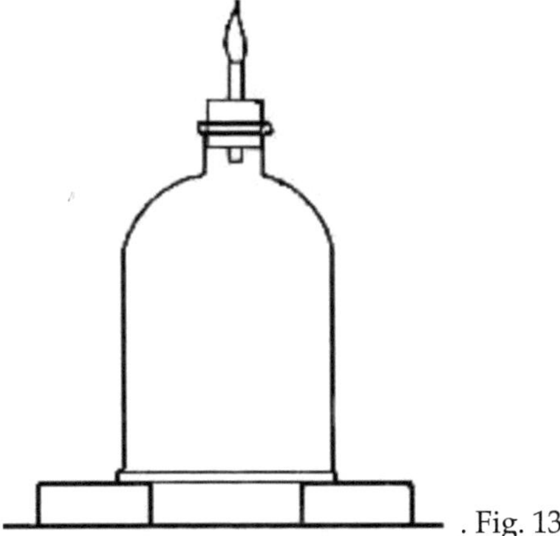

. Fig. 13

A mixture of hydrogen and oxygen is explosive. That a mixture of hydrogen and air is explosive may be shown safely as follows: A cork through which passes a short glass tube about 1 cm. in diameter is fitted air-tight into the tubule of a bell jar of 2 l. or 3 l. capacity. (A thick glass bottle with bottom removed may be used.) The tube is closed with a small rubber stopper and the bell jar filled with hydrogen, the gas being collected over water. When entirely filled with the gas the jar is removed from the water and supported by blocks of wood in order to leave the bottom of the jar open, as shown in Fig. 13. The stopper is now removed from the tube in the cork, and the hydrogen, which on account of its lightness escapes from the tube, is at once lighted. As the hydrogen escapes, the air flows in at the bottom of the jar and mixes with the remaining portion of the hydrogen, so that a mixture of the two soon forms, and a loud explosion results. The explosion is not dangerous, since the bottom of the jar is open, thus leaving room for the expansion of the hot gas.

Since air is only one fifth oxygen, the remainder being inert gases, it may readily be inferred that a mixture of hydrogen with pure oxygen would be far more explosive than a mixture of hydrogen with air. Such mixtures should not be made except in small quantities and by experienced workers. [Pg 36]

Hydrogen does not support combustion. While hydrogen is readily combustible, it is not a supporter of combustion. In other words, substances will not burn in it. This may be shown by bringing a lighted candle supported by a stiff wire into a bottle or cylinder of the pure gas, as shown in Fig. 14. The hydrogen is ignited by the flame of the candle and burns at the mouth of the bottle, where it comes in contact with the oxygen in the air. When the candle is thrust up into the gas, its flame is extinguished on account of the absence of oxygen. If slowly withdrawn, the candle is relighted as it passes through the layer of burning hydrogen.

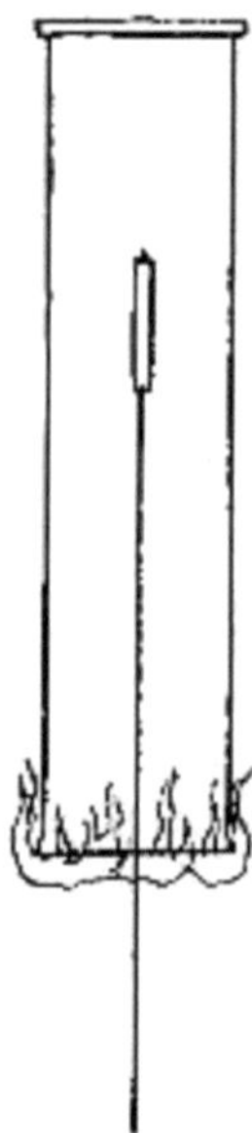

Fig. 14

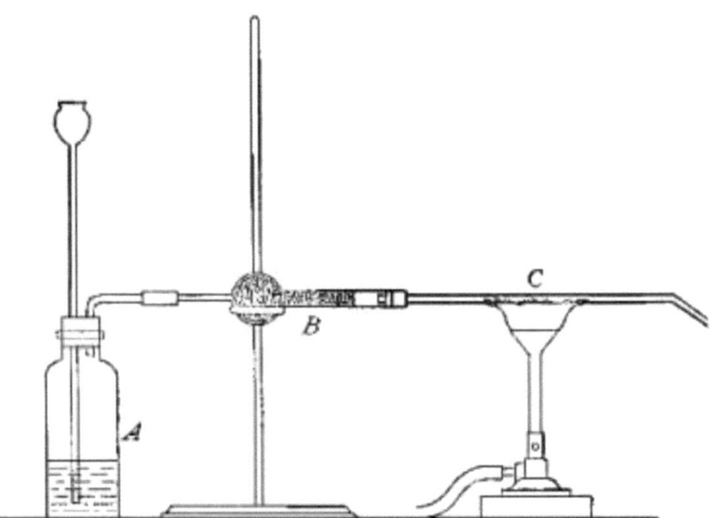

Fig. 15

Reduction. On account of its great affinity for oxygen, hydrogen has the power of abstracting it from many of its compounds. Thus, if a stream of hydrogen, dried by passing through the tube B (Fig. 15), filled with [Pg 37] calcium chloride, is conducted through the tube C containing some copper oxide, heated to a moderate temperature, the hydrogen abstracts the oxygen from the copper oxide. The change may be represented as follows:

hydrogen + {copper} {hydrogen}
{oxygen}(copper oxide) = {oxygen }(water) + copper

The water formed collects in the cold portions of the tube C near its end. In this experiment the copper oxide is said to undergo reduction. *Reduction may therefore be defined as the process of withdrawing oxygen from a compound.*

Relation of reduction to oxidation. At the same time that the copper oxide is reduced it is clear that the hydrogen is oxidized, for it combines with the oxygen given up by the copper oxide. The two processes are therefore very closely related, and it usually happens

that when one substance is oxidized some other substance is reduced. That substance which gives up its oxygen is called an *oxidizing agent,* while the substance which unites with the oxygen is called
a *reducing agent.*

The oxyhydrogen blowpipe. This is a form of apparatus used for
burning hydrogen in pure oxygen. As has been previously stated,
the flame produced by the combustion of hydrogen in the air is very
hot. It is evident that if pure oxygen is substituted for air, the temperature reached will be much higher, since there are no inert gases
to absorb the heat. The oxyhydrogen blowpipe, used to effect this
combination, consists of a small tube placed within a larger one, as
shown in Fig. 16.

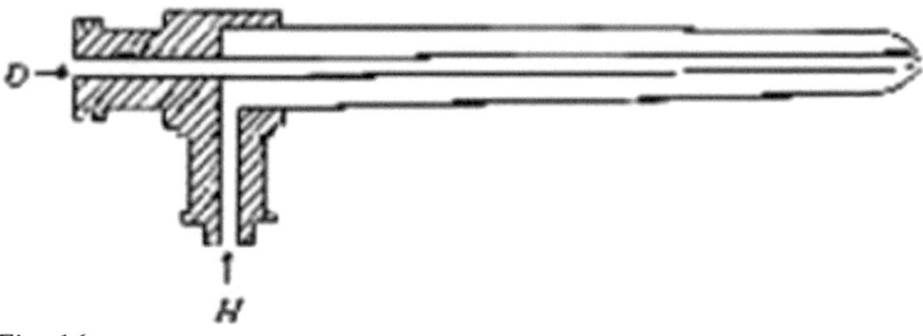

Fig. 16

[Pg 38]

The hydrogen, stored under pressure, generally in steel cylinders,
is first passed through the outer tube and ignited at the open end of
the tube. The oxygen from a similar cylinder is then conducted
through the inner tube, and mixes with the hydrogen at the end of
the tube. In order to produce the maximum heat, the hydrogen and
oxygen must be admitted to the blowpipe in the exact proportion in
which they combine, viz., 2 volumes of hydrogen to 1 of oxygen, or
by weight, 1 part of hydrogen to 7.94 parts of oxygen. The intensity
of the heat may be shown by bringing into the flame pieces of metal
such as iron wire or zinc. These burn with great brilliancy. Even
platinum, having a melting point of 1779°, may be melted by the
heat of the flame.

While the oxyhydrogen flame is intensely hot, it is almost non-
luminous. If directed against some infusible substance like ordinary

lime (calcium oxide), the heat is so intense that the lime becomes incandescent and glows with a brilliant light. This is sometimes used as a source of light, under the name of *Drummond* or *lime light*.

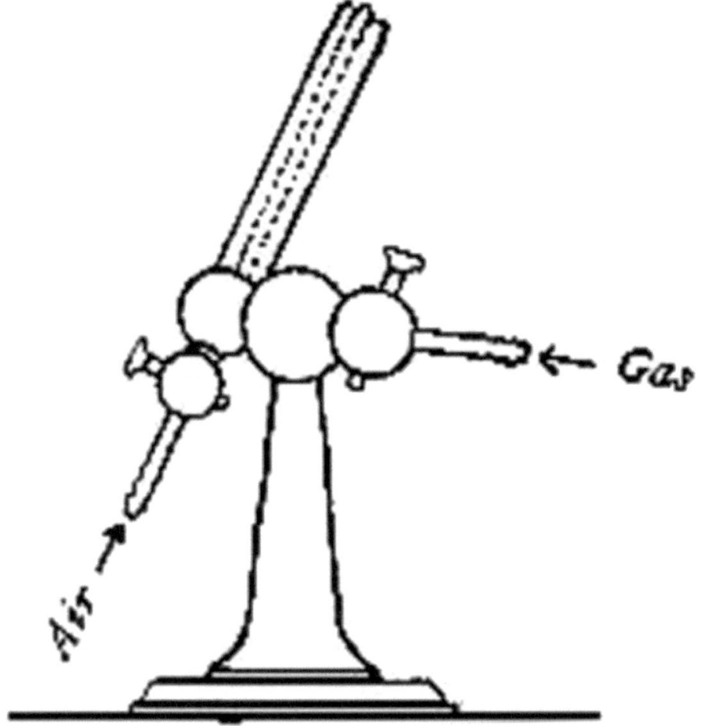

Fig. 17

The blast lamp. A similar form of apparatus is commonly used in the laboratory as a source of heat under the name *blast lamp* (Fig. 17). This differs from the oxyhydrogen blowpipe only in the size of the tubes. In place of the hydrogen and oxygen the more accessible coal gas and air are respectively used. The former is composed largely of a mixture of free hydrogen and gaseous compounds of carbon and hydrogen. While the temperature of the flame is not so high as that of the oxyhydrogen blowpipe, it nevertheless suffices for most chemical operations carried out in the laboratory.

Uses of hydrogen. On account of its cost, hydrogen is but little used for commercial purposes. It is sometimes used as a material for the inflation of balloons, but usually the much cheaper coal gas is substituted for it. Even hot air is often used when the duration of ascension is very short. It has been used also as a source of heat and light [Pg 39] in the oxyhydrogen blowpipe. Where the electric current is available, however, this form of apparatus has been displaced almost entirely by the electric light and electric furnace, which are much more economical and more powerful sources of light and heat.

EXERCISES

1. Will a definite weight of iron decompose an unlimited weight of steam?

2. Why is oxygen passed through the inner tube of the oxyhydrogen blowpipe rather than the outer?

3. In Fig. 14, will the flame remain at the mouth of the tube?

4. From Fig. 15, suggest a way for determining experimentally the quantity of water formed in the reaction.

5. Distinguish clearly between the following terms: oxidation, reduction, combustion, and kindling temperature.

6. Is oxidation always accompanied by reduction?

7. What is the source of heat in the lime light? What is the exact use of lime in this instrument?

8. In Fig. 12, why is it necessary to dry the hydrogen by means of the calcium chloride in the tube X?

9. At what pressure would the weight of 1 l. of hydrogen be equal to that of oxygen under standard conditions?

10. (*a*) What weight of hydrogen can be obtained from 150 g. of sulphuric acid? (*b*) What volume would this occupy under standard conditions? (*c*) The density of sulphuric acid is 1.84. What volume would the 150 g. of the acid occupy?

11. How many liters of hydrogen can be obtained from 50 cc. of sulphuric acid having a density of 1.84?

12. Suppose you wish to fill five liter bottles with hydrogen, the gas to be collected over water in your laboratory, how many cubic centimeters of sulphuric acid would be required?

[Pg 40]

CHAPTER IV

COMPOUNDS OF HYDROGEN AND OXYGEN; WATER AND HYDROGEN DIOXIDE

WATER

Historical. Water was long regarded as an element. In 1781 Cavendish showed that it is formed by the union of hydrogen and oxygen. Being a believer in the phlogiston theory, however, he failed to interpret his results correctly. A few years later Lavoisier repeated Cavendish's experiments and showed that water must be regarded as a compound of hydrogen and oxygen.

General methods employed for the determination of the composition of a compound. The composition of a compound may be determined by either of two general processes these are known as *analysis* and *synthesis*.

1. *Analysis* is the process of decomposing a compound into its constituents and determining what these constituents are. The analysis is *qualitative* when it results in merely determining what elements compose the compound; it is *quantitative* when the exact percentage of each constituent is determined. Qualitative analysis must therefore precede quantitative analysis, for it must be known what elements, are in a compound before a method can be devised for determining exactly how much of each is present.

2. *Synthesis* is the process of forming a compound from its constituent parts. It is therefore the reverse of analysis. Like analysis, it may be either qualitative or quantitative. [Pg 41]

Application of these methods to the determination of the composition of water. The determination of the composition of water is a matter of great interest not only because of the importance of the compound but also because the methods employed illustrate the general methods of analysis and synthesis.

Methods based on analysis. The methods based on analysis may be either qualitative or quantitative in character.

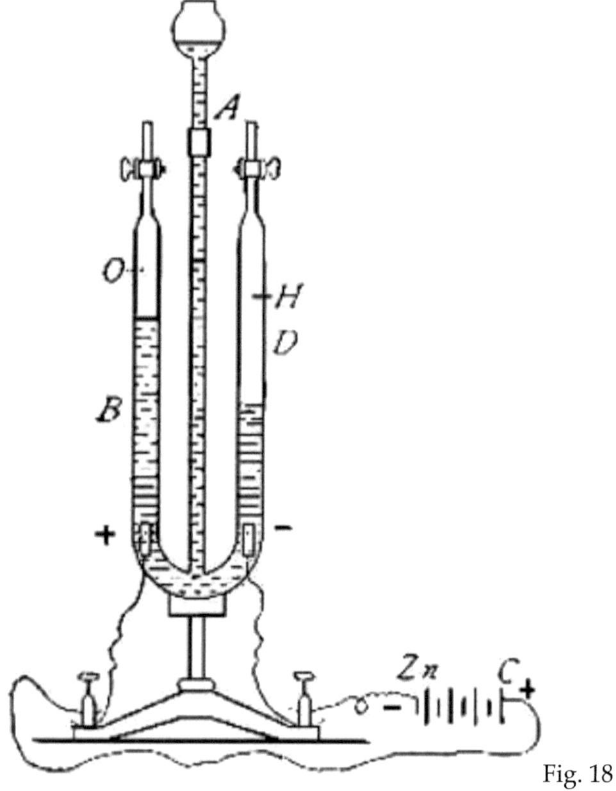

Fig. 18

1. *Qualitative analysis.* As was stated in the study of oxygen, water may be separated into its component parts by means of the electric current. The form of apparatus ordinarily used for effecting this analysis is shown in Fig. 18. A platinum wire, to the end of which is attached a small piece of platinum foil (about 15 mm. by 25 mm.), is fused through each of the tubes *B* and *D*, as shown in the figure. The stopcocks at the ends of these tubes are opened and water, to which has been added about one tenth of its volume of sulphuric

acid, is poured into the tube *A* until the side tubes *B* and *D* are completely filled. The stopcocks are then closed. The platinum wires extending into the tubes *B* and *D* are now connected with the wires leading from two or three dichromate cells joined in series. The pieces of platinum foil within the tubes thus become the electrodes, and the current flows from one to the other through the acidulated water. As soon as the current passes, bubbles of gas rise from each of the electrodes and collect in the upper part of the tubes. The gas [Pg 42] rising from the negative electrode is found to be hydrogen, while that from the positive electrode is oxygen. It will be seen that the volume of the hydrogen is approximately double that of the oxygen. Oxygen is more soluble in water than hydrogen, and a very little of it is also lost by being converted into ozone and other substances. It has been found that when the necessary corrections are made for the error due to these facts, the volume of the hydrogen is exactly double that of the oxygen.

Fig. 19 illustrates a simpler form of apparatus, which may be used in place of that shown in Fig. 18. A glass or porcelain dish is partially filled with water to which has been added the proper amount of acid. Two tubes filled with the same liquid are inverted over the electrodes. The gases resulting from the decomposition of the water collect in the tubes.

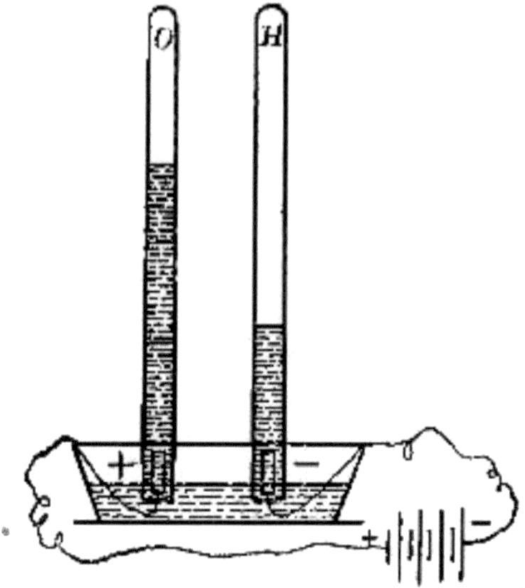

Fig. 19

2. *Quantitative analysis.* The analysis just described is purely qualitative and simply shows that water contains hydrogen and oxygen. It does not prove the absence of other elements; indeed it does not prove that the hydrogen and oxygen are present in the proportion in which they are liberated by the electric current. The method may be made quantitative, however, by weighing the water decomposed and also the hydrogen and oxygen obtained in its decomposition. If the combined weights of the hydrogen and oxygen exactly equal the weight of the water decomposed, then it would [Pg 43] be proved that the water consists of hydrogen and oxygen in the proportion in which they are liberated by the electric current. This experiment is difficult to carry out, however, so that the more accurate methods based on synthesis are used.

Methods based on synthesis. Two steps are necessary to ascertain the exact composition of water by synthesis: (1) to show by qualitative synthesis that water is formed by the union of oxygen with hydrogen; (2) to determine by quantitative synthesis in what proportion the two elements unite to form water. The fact that water is formed by the combination of oxygen with hydrogen was proved in the preceding chapter. The quantitative synthesis may be made as follows:

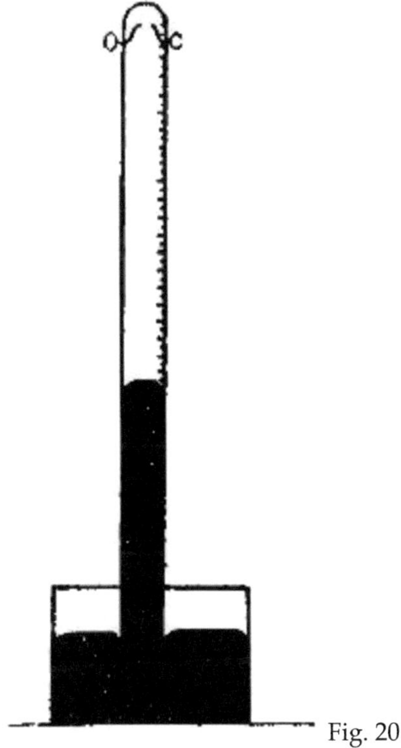

Fig. 20

The combination of the two gases is brought about in a tube called a eudiometer. This is a graduated tube about 60 cm. long and 2 cm. wide, closed at one end (Fig. 20). Near the closed end two platinum wires are fused through the glass, the ends of the wires within the tube being separated by a space of 2 mm or 3 mm. The tube is entirely filled with mercury and inverted in a vessel of the same liquid. Pure hydrogen is passed into the tube until it is about one fourth filled. The volume of the gas is then read off on the scale and reduced to standard conditions. Approximately an equal volume of pure oxygen is then introduced and the volume again read off and reduced to standard conditions. This gives the total volume of the two gases. From this the volume of the oxygen introduced may be determined by [Pg 44] subtracting from it the volume of the hydrogen. The combination of the two gases is now brought about by connecting the two platinum wires with an induction coil and passing a spark from one wire to the other. Immediately a slight explosion occurs. The mercury in the tube is at first depressed because of the expansion of the gases due to the heat generated, but at once rebounds, taking the place of the gases which have combined to form water. The volume of the water in the liquid state is so small that it may be disregarded in the calculations. In order that the temperature of the residual gas and the mercury may become uniform, the apparatus is allowed to stand for a few minutes. The volume of the gas is then read off and reduced to standard conditions, so that it may be compared with the volumes of the hydrogen and oxygen originally taken. The residual gas is then tested in order to ascertain whether it is hydrogen or oxygen, experiments having proved that it is never a mixture of the two. From the information thus obtained the composition of the water may be calculated. Thus, suppose the readings were as follows:

Volume of hydrogen taken	20.3 cc.
Volume of hydrogen and oxygen	38.7
Volume of oxygen	18.4
Volume of gas left after combination has taken place (oxygen)	8.3

The 20.3 cc. of hydrogen have combined with 18.4 cc. minus 8.3 cc. (or 10.1 cc.) of oxygen; or approximately 2 volumes of hydrogen have combined with 1 of oxygen. Since oxygen is 15.88 times as heavy as hydrogen, the proportion by weight in which the two gases combine is 1 part of hydrogen to 7.94 of oxygen. [Pg 45]

Precaution. If the two gases are introduced into the eudiometer in the exact proportions in which they combine, after the combination has taken place the liquid will rise and completely fill the tube. Under these conditions, however, the tube is very likely to be broken by the sudden upward rush of the liquid. Hence in performing the experiment care is taken to introduce an excess of one of the gases.

A more convenient form of eudiometer. A form of eudiometer (Fig. 21) different from that shown on page 43 is sometimes used to avoid the calculations necessary in reducing the volumes of the gases to the same conditions of temperature and pressure in order to make comparisons. With this apparatus it is possible to take the readings of the volumes under the same conditions of temperature and pressure, and thus compare them directly. The apparatus (Fig. 21) is filled with mercury and the gases introduced into the tube A. The experiment is carried out as in the preceding one, except that before taking the reading of the gas volumes, mercury is either added to the tube B or withdrawn from it by means of the stopcock C, until it stands at exactly the same height in both tubes. The gas inclosed in tube A is then under atmospheric pressure; and since but a few minutes are required for performing the experiment, the conditions of temperature and pressure may be regarded as constant. Hence the volumes of the hydrogen and oxygen and of the residual gas may be read off from the tube and directly compared.

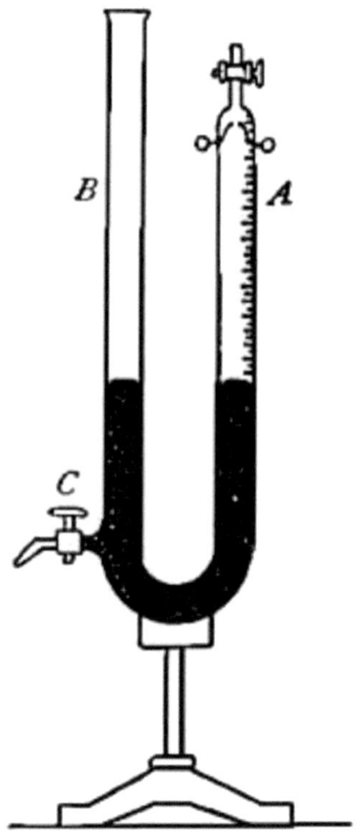

Fig. 21

Method used by Berzelius and Dumas. The method used by
these investigators enables us to determine directly the proportion
by weight in which the hydrogen and oxygen combine. Fig. 22 illustrates the apparatus used in making this determination. *B* is a glass
tube containing copper oxide. *C* and *D* are glass tubes filled with
calcium chloride, a substance which has great affinity for water. [Pg
46] The tubes *B* and *C*, including their contents, are carefully
weighed, and the apparatus connected as shown in the figure. A
slow current of pure hydrogen is then passed through *A*, and that
part of the tube *B* which contains copper oxide is carefully heated.
The hydrogen combines with the oxygen present in the copper oxide to form water, which is absorbed by the calcium chloride in tube
C. The calcium chloride in tube *D* prevents any moisture entering

tube C from the air. The operation is continued until an appreciable amount of water has been formed. The tubes B and C are then weighed once more. The loss of weight in the tube B will exactly equal the weight of oxygen taken up from the copper oxide in the formation of the water. The gain in weight in the tube C will exactly equal the weight of the water formed. The difference in these weights will of course equal the weight of the hydrogen present in the water formed.

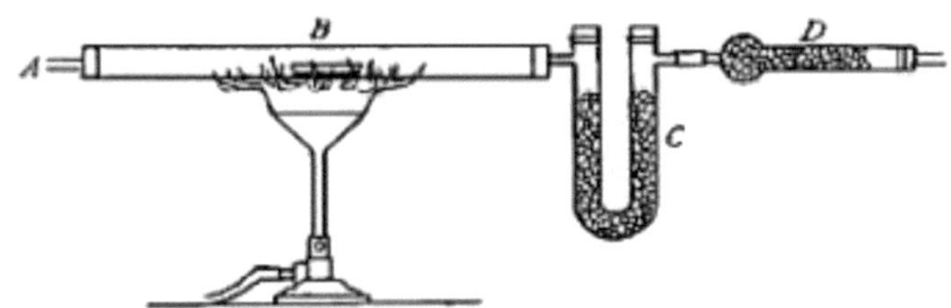

Fig. 22

Dumas' results. The above method for the determination of the composition of water was first used by Berzelius in 1820. The work was repeated in 1843 by Dumas, the average of whose results is as follows:

Weight of water formed	236.36 g.
Oxygen given up by the copper oxide	210.04
	— — —
Weight of hydrogen present in water	26.32

[Pg 47]

According to this experiment the ratio of hydrogen to oxygen in water is therefore 26.32 to 210.04, or as l to 7.98

Morley's results. The American chemist Morley has recently determined the composition of water, extreme precautions being taken to use pure materials and to eliminate all sources of error. The hydrogen and oxygen which combined, as well as the water formed, were all accurately weighed. According to Morley's results, 1 part of hydrogen by weight combines with 7.94 parts of oxygen to form water.

Comparison of results obtained. From the above discussions it is easy to see that it is by experiment alone that the composition of a compound can be determined. Different methods may lead to slightly different results. The more accurate the method chosen and the greater the skill with which the experiment is carried out, the more accurate will be the results. It is generally conceded by chemists that the results obtained by Morley in reference to the composition of water are the most accurate ones. In accordance with these results, then, *water must be regarded as a compound containing hydrogen and oxygen in the proportion of 1 part by weight of hydrogen to 7.94 parts by weight of oxygen.*

Relation between the volume of aqueous vapor and the volumes of the hydrogen and oxygen which combine to form it. When the quantitative synthesis of water is carried out in the eudiometer as described above, the water vapor formed by the union of the hydrogen and oxygen at once condenses. The volume of the resulting liquid is so small that it may be disregarded in making the calculations. If, however, the experiment is carried out at a temperature of 100° or above, the water-vapor formed is not condensed and it thus becomes possible to compare the volume of the [Pg 48] vapor with the volumes of hydrogen and oxygen which combined to form it. This can be accomplished by surrounding the arm *A* of the eudiometer (Fig. 23) with the tube *B* through which is passed the vapor obtained by boiling some liquid which has a boiling point above 100°. In this way it has been proved that 2 volumes of hydrogen and 1 volume of oxygen combine to form exactly 2 volumes of water vapor, the volumes all being measured under the same conditions of temperature and pressure. It will be noted that the relation between these volumes may be expressed by whole numbers. The significance of this very important fact will be discussed in a subsequent chapter.

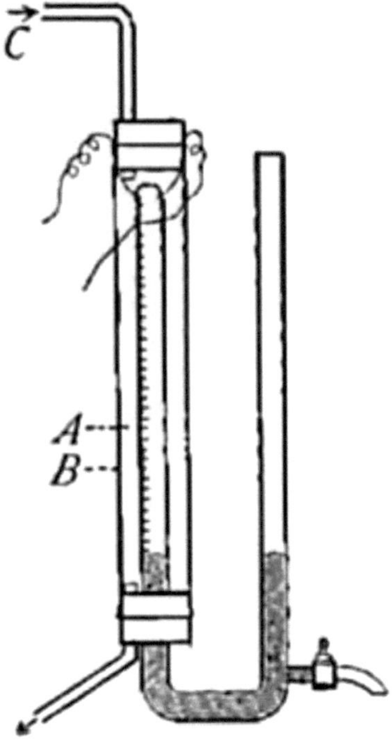

Fig. 23

Occurrence of water. Water not only covers about three fourths of the surface of the earth, and is present in the atmosphere in the form of moisture, but it is also a common constituent of the soil and rocks and of almost every form of animal and vegetable organism. The human body is nearly 70% water. This is derived not only from the water which we drink but also from the food which we eat, most of which contains a large percentage of water. Thus potatoes contain about 78% of water, milk 85%, beef over 50%, apples 84%, tomatoes 94%.

Impurities in water. Chemically pure water contains only hydrogen and oxygen. Such a water never occurs in nature, however, for being a good solvent, it takes up certain substances from the rocks and soil with which it comes in contact. When such waters are evaporated these [Pg 49] substances are deposited in the form of a residue. Even rain water, which is the purest form occurring in nature, contains dust particles and gases dissolved from the atmosphere. The foreign matter in water is of two kinds, namely, *mineral*, such as common salt and limestone, and *organic*, that is the products of animal and vegetable life.

Mineral matter in water. The amount and nature of the mineral matter present in different waters vary greatly, depending on the character of the rocks and soil with which the waters come in contact. The more common of the substances present are common salt and compounds of calcium, magnesium, and iron. One liter of the average river water contains about 175 mg. of mineral matter. Water from deep wells naturally contains more mineral matter than river water, generally two or three times as much, while sea water contains as much as 35,000 mg. to the liter.

Effect of impurities on health. The mineral matter in water does not, save in very exceptional cases, render the water injurious to the human system. In fact the presence of a certain amount of such matter is advantageous, supplying the mineral constituents necessary for the formation of the solid tissues of the body. The presence of organic matter, on the other hand, must always be regarded with suspicion. This organic matter may consist not only of the products of animal and vegetable life but also of certain microscopic forms of living organisms which are likely to accompany such products. Contagious diseases are known to be due to the presence in the body of minute living organisms or germs. Each disease is caused by its own particular kind of germ. Through sewage these germs may find their way from persons afflicted with disease into the water supply, and it is principally through the drinking water that certain of these diseases, especially typhoid fever, are spread. It becomes of great importance, therefore, to be [Pg 50] able to detect such matter when present in drinking water as well as to devise methods whereby it can be removed or at least rendered harmless.

Analysis of water. The mineral analysis of a water is, as the name suggests, simply the determination of the mineral matter present. Sanitary analysis, on the other hand, is the determination of the organic matter present. The physical properties of a water give no conclusive evidence as to its purity, since a water may be unfit for drinking purposes and yet be perfectly clear and odorless. Neither can any reliance be placed on the simple methods often given for testing the purity of water. Only the trained chemist can carry out such methods of analysis as can be relied upon.

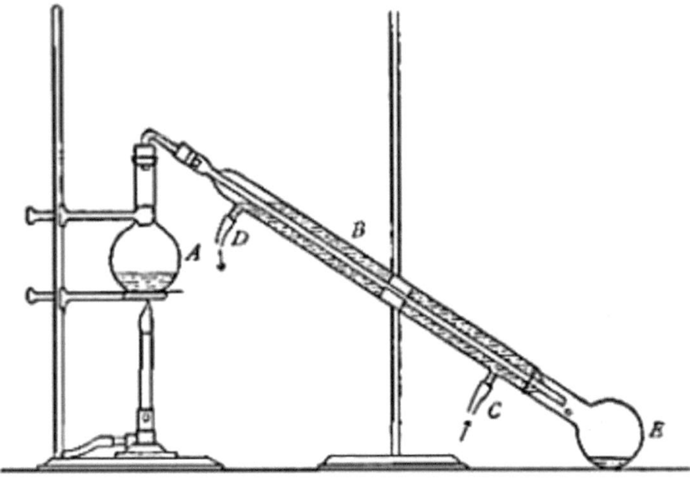

Fig. 24

Purification of water. Three general methods are used for the purification of water, namely, *distillation, filtration,* and *boiling.*

1. *Distillation.* The most effective way of purifying natural waters is by the process of distillation. This consists in boiling the water and condensing the steam. Fig. 24 illustrates the process of distillation, as commonly conducted [Pg 51] in the laboratory. Ordinary water is poured into the flask A and boiled. The steam is conducted through the condenser B, which consists essentially of a narrow glass tube sealed within a larger one, the space between the two being filled with cold water, which is admitted at C and escapes at D. The inner tube is thus kept cool and the steam in passing through it is condensed. The water formed by the condensation of the steam collects in the receiver E and is known as *distilled* water. Such water

is practically pure, since the impurities are nonvolatile and remain in the flask *A*.

Commercial distillation. In preparing distilled water on a large scale, the steam is generated in a boiler or other metal container and condensed by passing it through a pipe made of metal, generally tin. This pipe is wound into a spiral and is surrounded by a current of cold water. Distilled water is used by the chemist in almost all of his work. It is also used in the manufacture of artificial ice and for drinking water.

Fractional distillation. In preparing distilled water, it is evident that if the natural water contains some substance which is volatile its vapor will pass over and be condensed with the steam, so that the distillate will not be pure water. Even such mixtures, however, may generally be separated by repeated distillation. Thus, if a mixture of water (boiling point 100°) and alcohol (boiling point 78°) is distilled, the alcohol, having the lower boiling point, tends to distill first, followed by the water. The separation of the two is not perfect, however, but may be made nearly so by repeated distillations. The process of separating a mixture of volatile substances by distillation is known as *fractional distillation.*

2. *Filtration.* The process of distillation practically removes all nonvolatile foreign matter, mineral as well as organic. In purifying water for drinking purposes, however, it is only necessary to eliminate the latter or to render it harmless. This is ordinarily done either by filtration or [Pg 52] boiling. In filtration the water is passed through some medium which will retain the organic matter. Ordinary charcoal is a porous substance and will condense within its pores the organic matter in water if brought in contact with it. It is therefore well adapted to the construction of filters. Such filters to be effective must be kept clean, since it is evident that the charcoal is useless after its pores are filled. A more effective type of filter is the Chamberlain-Pasteur filter. In this the water is forced through a porous cylindrical cup, the pores being so minute as to strain out the organic matter.

City filtration beds. For purifying the water supply of cities, large filtration beds are prepared from sand and gravel, and the water is allowed to filter through these. Some of the impurities are strained out by the filter, while others are decomposed by the action of certain kinds of bacteria present in the sand. Fig. 25 shows a cross section of a portion of the filter used in purifying the water supply of Philadelphia. The water filters through the sand and gravel and passes into the porous pipe *A*, from which it is pumped into the city mains. The filters are covered to prevent the water from freezing in cold weather.

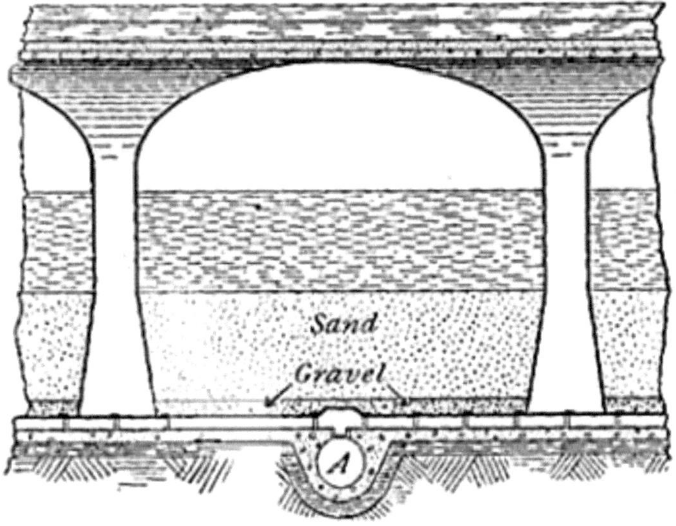

Fig. 25

3. *Boiling.* A simpler and equally efficient method for purifying water for drinking purposes consists in boiling the water. It is the germs in water that render it dangerous to health. These germs are living forms of matter. If the [Pg 53] water is boiled, the germs are killed and the water rendered safe. While these germs are destroyed by heat, cold has little effect upon them. Thus Dewar, in working with liquid hydrogen, exposed some of these minute forms of life to the temperature of boiling hydrogen (-252°) without killing them.

Self-purification of water. It has long been known that water contaminated with organic matter tends to purify itself when exposed to the air. This is due to the fact that the water takes up a small amount of oxygen from the air, which gradually oxidizes the organic matter present in the water. While water is undoubtedly purified in this way, the method cannot be relied upon to purify a contaminated water so as to render it safe for drinking purposes.

Physical properties. Pure water is an odorless and tasteless liquid, colorless in thin layers, but having a bluish tinge when observed through a considerable thickness. It solidifies at 0° and boils at 100° under the normal pressure of one atmosphere. If the pressure is increased, the boiling point is raised. When water is cooled it steadily contracts until the temperature of 4° is reached: it then expands. Water is remarkable for its ability to dissolve other substances, and is the best solvent known. Solutions of solids in water are more frequently employed in chemical work than are the solid substances, for chemical action between substances goes on more readily when they are in solution than it does when they are in the solid state.

Chemical properties. Water is a very stable substance, or, in other words, it does not undergo decomposition readily. To decompose it into its elements by heat alone requires a very high temperature; at 2500°, for example, only about 5% of the entire amount is decomposed. Though very [Pg 54] stable towards heat, water can be decomposed in other ways, as by the action of the electrical current or by certain metals.

Heat of formation and heat of decomposition are equal. The fact that a very high temperature is necessary to decompose water into hydrogen and oxygen is in accord with the fact that a great deal of heat is evolved by the union of hydrogen and oxygen; for it has been proved that the heat necessary to decompose a compound into its elements (heat of decomposition) is equal to the heat evolved in the formation of a compound from its elements (heat of formation).

Water of crystallization. When a solid is dissolved in water and the resulting solution is allowed to evaporate, the solid separates out, often in the form of crystals. It has been found that the crystals of many compounds, although perfectly dry, give up a definite amount of water when heated, the substance at the same time losing its crystalline form. Such water is called *water of crystallization.* This varies in amount with different compounds, but is perfectly definite for the same compound. Thus, if a perfectly dry crystal of copper sulphate is strongly heated in a tube, water is evolved and condenses on the sides of the tube, the crystal crumbling to a light powder. The weight of the water evolved is always equal to exactly 36.07% of the weight of copper sulphate crystals heated. The water must therefore be in chemical combination with the substance composing the crystal; for if simply mixed with it or adhering to it, not only would the substance appear moist but the amount present would undoubtedly vary. The combination, however, must be a very weak one, since the water is often expelled by even a gentle heat. Indeed, in some cases the water is given up on simple exposure to air. Such compounds are said to be *efflorescent*. Thus a crystal of sodium sulphate [Pg 55] (Glauber's salt) on exposure to air crumbles to a fine powder, owing to the escape of its water of crystallization. Other substances have just the opposite property: they absorb moisture when exposed to the air. For example, if a bit of dry calcium chloride is placed in moist air, in the course of a few hours it will have absorbed sufficient moisture to dissolve it. Such substances are said to be *deliquescent*. A deliquescent body serves as a good drying or *desiccating* agent. We have already employed calcium chloride as an agent for absorbing the moisture from hydrogen. Many substances, as for example quartz, form crystals which contain no water of crystallization.

Mechanically inclosed water. Water of crystallization must be carefully distinguished from water which is mechanically inclosed in a crystal and which can be removed by powdering the crystal and drying. Thus, when crystals of common salt are heated, the water inclosed in the crystal is changed into steam and bursts the crystal with a crackling sound. Such crystals are said to *decrepitate*. That this water is not combined is proved by the fact that the amount present varies and that it has all the properties of water.

Uses of water. The importance of water in its relation to life and commerce is too well known to require comment. Its importance to the chemist has also been pointed out. It remains to call attention to the fact that it is used as a standard in many physical measurements. Thus 0° and 100° on the centigrade scale are respectively the freezing and the boiling points of water under normal pressure. The weight of 1 cc. of water at its point of greatest density is the unit of weight in the metric system, namely, the gram. It is also taken as the unit for the determination of the density of liquids and solids as well as for the measurement of amounts of heat. [Pg 56]

HYDROGEN DIOXIDE

Composition. As has been shown, 1 part by weight of hydrogen combines with 7.94 parts by weight of oxygen to form water. It is possible, however, to obtain a second compound of hydrogen and oxygen differing from water in composition in that 1 part by weight of hydrogen is combined with 2 × 7.94, or 15.88 parts, of oxygen. This compound is called *hydrogen dioxide* or *hydrogen peroxide*, the prefixes *di-* and *per-* signifying that it contains more oxygen than hydrogen oxide, which is the chemical name for water.

Preparation. Hydrogen dioxide cannot be prepared cheaply by the direct union of hydrogen and oxygen, and indirect methods must therefore be used. It is commonly prepared by the action of a solution of sulphuric acid on barium dioxide. The change which takes place may be indicated as follows:

sulphuric acid	+ barium dioxide	= barium sulphate	+ hydrogen dioxide
- - - - - - -	- - - - - - -	- - - - - - -	- - - - - - -
hydrogen	barium	barium	hydrogen
sulphur	oxygen	sulphur	oxygen
oxygen	oxygen		

In other words, the barium and hydrogen in the two compounds exchange places. By this method a dilute solution of the dioxide in water is obtained. It is possible to separate the dioxide from the water by fractional distillation. This is attended with great difficulties, however, since the pure dioxide is explosive. The distillation is carried on under diminished pressure so as to lower the boiling points as much as possible; otherwise the high temperature would decompose the dioxide. [Pg 57]

Properties. Pure hydrogen dioxide is a colorless sirupy liquid having a density of 1.49. Its most characteristic property is the ease with which it decomposes into water and oxygen. One part by weight of hydrogen is capable of holding firmly only 7.94 parts of oxygen. The additional 7.94 parts of oxygen present in hydrogen dioxide are therefore easily evolved, the compound breaking down into water and oxygen. This decomposition is attended by the generation of considerable heat. In dilute solution hydrogen dioxide is fairly stable, although such a solution should be kept in a dark, cool place, since both heat and light aid in the decomposition of the dioxide.

Uses. Solutions of hydrogen dioxide are used largely as oxidizing agents. The solution sold by druggists contains 3% of the dioxide and is used in medicine as an antiseptic. Its use as an antiseptic depends upon its oxidizing properties.

EXERCISES

1. Why does the chemist use distilled water in making solutions, rather than filtered water?

2. How could you determine the total amount of solid matter dissolved in a sample of water?

3. How could you determine whether a given sample of water is distilled water?

4. How could the presence of air dissolved in water be detected?

5. How could the amount of water in a food such as bread or potato be determined?

6. Would ice frozen from impure water necessarily be free from disease germs?

7. Suppose that the maximum density of water were at 0° in place of 4°; what effect would this have on the formation of ice on bodies of water?

8. Is it possible for a substance to contain both mechanically inclosed water and water of crystallization? [Pg 58]

9. If steam is heated to 2000° and again cooled, has any chemical change taken place in the steam?

10. Why is cold water passed into *C* instead of *D* (Fig. 24)?

11. Mention at least two advantages that a metal condenser has over a glass condenser.

12. Draw a diagram of the apparatus used in your laboratory for supplying distilled water.

13. 20 cc. of hydrogen and 7 cc. of oxygen are placed in a eudiometer and the mixture exploded. (*a*) How many cubic centimeters of aqueous vapor are formed? (*b*) What gas and how much of it remains in excess?

14. (*a*) What weight of water can be formed by the combustion of 100 L of hydrogen, measured under standard conditions? (*b*)What volume of oxygen would be required in (*a*)? (*c*)What weight of potassium chlorate is necessary to prepare this amount of oxygen?

15. What weight of oxygen is present in 1 kg. of the ordinary hydrogen dioxide solution? In the decomposition of this weight of the dioxide into water and oxygen, what volume of oxygen (measured under standard conditions) is evolved?

[Pg 59]

CHAPTER V

THE ATOMIC THEORY

Three fundamental laws of matter. Before we can gain any very definite idea in regard to the structure of matter, and the way in which different kinds of substances act chemically upon each other, it is necessary to have clearly in view three fundamental laws of matter. These laws have been established by experiment, and any conception which may be formed concerning matter must therefore be in harmony with them. The laws are as follows:

Law of conservation of matter. This law has already been touched upon in the introductory chapter, and needs no further discussion. It will be recalled that it may be stated thus: *Matter can neither be created nor destroyed, though it can be changed from one form into another.*

Law of definite composition. In the earlier days of chemistry there was much discussion as to whether the composition of a given compound is always precisely the same or whether it is subject to some variation. Two Frenchmen, Berthollet and Proust, were the leaders in this discussion, and a great deal of most useful experimenting was done to decide the question. Their experiments, as well as all succeeding ones, have shown that the composition of a pure chemical compound is always exactly the same. Water obtained by melting pure ice, condensing steam, burning hydrogen in oxygen, has always 11.18% hydrogen and 88.82% oxygen in it. Red oxide of mercury, from whatever source it is obtained, contains 92.6% [Pg 60] mercury and 7.4% oxygen. This truth is known as *the law of definite composition*, and may be stated thus: *The composition of a chemical compound never varies.*

Law of multiple proportion. It has already been noted, however, that hydrogen and oxygen combine in two different ratios to form water and hydrogen dioxide respectively. It will be observed that this fact does not contradict the law of definite composition, for entirely different substances are formed. These compounds differ from each other in composition, but the composition of each one is always constant. This ability of two elements to unite in more than one ratio is very frequently observed. Carbon and oxygen combine in two different ratios; nitrogen and oxygen combine to form as many as five distinct compounds, each with its own precise composition.

In the first decade of the last century John Dalton, an English school-teacher and philosopher, endeavored to find some rule which holds between the ratios in which two given substances combine. His studies brought to light a very simple relation, which the following examples will make clear. In water the hydrogen and oxygen are combined in the ratio of 1 part by weight of hydrogen to 7.94 parts by weight of oxygen. In hydrogen dioxide the 1 part by weight of hydrogen is combined with 15.88 parts by weight of oxygen. The ratio between the amounts of oxygen which combine with the same amount of hydrogen to form water and hydrogen dioxide respectively is therefore 7.94: 15.88, or 1: 2.

JOHN DALTON (English) (1766-1844)

Developed the atomic theory; made many studies on the properties and the composition of gases. His book entitled "A New System of Chemical Philosophy" had a large influence on the development of chemistry

[Pg 61]

Similarly, the element iron combines with oxygen to form two oxides, one of which is black and the other red. By analysis it has been shown that the former contains 1 part by weight of iron combined with 0.286 parts by weight of oxygen, while the latter contains 1 part by weight of iron combined with 0.429 parts by weight of oxygen. Here again we find that the amounts of oxygen which combine with the same fixed amount of iron to form the two compounds are in the ratio of small whole numbers, viz., 2:3.

Many other examples of this simple relation might be given, since it has been found to hold true in all cases where more than one compound is, formed from the same elements. Dalton's law of multiple proportion states these facts as follows: *When any two elements, A and B, combine to form more than one compound, the amounts of B which unite with any fixed amount of A bear the ratio of small whole numbers to each other.*

Hypothesis necessary to explain the laws of matter. These three generalizations are called *laws*, because they express in concise language truths which are found by careful experiment to hold good in all cases. They do not offer any explanation of the facts, but merely state them. The human mind, however, does not rest content with the mere bare facts, but seeks ever to learn the explanation of the facts. A suggestion which is offered to explain such a set of facts is called an *hypothesis*. The suggestion which Dalton offered to explain the three laws of matter, called the *atomic hypothesis*, was prompted by his view of the constitution of matter, and it involves three distinct assumptions in regard to the nature of matter and chemical action. Dalton could not prove these assumptions to be true, but he saw that if they were true the laws of matter become very easy to understand.

Dalton's atomic hypothesis. The three assumptions which Dalton made in regard to the nature of matter, and which together constitute the atomic hypothesis, are these: [Pg 62]

1. All elements are made up of minute, independent particles which Dalton designated as *atoms*.

2. All atoms of the same element have equal masses; those of different elements have different masses; in any change to which an atom is subjected its mass does not change.

3. When two or more elements unite to form a compound, the action consists in the union of a definite small number of atoms of each element to form a small particle of the compound. The smallest particles of a given compound are therefore exactly alike in the number and kinds of atoms which they contain, and larger masses of the substances are simply aggregations of these least particles.

Molecules and atoms. Dalton applied the name atom not only to the minute particles of the elements but also to the least particles of compounds. Later Avogadro, an Italian scientist, pointed out the fact that the two are different, since the smallest particle of an element is a unit, while that of a compound must have at least two units in it. He suggested the name *molecule* for the least particle of a compound which can exist, retaining the name *atom* for the smallest particle of an element. In accordance with this distinction, we may define the atom and the molecule as follows: *An atom is the smallest particle of an element which can exist. A molecule is the smallest particle of a compound which can exist.* It will be shown in a subsequent chapter that sometimes two or more atoms of the same element unite with each other to form molecules of the element. While the term atom, therefore, is applicable only to elements, the term molecule is applicable both to elements and compounds. [Pg 63]

The atomic hypothesis and the laws of matter. Supposing the atomic hypothesis to be true, let us now see if it is in harmony with the laws of matter.

1. *The atomic hypothesis and the law of conservation of matter.* It is evident that if the atoms never change their masses in any change which they undergo, the total quantity of matter can never change and the law of conservation of matter must follow.

2. *The atomic hypothesis and the law of definite composition.* According to the third supposition, when iron combines with sulphur the union is between definite numbers of the two kinds of atoms. In the simplest case one atom of the one element combines with one atom of the other. If the sulphur and the iron atoms never change their respective masses when they unite to form a molecule of iron sulphide, all iron sulphide molecules will have equal amounts of iron in them and also of sulphur. Consequently any mass made up of iron sulphide molecules will have the same fraction of iron by

weight as do the individual iron sulphide molecules. Iron sulphide, from whatever source, will have the same composition, which is in accordance with the law of definite composition.

3. *The atomic hypothesis and the law of multiple proportion.* But this simplest case may not always be the only one. Under other conditions one atom of iron might combine with two of sulphur to form a molecule of a second compound. In such a case the one atom of iron would be in combination with twice the mass of sulphur that is in the first compound, since the sulphur atoms all have equal masses. What is true for one molecule will be true for any number of them; consequently when such quantities of these two compounds are selected as are found to contain [Pg 64] the same amount of iron, the one will contain twice as much sulphur as the other.

The combination between the atoms may of course take place in other simple ratios. For example, two atoms of one element might combine with three or with five of the other. In all such cases it is clear that the law of multiple proportion must hold true. For on selecting such numbers of the two kinds of molecules as have the same number of the one kind of atoms, the numbers of the other kind of atoms will stand in some simple ratio to each other, and their weights will therefore stand in the same simple ratio.

Testing the hypothesis. Efforts have been made to find compounds which do not conform to these laws, but all such attempts have resulted in failure. If such compounds should be found, the laws would be no longer true, and the hypothesis of Dalton would cease to possess value. When an hypothesis has been tested in every way in which experiment can test it, and is still found to be in harmony with the facts in the case, it is termed a *theory*. We now speak of the atomic theory rather than of the atomic hypothesis.

Value of a theory. The value of a theory is twofold. It aids in the clear understanding of the laws of nature because it gives an intelligent idea as to why these laws should be in operation.

A theory also leads to discoveries. It usually happens that in testing a theory much valuable work is done, and many new facts are discovered. Almost any theory in explaining given laws will involve a number of consequences apart from the laws it seeks to explain. Experiment will soon show whether these facts are as the theory predicts they will be. Thus Dalton's atomic theory predicted many properties of gases which experiment has since verified. [Pg 65]

Atomic weights. It would be of great advantage in the study of chemistry if we could determine the weights of the different kinds of atoms. It is evident that this cannot be done directly. They are so small that they cannot be seen even with a most powerful microscope. It is calculated that it would take 200,000,000 hydrogen atoms placed side by side to make a row one centimeter long. No balance can weigh such minute objects. It is possible, however, to determine their relative weights, — that is, how much heavier one is than another. *These relative weights of the atoms are spoken of as the atomic weights of the elements.*

If elements were able to combine in only one way, — one atom of one with one atom of another, — the problem of determining the atomic weights would be very simple. We should merely have to take some one convenient element as a standard, and find by experiment how much of each other element would combine with a fixed weight of it. The ratios thus found would be the same ratios as those between the atoms of the elements, and thus we should have their relative atomic weights. The law of multiple proportion calls attention to the fact that the atoms combine in other ratios than 1: 1, and there is no direct way of telling which one, if any, of the several compounds in a given case is the one consisting of a single atom of each element.

If some way were to be found of telling how much heavier the entire molecule of a compound is than the atom chosen as a standard, — that is, of determining the molecular weights of compounds, — the problem could be solved, though its solution would not be an entirely simple matter. There are ways of determining the molecular weights of [Pg 66] compounds, and there are other experiments which throw light directly upon the relative weights of the atoms. These methods cannot be described until the facts upon which they rest have been studied. It will be sufficient for the present to assume that these methods are trustworthy.

Standard for atomic weights. Since the atomic weights are merely relative to some one element chosen as a standard, it is evident that any one of the elements may serve as this standard and that

84

any convenient value may be assigned to its atom. At one time oxygen was taken as this standard, with the value 100, and the atomic weights of the other elements were expressed in terms of this standard. It would seem more rational to take the element of smallest atomic weight as the standard and give it unit value; accordingly hydrogen was taken as the standard with an atomic weight of 1. Very recently, however, this unit has been replaced by oxygen, with an atomic weight of 16.

Why oxygen is chosen as the standard for atomic weights. In the determination of the atomic weight of an element it is necessary to find the weight of the element which combines with a definite weight of another element, preferably the element chosen as the standard. Since oxygen combines with the elements far more readily than does hydrogen to form definite compounds, it is far better adapted for the standard element, and has accordingly replaced hydrogen as the standard. Any definite value might be given to the weight of the oxygen atom. In assigning a value to it, however, it is convenient to choose a whole number, and as small a number as possible without making the atomic weight of any other element less than unity. For these reasons the number 16 has been chosen as the atomic [Pg 67] weight of oxygen. This makes the atomic weight of hydrogen equal to 1.008, so that there is but little difference between taking oxygen as 16 and hydrogen as 1 for the unit.

The atomic weights of the elements are given in the Appendix.

EXERCISES

1. Two compounds were found to have the following compositions: (*a*) oxygen = 69.53%, nitrogen = 30.47%; (*b*) oxygen = 53.27%, nitrogen = 46.73%. Show that the law of multiple proportion holds in this case.

2. Two compounds were found to have the following compositions: (*a*) oxygen = 43.64%, phosphorus = 56.36%; (*b*) oxygen = 56.35%, phosphorus = 43.65%. Show that the law of multiple proportion holds in this case.

3. Why did Dalton assume that all the atoms of a given element have the same weight?

CHAPTER VI

CHEMICAL EQUATIONS AND CALCULATIONS

Formulas. Since the molecule of any chemical compound consists of a definite number of atoms, and this number never changes without destroying the identity of the compound, it is very convenient to represent the composition of a compound by indicating the composition of its molecules. This can be done very easily by using the symbols of the atoms to indicate the number and the kind of the atoms which constitute the molecule. HgO will in this way represent mercuric oxide, a molecule of which has been found to contain 1 atom each of mercury and oxygen. H_2O will represent water, the molecules of which consist of 1 atom of oxygen and 2 of hydrogen, the subscript figure indicating the number of the atoms of the element whose symbol precedes it. H_2SO_4 will stand for sulphuric acid, the molecules of which contain 2 atoms of hydrogen, 1 of sulphur, and 4 of oxygen. The combination of symbols which represents the molecule of a substance is called its *formula*.

Equations. When a given substance undergoes a chemical change it is possible to represent this change by the use of such symbols and formulas. In a former chapter it was shown that mercuric oxide decomposes when heated to form mercury and oxygen. This may be expressed very briefly in the form of the equation

$$(1) \quad HgO = Hg + O.$$

When water is electrolyzed two new substances, hydrogen and oxygen, are formed from it. This statement in the form of an equation is

$$(2) \quad H_2O = 2H + O.$$

The coefficient before the symbol for hydrogen indicates that a single molecule of water yields two atoms of hydrogen on decomposition.

In like manner the combination of sulphur with iron is expressed by the equation

(3) $Fe + S = FeS$.

The decomposition of potassium chlorate by heat takes place as represented by the equation

(4) $KClO_3 = KCl + 3O$.

Reading of equations. Since equations are simply a kind of shorthand way of indicating chemical changes which occur under certain conditions, in reading an equation the full statement for which it stands should be given. Equation (1) should be read, "Mercuric oxide when heated gives mercury and oxygen"; equation (2) is equivalent to the statement, "When electrolyzed, water produces hydrogen and oxygen"; equation (3), "When heated together iron and sulphur unite to form iron sulphide"; equation (4), "Potassium chlorate when heated yields potassium chloride and oxygen."

Knowledge required for writing equations. In order to write such equations correctly, a considerable amount of exact knowledge is required. Thus, in equation (1) the fact that red oxide of mercury has the composition represented by the formula HgO, that it is decomposed by heat, that in this decomposition mercury and oxygen are formed and [Pg 70] no other products,—all these facts must be ascertained by exact experiment before the equation can be written. An equation expressing these facts will then have much value.

Having obtained an equation describing the conduct of mercuric oxide on being heated, it will not do to assume that other oxides will behave in like manner. Iron oxide (FeO) resembles mercuric oxide in many respects, but it undergoes no change at all when heated. Manganese dioxide, the black substance used in the preparation of oxygen, has the formula MnO_2. When this substance is heated oxygen is set free, but the metal manganese is not liberated; instead, a different oxide of manganese containing less oxygen is produced. The equation representing the reaction is

$$3MnO_2 = Mn_3O_4 + 2O.$$

Classes of reactions. When a chemical change takes place in a substance the substance is said to undergo a reaction. Although a great many different reactions will be met in the study of chemistry, they may all be grouped under the following heads.

1. *Addition.* This is the simplest kind of chemical action. It consists in the union of two or more substances to produce a new substance. The combination of iron with sulphur is an example:

$$Fe + S = FeS.$$

2. *Decomposition.* This is the reverse of addition, the substance undergoing reaction being parted into its constituents. The decomposition of mercuric oxide is an example: $HgO = Hg + O$.

3. *Substitution.* It is sometimes possible for an element in the free state to act upon a compound in such a way that [Pg 71] it takes the place of one of the elements of the compound, liberating it in turn. In the study of the element hydrogen it was pointed out that hydrogen is most conveniently prepared by the action of sulphuric or hydrochloric acid upon zinc. When sulphuric acid is used a substance called zinc sulphate, having the composition represented by the formula $ZnSO_4$, is formed together with hydrogen. The equation is

$$Zn + H_2SO_4 = ZnSO_4 + 2H.$$

When hydrochloric acid is used zinc chloride and hydrogen are the products of reaction:

$$Zn + 2HCl = ZnCl_2 + 2H.$$

When iron is used in place of zinc the equation is

$$Fe + H_2SO_4 = FeSO_4 + 2H.$$

These reactions are quite similar, as is apparent from an examination of the equations. In each case 1 atom of the metal replaces 2 atoms of hydrogen in the acid, and the hydrogen escapes as a gas. When an element in the free state, such as the zinc in the equations just given, takes the place of some one element in a compound, setting it free from chemical combination, the act is called *substitution*.

Other reactions illustrating substitution are the action of sodium on water,

$$Na + H_2O = NaOH + H;$$

and the action of heated iron upon water,

$$3Fe + 4H_2O = Fe_3O_4 + 8H.$$

4. *Double decomposition.* When barium dioxide (BaO_2) is treated with sulphuric acid two compounds are formed, namely, hydrogen dioxide (H_2O_2) and barium sulphate ($BaSO_4$). The equation is

$$BaO_2 + H_2SO_4 = BaSO_4 + H_2O_2.$$

[Pg 72]

In this reaction it will be seen that the two elements barium and hydrogen simply exchange places. Such a reaction is called a *double decomposition*. We shall meet with many examples of this kind of chemical reactions.

Chemical equations are quantitative. The use of symbols and formulas in expressing chemical changes has another great advantage. Thus, according to the equation

$$H_2O = 2H + O,$$

1 molecule of water is decomposed into 2 atoms of hydrogen and 1 atom of oxygen. But, as we have seen, the relative weights of the atoms are known, that of hydrogen being 1.008, while that of oxygen is 16. The molecule of water, being composed of 2 atoms of hydrogen and 1 atom of oxygen, must therefore weigh relatively 2.016 + 16, or 18.016. The amount of hydrogen in this molecule must be 2.016/18.016, or 11.18% of the whole, while the amount of oxygen must be 16/18.018, or 88.82% of the whole. Now, since any definite quantity of water is simply the sum of a great many molecules of water, it is plain that the fractions representing the relative amounts of hydrogen and oxygen present in a molecule must likewise express the relative amounts of hydrogen and oxygen present in any quantity of water. Thus, for example, in 20 g. of water there are 2.016/18.016 × 20, or 2.238 g. of hydrogen, and 16/18.016 × 20, or 17.762 g. of oxygen. These results in reference to the composition of water of course agree exactly with the facts obtained by the experiments described in the chapter on water, for it is because of those experiments that the values 1.008 and 16 are given to hydrogen and oxygen respectively.

It is often easier to make calculations of this kind in the form of a proportion rather than by fractions. Since the [Pg 73] molecule of water and the two atoms of hydrogen which it contains have the ratio by weight of 18.016: 2.016, any mass of water has the same ratio between its total weight and the weight of the hydrogen in it. Hence, to find the number of grams (x) of hydrogen in 20 g. of water, we have the proportion

$$18.016 : 2.016 :: 20 \text{ g.} : x \text{ (grams of hydrogen)}.$$

Solving for x, we get 2.238 for the number of grams of hydrogen. Similarly, to find the amount (x) of oxygen present in the 20 g. of water, we have the proportion

$$18.016 : 16 :: 20 : x$$

from which we find that $x = 17.762$ g.

Again, suppose we wish to find what weight of oxygen can be obtained from 15 g. of mercuric oxide. The equation representing the decomposition of mercuric oxide is

$$HgO = Hg + O.$$

The relative weights of the mercury and oxygen atoms are respectively 200 and 16. The relative weight of the mercuric oxide molecule must therefore be the sum of these, or 216. The molecule of mercuric oxide and the atom of oxygen which it contains have the ratio 216: 16. This same ratio must therefore hold between the weight of any given quantity of mercuric oxide and that of the oxygen which it contains. Hence, to find the weight of oxygen in 15 g. of mercuric oxide, we have the proportion

$$216 : 16 :: 15 : x \text{ (grams of oxygen)}.$$

On the other hand, suppose we wish to prepare, say, 20 g. of oxygen. The problem is to find out what weight of mercuric oxide will yield 20 g. of oxygen. The following proportion evidently holds

$$216 : 16 :: x \text{ (grams of mercuric oxide)} : 20;$$

from which we get $x = 270$.

In the preparation of hydrogen by the action of sulphuric acid upon zinc, according to the equation,

$$Zn + H_2SO_4 = ZnSO_4 + 2\,H,$$

[Pg 74]

suppose that 50 g. of zinc are available; let it be required to calculate the weight of hydrogen which can be obtained. It will be seen that 1 atom of zinc will liberate 2 atoms of hydrogen. The ratio by weight of a zinc to an hydrogen atom is 65.4 : 1.008; of 1 zinc atom to 2 hydrogen atoms, 65.4 : 2.016. Zinc and hydrogen will be related in this reaction in this same ratio, however many atoms of zinc are concerned. Consequently in the proportion

$$65.4 : 2.016 :: 50 : x,$$

x will be the weight of hydrogen set free by 50 g. of zinc. The weight of zinc sulphate produced at the same time can be found from the proportion

$$65.4 : 161.46 :: 50 : x;$$

where 161.46 is the molecular weight of the zinc sulphate, and x the weight of zinc sulphate formed. In like manner, the weight of sulphuric acid used up can be calculated from the proportion

$$65.4 : 98.076 :: 50 : x.$$

These simple calculations are possible because the symbols and formulas in the equations represent the relative weights of the substances concerned in a chemical reaction. When once the relative weights of the atoms have been determined, and it has been agreed to allow the symbols to stand for these relative weights, an equation or formula making use of the symbols becomes a statement of a definite numerical fact, and calculations can be based on it.

Chemical equations not algebraic. Although chemical equations are quantitative, it must be clearly understood that they are not algebraic. A glance at the equations

$$7 + 4 = 11, 8 + 5 = 9 + 4$$

will show at once that they are true. The equations

$$HgO = Hg + O, FeO = Fe + O$$

are equally true in an algebraic sense, but experiment shows that only the first is true chemically, for iron oxide (FeO) [Pg 75] cannot be directly decomposed into iron and oxygen. Only such equations as have been found by careful experiment to express a real chemical transformation, true both for the kinds of substances as well as for the weights, have any value.

Chemical formulas and equations, therefore, are a concise way of representing qualitatively and quantitatively facts which have been found by experiment to be true in reference to the composition of substances and the changes which they undergo.

Formulas representing water of crystallization. An examination of substances containing water of crystallization has shown that in every case the water is present in such proportion by weight as can readily be represented by a formula. For example, copper sulphate ($CuSO_4$) and water combine in the ratio of 1 molecule of the sulphate to 5 of water; calcium sulphate ($CaSO_4$) and water combine in the ratio 1: 2 to form gypsum. These facts are expressed by writing the formulas for the two substances with a period between them. Thus the formula for crystallized copper sulphate is $CuSO_4 \cdot 5H_2O$; that of gypsum is $CaSO_4 \cdot 2H_2O$.

Heat of reaction. Attention has frequently been directed to the fact that chemical changes are usually accompanied by heat changes. In general it has been found that in every chemical action heat is either absorbed or given off. By adopting a suitable unit for the measurement of heat, the heat change during a chemical reaction can be expressed in the equation for the reaction.

Heat cannot be measured by the use of a thermometer alone, since the thermometer measures the intensity of heat, not its quantity. The easiest way to measure a quantity of heat is to note how warm it will make a definite amount of [Pg 76] a given substance chosen as a standard. Water has been chosen as the standard, and the unit of heat is called a *calorie. A calorie is defined as the amount of heat required to raise the temperature of one gram of water one degree.*

By means of this unit it is easy to indicate the heat changes in a given chemical reaction. The equation

$$2H + O = H_2O + 68{,}300 \text{ cal.}$$

means that when 2.016 g. of hydrogen combine with 16 g. of oxygen, 18.016 g. of water are formed and 68,300 cal. are set free.

$$C + 2S = CS_2 - 19{,}000 \text{ cal.}$$

means that an expenditure of 19,000 cal. is required to cause 12 g. of carbon to unite with 64.12 g. of sulphur to form 76.12 g. of carbon disulphide. In these equations it will be noted that the symbols stand for as many grams of the substance as there are units in the weights of the atoms represented by the symbols. This is always understood to be the case in equations where the heat of reaction is given.

Conditions of a chemical action are not indicated by equations. Equations do not tell the conditions under which a reaction will take place. The equation

$$HgO = Hg + O$$

does not tell us that it is necessary to keep the mercuric oxide at a high temperature in order that the decomposition may go on. The equation

$$Zn + 2HCl = ZnCl_2 + 2H$$

in no way indicates the fact that the hydrochloric acid must be dissolved in water before it will act upon the zinc. From the equation

$$H + Cl = HCl$$

[Pg 77]

it would not be suspected that the two gases hydrogen and chlorine will unite instantly in the sunlight, but will stand mixed in the dark a long time without change. It will therefore be necessary to pay much attention to the details of the conditions under which a given reaction occurs, as well as to the expression of the reaction in the form of an equation.

EXERCISES

1. Calculate the percentage composition of the following substances: (*a*) mercuric oxide; (*b*) potassium chlorate; (*c*) hydrochloric acid; (*d*) sulphuric acid. Compare the results obtained with the compositions as given in Chapters II and III.

2. Determine the percentage of copper, sulphur, oxygen, and water in copper sulphate crystals. What weight of water can be obtained from 150 g. of this substance?

3. What weight of zinc can be dissolved in 10 g. of sulphuric acid? How much zinc sulphate will be formed?

4. How many liters of hydrogen measured under standard conditions can be obtained from the action of 8 g. of iron on 10 g. of sulphuric acid? How much iron sulphate ($FeSO_4$) will be formed?

5. 10 g. of zinc were used in the preparation of hydrogen; what weight of iron will be required to prepare an equal volume?

6. How many grams of barium dioxide will be required to prepare 1 kg. of common hydrogen dioxide solution? What weight of barium sulphate will be formed at the same time?

7. What weight of the compound Mn_3O_4 will be formed by strongly heating 25 g. of manganese dioxide? What volume of oxygen will be given off at the same time, measured under standard conditions?

8. (*a*) What is the weight of 100 l. of hydrogen measured in a laboratory in which the temperature is 20° and pressure 750 mm.? (*b*) What weight of sulphuric acid is necessary to prepare this amount of hydrogen? (*c*) The density of sulphuric acid is 1.84. Express the acid required in (*b*) in cubic centimeters.

9. What weight of potassium chlorate is necessary to furnish sufficient oxygen to fill four 200 cc. bottles in your laboratory (the gas to be collected over water)?

[Pg 78]

CHAPTER VII

NITROGEN AND THE RARE ELEMENTS: ARGON, HELIUM, NEON, KRYPTON, XENON

Historical. Nitrogen was discovered by the English chemist Rutherford in 1772. A little later Scheele showed it to be a constituent of air, and Lavoisier gave it the name *azote*, signifying that it would not support life. The name *nitrogen* was afterwards given it because of its presence in saltpeter or niter. The term azote and symbol Az are still retained by the French chemists.

Occurrence. Air is composed principally of oxygen and nitrogen in the free state, about 78 parts by volume out of every 100 parts being nitrogen. Nitrogen also occurs in nature in the form of potassium nitrate (KNO$_3$)—commonly called saltpeter or niter—as well as in sodium nitrate (NaNO$_3$). Nitrogen is also an essential constituent of all living organisms; for example, the human body contains about 2.4% of nitrogen.

Preparation from air. Nitrogen can be prepared from air by the action of some substance which will combine with the oxygen, leaving the nitrogen free. Such a substance must be chosen, however, as will combine with the oxygen to form a product which is not a gas, and which can be readily separated from the nitrogen. The substances most commonly used for this purpose are phosphorus and copper.

1. *By the action of phosphorus.* The method used for the preparation of nitrogen by the action of phosphorus is as follows: [Pg 79]

The phosphorus is placed in a little porcelain dish, supported on a cork and floated on water (Fig. 26). It is then ignited by contact with a hot wire, and immediately a bell jar or bottle is brought over it so as to confine a portion of the air. The phosphorus combines with the oxygen to form an oxide of phosphorus, known as phosphorus pentoxide. This is a white solid which floats about in the bell jar, but in a short time it is all absorbed by the water, leaving the nitrogen. The withdrawal of the oxygen is indicated by the rising of the water in the bell jar.

Fig. 26

2. *By the action of copper.* The oxygen present in the air may also be removed by passing air slowly through a heated tube containing copper. The copper combines with the oxygen to form copper oxide, which is a solid. The nitrogen passes on and may be collected over water.

Nitrogen obtained from air is not pure. Inasmuch as air, in addition to oxygen and nitrogen, contains small amounts of other gases, and since the phosphorus as well as the copper removes only the oxygen, it is evident that the nitrogen obtained by these methods is never quite pure. About 1% of the product is composed of other gases, from which it is very difficult to separate the nitrogen. The impure nitrogen so obtained may, however, be used for a study of most of the properties of nitrogen, since these are not materially affected by the presence of the other gases.

Preparation from compounds of nitrogen. Pure nitrogen may be obtained from certain compounds of the element. Thus, if heat is applied to the compound ammonium nitrite (NH_4NO_2), the change represented in the following equation takes place:

$$NH_4NO_2 = 2H_2O + 2N.$$
[Pg 80]

Physical properties. Nitrogen is similar to oxygen and hydrogen in that it is a colorless, odorless, and tasteless gas. One liter of nitrogen weighs 1.2501 g. It is almost insoluble in water. It can be obtained in the form of a colorless liquid having a boiling point of -195° at ordinary pressure. At -214° it solidifies.

Chemical properties. Nitrogen is characterized by its inertness. It is neither combustible nor a supporter of combustion. At ordinary temperatures it will not combine directly with any of the elements except under rare conditions. At higher temperatures it combines with magnesium, lithium, titanium, and a number of other elements. The compounds formed are called *nitrides*, just as compounds of an element with oxygen are called *oxides*. When it is mixed with oxygen and subjected to the action of electric sparks, the two gases slowly combine forming oxides of nitrogen. A mixture of nitrogen and hydrogen when treated similarly forms ammonia, a gaseous compound of nitrogen and hydrogen. Since we are constantly inhaling nitrogen, it is evident that it is not poisonous. Nevertheless life would be impossible in an atmosphere of pure nitrogen on account of the exclusion of the necessary oxygen.

Argon, helium, neon, krypton, xenon. These are all rare elements occurring in the air in very small quantities. Argon, discovered in 1894, was the first one obtained. Lord Rayleigh, an English scientist, while engaged in determining the exact weights of various gases, observed that the nitrogen obtained from the air is slightly heavier than pure nitrogen obtained from its compounds. After repeating his experiments many times, always with the same results, Rayleigh finally concluded that the nitrogen which he had obtained from the air was not pure, but was mixed with a small amount of some unknown gas, the density of which is greater than that of nitrogen. Acting on this assumption, Rayleigh, together with the [Pg 81] English chemist Ramsay, attempted to separate the nitrogen from the unknown gas. Knowing that nitrogen would combine with magnesium, they passed the nitrogen obtained from the air and freed from all known substances through tubes containing magnesium heated to the necessary temperature. After repeating this operation, they finally succeeded in obtaining from the atmospheric nitrogen a small volume of gas which would not combine with magnesium and hence could not be nitrogen. This proved to be a new element, to which they gave the name *argon*. As predicted, this new element was found to be heavier than nitrogen, its density as compared with hydrogen as a standard being approximately 20, that of nitrogen being only 14. About 1% of the atmospheric nitrogen proved to be argon. The new element is characterized by having no affinity for other elements. Even under the most favorable conditions it has not been made to combine with any other element. On this account it was given the name argon, signifying lazy or idle. Like nitrogen, it is colorless, odorless, and tasteless. It has been liquefied and solidified. Its boiling point is -187°.

Helium was first found in the gases expelled from certain minerals by heating. Through the agency of the spectroscope it had been known to exist in the sun long before its presence on the earth had been demonstrated,—a fact suggested by the name helium, signifying the sun. Its existence in traces in the atmosphere has also been proven. It was first liquefied by Onnes in July, 1908. Its boiling point, namely -269°, is the lowest temperature yet reached.

The remaining elements of this group—neon, krypton, and xenon—have been obtained from liquid air. When liquid air is allowed to boil, the constituents which are the most difficult to liquefy, and which therefore have the lowest boiling points, vaporize first, followed by the others in the order of their boiling points. It is possible in this way to make at least a partial separation of the air into its constituents, and Ramsay thus succeeded in obtaining from liquid air not only the known constituents, including argon and helium, but also the new elements, neon, krypton, and xenon. These elements, as well as helium, all proved to be similar to argon in that they are without chemical activity, apparently forming no compounds whatever. The percentages present in the air are very small. The names, neon, krypton, xenon, signify respectively, new, hidden, stranger.

[Pg 82]

100

EXERCISES

1. How could you distinguish between oxygen, hydrogen, and nitrogen?

2. Calculate the relative weights of nitrogen and oxygen; of nitrogen and hydrogen.

3. In the preparation of nitrogen from the air, how would hydrogen do as a substance for the removal of the oxygen?

4. What weight of nitrogen can be obtained from 10 l. of air measured under the conditions of temperature and pressure which prevail in your laboratory?

5. How many grams of ammonium nitrite are necessary in the preparation of 20 l. of nitrogen measured over water under the conditions of temperature and pressure which prevail in your laboratory?

6. If 10 l. of air, measured under standard conditions, is passed over 100 g. of hot copper, how much will the copper gain in weight?

WILLIAM RAMSAY (Scotch) (1855-)

Has made many studies in the physical properties of substances;
discovered helium; together with Lord Rayleigh and others he discovered argon, krypton, xenon, and neon; has contributed largely to
the knowledge of radio-active substances, showing that radium
gradually gives rise to helium; professor at University College,
London

CHAPTER VIII

THE ATMOSPHERE

Atmosphere and air. The term *atmosphere* is applied to the gaseous envelope surrounding the earth. The term *air* is generally applied to a limited portion of this envelope, although the two words are often used interchangeably. Many references have already been made to the composition and properties of the atmosphere. These statements must now be collected and discussed somewhat more in detail.

Air formerly regarded as an element. Like water, air was at first regarded as elementary in character. Near the close of the eighteenth century Scheele, Priestley, and Lavoisier showed by their experiments that it is a mixture of at least two gases,—those which we now call oxygen and nitrogen. By burning substances in an inclosed volume of air and noting the contraction in volume due to the removal of the oxygen, they were able to determine with some accuracy the relative volumes of oxygen and nitrogen present in the air.

The constituents of the atmosphere. The constituents of the atmosphere may be divided into two general groups: those which are essential to life and those which are not essential.

1. *Constituents essential to life.* In addition to oxygen and nitrogen at least two other substances, namely, carbon dioxide and water vapor, must be present in the atmosphere in order that life may exist. The former of these is a [Pg 84] gaseous compound of carbon and oxygen having the formula CO_2. Its properties will be discussed in detail in the chapter on the compounds of carbon. Its presence in the air may be shown by causing the air to bubble through a solution of calcium hydroxide ($Ca(OH)_2$), commonly called lime water. The carbon dioxide combines with the calcium hydroxide in accordance with the following equation:

$$Ca(OH)_2 + CO_2 = CaCO_3 + H_2O.$$

The resulting calcium carbonate ($CaCO_3$) is insoluble in water and separates in the form of a white powder, which causes the solution to appear milky.

The presence of water vapor is readily shown by its condensation on cold objects as well as by the fact that a bit of calcium chloride when exposed to the air becomes moist, and may even dissolve in the water absorbed from the air.

2. *Constituents not essential to life.* In addition to the essential constituents, the air contains small percentages of various other gases, the presence of which so far as is known is not essential to life. This list includes the rare elements, argon, helium, neon, krypton, and xenon; also hydrogen, ammonia, hydrogen dioxide, and probably ozone. Certain minute forms of life (germs) are also present, the decay of organic matter being due to their presence.

Function of each of the essential constituents. (1) The oxygen directly supports life through respiration. (2) The nitrogen, on account of its inactivity, serves to dilute the oxygen, and while contrary to the older views, it is possible that life might continue to exist in the absence of the atmospheric nitrogen, yet the conditions of life would be entirely changed. Moreover, nitrogen is an essential constituent of all animal and plant life. It was formerly supposed that neither animals nor plants could assimilate the free nitrogen, but it has been shown recently that the plants of at least one natural [Pg 85] order, the Leguminosæ, to which belong the beans, peas, and clover, have the power of directly assimilating the free nitrogen from the atmosphere. This is accomplished through the agency of groups of bacteria, which form colonies in little tubercles on the roots of the plants. These bacteria probably assist in the absorption of nitrogen by changing the free nitrogen into compounds which can be assimilated by the plant. Fig. 27 shows the tubercles on the roots of a variety of bean. (3) The presence of water vapor in the air is necessary to prevent excessive evaporation from both plants and animals. (4) Carbon dioxide is an essential plant food.

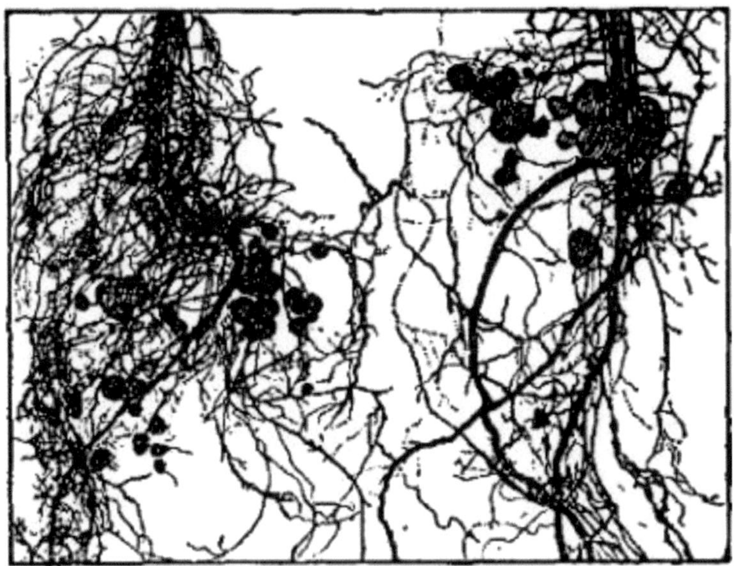

Fig. 27

The quantitative analysis of air. A number of different methods have been devised for the determination of the percentages of the constituents present in the atmosphere. Among these are the following.

1. *Determination of oxygen.* (1) The oxygen is withdrawn from a measured volume of air inclosed in a tube, by means of phosphorus.

To make the determination, a graduated tube is filled with water and inverted in a vessel of water. Air is introduced into the tube until it is partially filled with the gas. The volume of the inclosed air is carefully noted and reduced to standard conditions. A small piece of phosphorus is attached to a wire and brought within the tube as shown in Fig. 28. After a few hours the oxygen in the inclosed air will have combined with the phosphorus, the water rising to take its place. The phosphorus is removed and the volume is again noted and reduced to standard conditions. The contraction in the volume of the air is equal to the volume of oxygen absorbed.

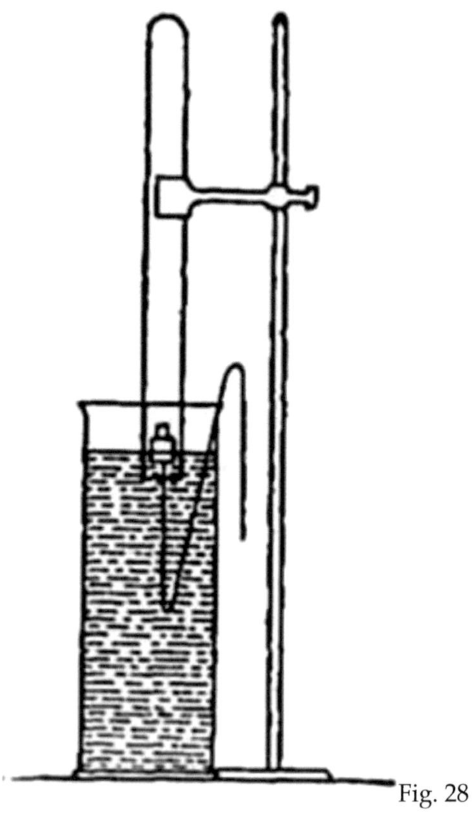

Fig. 28

[Pg 86]

(2) The oxygen may also be estimated by passing a measured volume of air through a tube containing copper heated to a high temperature. The oxygen in the air combines with the copper to form copper oxide (CuO). Hence the increase in the weight of the copper equals the weight of the oxygen in the volume of air taken.

(3) A more accurate method is the following. A eudiometer tube is filled with mercury and inverted in a vessel of the same liquid. A convenient amount of air is then introduced into the tube and its volume accurately noted. There is then introduced more than sufficient hydrogen to combine with the oxygen present in the inclosed air, and the volume is again accurately noted. The mixture is then

exploded by an electric spark, and the volume is once more taken. By subtracting this volume from the total volume of the air and hydrogen there is obtained the contraction in volume due to the union of the oxygen and hydrogen. The volume occupied by the water formed by the union of the two gases is so small that it may be disregarded in the calculation. Since oxygen and hydrogen combine in the ratio 1: 2 by volume, it is evident that the contraction in volume due to the combination is equal to the volume occupied by the oxygen in the air contained in the tube, plus twice this volume of hydrogen. In other words, one third of the total contraction is equal to the volume occupied by the oxygen in the inclosed air. The following example will make this clear:

Volume of air in tube	50.0 cc.
Volume after introducing hydrogen	80.0
Volume after combination of oxygen and hydrogen	48.5
Contraction in volume due to combination (80 cc.-48.5 cc.)	31.5
Volume of oxygen in 50 cc. of air (1/3 of 31.5)	10.5

[Pg 87]

All these methods agree in showing that 100 volumes of dry air contain approximately 21 volumes of oxygen.

2. *Determination of nitrogen.* If the gas left after the removal of oxygen from a portion of air is passed over heated magnesium, the nitrogen is withdrawn, argon and the other rare elements being left. It may thus be shown that of the 79 volumes of gas left after the removal of the oxygen from 100 volumes of air, approximately 78 are nitrogen and 0.93 argon. The other elements are present in such small quantities that they may be neglected.

3. *Determination of carbon dioxide.* The percentage of carbon dioxide in any given volume of air may be determined by passing the air over calcium hydroxide or some other compound which will combine with the carbon dioxide. The increase in the weight of the hydroxide equals the weight of the carbon dioxide absorbed. The amount present in the open normal air is from 3 to 4 parts by volume in 10,000 volumes of air, or about 0.04%.

4. *Determination of water vapor.* The water vapor present in a given volume of air may be determined by passing the air over calcium chloride (or some other compound which has a strong affinity for water), and noting the increase in the weight of the chloride. The amount present varies not only with the locality, but there is a wide variation from day to day in the same locality because of the winds and changes in temperature.

Processes affecting the composition of the air. The most important of these processes are the following.

1. *Respiration.* In the process of respiration some of the oxygen in the inhaled air is absorbed by the blood and carried to all parts of the body, where it combines with the carbon of the worn-out tissues. The products of oxidation [Pg 88] are carried back to the lungs and exhaled in the form of carbon dioxide. The amount exhaled by an adult averages about 20 l. per hour. Hence in a poorly ventilated room occupied by a number of people the amount of carbon dioxide rapidly increases. While this gas is not poisonous unless present in large amounts, nevertheless air containing more than 15 parts in 10,000 is not fit for respiration.

2. *Combustion.* All of the ordinary forms of fuel contain large percentages of carbon. On burning, this carbon combines with oxygen in the air, forming carbon dioxide. Combustion and respiration, therefore, tend to diminish the amount of oxygen in the air and to increase the amount of carbon dioxide.

3. *Action of plants.* Plants have the power, when in the sunlight, of absorbing carbon dioxide from the air, retaining the carbon and returning at least a portion of the oxygen to the air. It will be observed that these changes are just the opposite of those brought about by the processes of respiration and combustion.

Poisonous effect of exhaled air. The differences in the percentages of oxygen, carbon dioxide, and moisture present in inhaled air and exhaled air are shown in the following analyses.

	INHALED AIR	EXHALED AIR
Oxygen	21.00%	16.00%
Carbon dioxide	0.04	4.38

| Moisture | variable | saturated |

The foul odor of respired air is due to the presence of a certain amount of organic matter. It is possible that this organic matter rather than the carbon dioxide is responsible for the injurious effects which follow the respiration of impure air. The extent of such organic impurities present may be judged, however, by the amount of carbon dioxide present, since the two are exhaled together.

The cycle of carbon in nature. Under the influence of sunlight, the carbon dioxide absorbed from the air by plants reacts with water [Pg 89] and small amounts of other substances absorbed from the soil to form complex compounds of carbon which constitute the essential part of the plant tissue. This reaction is attended by the evolution of oxygen, which is restored to the air. The compounds resulting from these changes are much richer in their energy content than are the substances from which they are formed; hence a certain amount of energy must have been absorbed in their formation. The source of this energy is the sun's rays.

If the plant is burned, the changes which took place in the formation of the compounds present are largely reversed. The carbon and hydrogen present combine with oxygen taken from the air to form carbon dioxide and water, while the energy absorbed from the sun's rays is liberated in the form of energy of heat. If, on the other hand, the plant is used as food, the compounds present are used in building up the tissues of the body. When this tissue breaks down, the changes which it undergoes are very similar to those which take place when the plant is burned. The carbon and hydrogen combine with the inhaled oxygen to form carbon dioxide and water, which are exhaled. The energy possessed by the complex substances is liberated partly in the form of energy of heat, which maintains the heat of the body, and partly in the various forms of muscular energy. The carbon originally absorbed from the air by the plant in the form of carbon dioxide is thus restored to the air and is ready to repeat the cycle of changes.

The composition of the air is constant. Notwithstanding the changes constantly taking place which tend to alter the composition of the air, the results of a great many analyses of air collected in the open fields show that the percentages of oxygen and nitrogen as well as of carbon dioxide are very nearly constant. Indeed, so constant are the percentages of oxygen and nitrogen that the question has arisen, whether these two elements are not combined in the air, forming a definite chemical compound. That the two are not combined but are simply mixed together can be shown in a number of ways, among which are the following. [Pg 90]

1. When air dissolves in water it has been found that the ratio of oxygen to nitrogen in the dissolved air is no longer 21: 78, but more nearly 35: 65. If it were a chemical compound, the ratio of oxygen to nitrogen would not be changed by solution in water.

2. A chemical compound in the form of a liquid has a definite boiling point. Water, for example, boils at 100°. Moreover the steam which is thus formed has the same composition as the water. The boiling point of liquid air, on the other hand, gradually rises as the liquid boils, the nitrogen escaping first followed by the oxygen. If the two were combined, they would pass off together in the ratio in which they are found in the air.

Why the air has a constant composition. If air is a mixture and changes are constantly taking place which tend to modify its composition, how, then, do we account for the constancy of composition which the analyses reveal? This is explained by several facts. (1) The changes which are caused by the processes of combustion and respiration, on the one hand, and the action of plants, on the other, tend to equalize each other. (2) The winds keep the air in constant motion and so prevent local changes. (3) The volume of the air is so vast and the changes which occur are so small compared with the total amount of air that they cannot be readily detected. (4) Finally it must be noted that only air collected in the open fields shows this constancy in composition. The air in a poorly ventilated room occupied by a number of people rapidly changes in composition.

The properties of the air. Inasmuch as air is composed principally of a mixture of oxygen and nitrogen, which elements have already been discussed, its properties may be inferred largely from those of the two gases. [Pg 91] One liter weighs 1.2923 g. It is thus 14.38 times as heavy as hydrogen. At the sea level it exerts an average pressure sufficient to sustain a column of mercury 760 mm. in height. This is taken as the standard pressure in determining the volumes of gases as well as the boiling points of liquids. Water may be made to boil at any temperature between 0° and considerably above 100° by simply varying the pressure. It is only when the pressure upon it is equal to the normal pressure of the atmosphere at the sea level, as indicated by a barometric reading of 760 mm., that it boils at 100°.

Preparation of liquid air. Attention has been called to the fact that both oxygen and nitrogen can be obtained in the liquid state by strongly cooling the gases and applying great pressure to them. Since air is largely a mixture of these two gases, it can be liquefied by the same methods.

The methods for liquefying air have been simplified greatly in that the low temperature required is obtained by allowing a portion of the compressed air to expand. The expansion of a gas is always attended by the absorption of heat. In liquefying air the apparatus is so constructed that the heat absorbed is withdrawn from air already under great pressure. This process is continued until the temperature is lowered to the point of liquefaction.

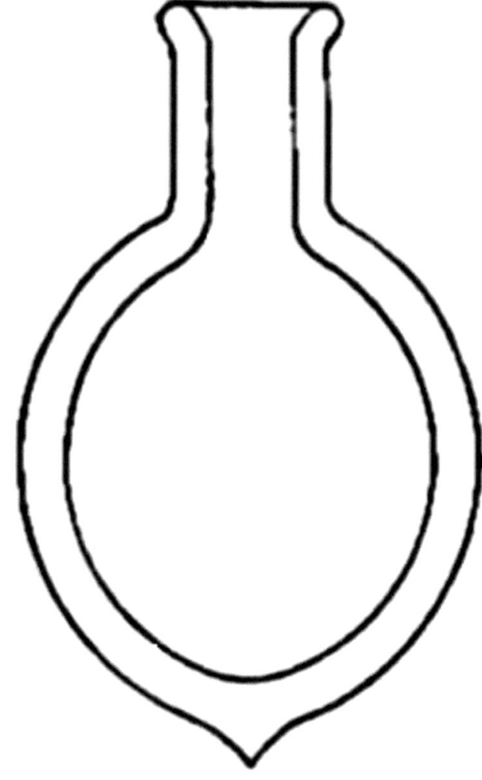

Fig. 29

The Dewar bulb. It is not possible to preserve air in the liquid state in a closed vessel, on account of the enormous pressure exerted by it in its tendency to pass into the gaseous state. It may however be preserved for some hours or even days before it will completely evaporate, by simply placing it in an open vessel surrounded by a nonconducting material. The most efficient vessel for this purpose is the *Dewar bulb* shown in Fig. 29. [Pg 92] The air is withdrawn from the space between the two walls, thus making it nonconducting.

Properties and uses of liquid air. When first prepared, liquid air is cloudy because of the presence of particles of solid carbon dioxide. These may be filtered off, leaving a liquid of slightly bluish color. It begins to boil at about -190°, the nitrogen passing off first, gradually followed by the oxygen, the last portions being nearly pure oxygen. To a certain extent oxygen is now prepared in this way for commercial purposes.

The extremely low temperature of liquid air may be inferred from the fact that mercury when cooled by it is frozen to a mass so hard that it may be used for driving nails.

Liquid air is used in the preparation of oxygen and as a cooling agent in the study of the properties of matter at low temperatures. It has thus been found that elements at extremely low temperatures largely lose their chemical activity.

EXERCISES

1. When oxygen and nitrogen are mixed in the proportion in which they exist in the atmosphere, heat is neither evolved nor absorbed by the process. What important point does this suggest?

2. What essential constituent of the air is found in larger amount in manufacturing districts than in the open country?

3. Can you suggest any reason why the growth of clover in a field improves the soil?

4. Why are the inner walls of a Dewar bulb sometimes coated with a film of silver?

5. To what is the blue color of liquid air due? Does this color increase in intensity on standing?

6. When ice is placed in a vessel containing liquid air, the latter boils violently. Explain. [Pg 93]

7. Taking the volumes of the oxygen and nitrogen in 100 volumes of air as 21 and 78 respectively, calculate the percentages of these elements present by weight.

8. Would combustion be more intense in liquid air than in the gaseous substance?

9. A tube containing calcium chloride was found to weigh 30.1293 g. A volume of air which weighed 15.2134 g. was passed through, after which the weight of the tube was found to be 30.3405 g. What was the percentage amount of moisture present in the air?

10. 10 l. of air measured at 20° and 740 mm. passed through lime water caused the precipitation of 0.0102 g. of CaCO₃. Find the number of volumes of carbon dioxide in 10,000 volumes of the air.

[Pg 94]

CHAPTER IX

SOLUTIONS

Definitions. When a substance disappears in a liquid in such a way as to thoroughly mix with it and to be lost to sight as an individual body, the resulting liquid is called a *solution*. The liquid in which the substance dissolves is called the *solvent*, while the dissolved substance is called the *solute*.

Classes of solutions. Matter in any one of its physical states may dissolve in a liquid, so that we may have solutions of gases, of liquids, and of solids. Solutions of liquids in liquids are not often mentioned in the following pages, but the other two classes will become very familiar in the course of our study, and deserve special attention.

SOLUTION OF GASES IN LIQUIDS

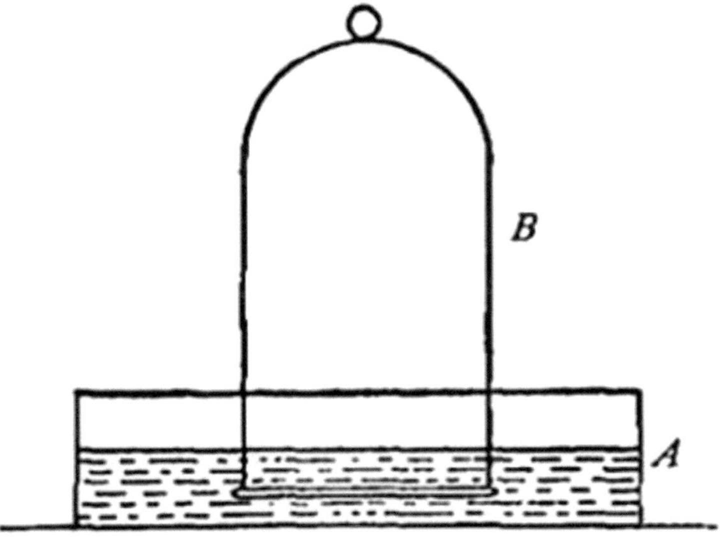

Fig. 30

It has already been stated that oxygen, hydrogen, and nitrogen are slightly soluble in water. Accurate study has led to the conclusion that all gases are soluble to some extent not only in water but in many other liquids. The amount of a gas which will dissolve in a liquid depends upon a number of conditions, and these can best be understood by [Pg 95] supposing a vessel B (Fig. 30), to be filled with the gas and inverted over the liquid. Under these circumstances the gas cannot escape or become mixed with another gas.

Circumstances affecting the solubility of gases. A number of circumstances affect the solubility of a gas in a liquid.

1. *Nature of the gas.* Other conditions being equal, each gas has its own peculiar solubility, just as it has its own special taste or odor. The solubility of gases varies between wide limits, as will be seen from the following table, but as a rule a given volume of a liquid will not dissolve more than two or three times its own volume of a gas.

1 l. of water at 760 mm. pressure and at 0° will dissolve:

Ammonia	1148.00 l.
Hydrochloric acid	503.00
Sulphur dioxide	79.79
Carbon dioxide	1.80
Oxygen	41.14 cc.
Hydrogen	21.15
Nitrogen	20.03

In the case of very soluble gases, such as the first three in the table, it is probable that chemical combination between the liquid and the gas takes place.

2. *Nature of the liquid.* The character of the liquid has much influence upon the solubility of a gas. Water, alcohol, and ether have each its own peculiar solvent power. From the solubility of a gas in water, no prediction can be made as to its solubility in other liquids.

3. *Influence of pressure.* It has been found that the weight of gas which dissolves in a given case is proportional to the pressure exerted upon the gas. If the [Pg 96] pressure is doubled, the weight of gas going into solution is doubled; if the pressure is diminished to one half of its original value, half of the dissolved gas will escape. Under high pressure, large quantities of gas can be dissolved in a liquid, and when the pressure is removed the gas escapes, causing the liquid to foam or *effervesce.*

4. *Influence of temperature.* In general, the lower the temperature of the liquid, the larger the quantity of gas which it can dissolve. 1000 volumes of water at 0° will dissolve 41.14 volumes of oxygen; at 50°, 18.37 volumes; at 100° none at all. While most gases can be expelled from a liquid by boiling the solution, some cannot. For example, it is

not possible to expel hydrochloric acid gas completely from its solution by boiling.

SOLUTION OF SOLIDS IN LIQUIDS

This is the most familiar class of solutions, since in the laboratory substances are much more frequently used in the form of solutions than in the solid state.

Circumstances affecting the solubility of a solid. The solubility of a solid in a liquid depends upon several factors.

1. *Nature of the solid.* Other conditions being the same, solids vary greatly in their solubility in liquids. This is illustrated in the following table:

Table of Solubility of Solids at 18°

100 cc. of water will dissolve:

Calcium chloride	71.0 g.
Sodium chloride	35.9
Potassium nitrate	29.1
Copper sulphate	21.4
Calcium sulphate	0.207

[Pg 97]

No solids are absolutely insoluble, but the amount dissolved may be so small as to be of no significance for most purposes. Thus barium sulphate, one of the most insoluble of common substances, dissolves in water to the extent of 1 part in 400,000.

2. *Nature of the solvent.* Liquids vary much in their power to dissolve solids. Some are said to be good solvents, since they dissolve a great variety of substances and considerable quantities of them. Others have small solvent power, dissolving few substances, and those to a slight extent only. Broadly speaking, water is the most general solvent, and alcohol is perhaps second in solvent power.

3. *Temperature.* The weight of a solid which a given liquid can dissolve varies with the temperature. Usually it increases rapidly as the temperature rises, so that the boiling liquid dissolves several times the weight which the cold liquid will dissolve. In some instances, as in the case of common salt dissolved in water, the temperature has little influence upon the solubility, and a few solids are more soluble in cold water than in hot. The following examples will serve as illustrations:

Table of Solubility at 0° and at 100°

100 cc. of water will dissolve:

	At 0°	At 100°
Calcium chloride	49.6 g.	155.0 g.
Sodium chloride	35.7	39.8
Potassium nitrate	13.3	247.0
Copper sulphate	15.5	73.5
Calcium sulphate	0.205	0.217
Calcium hydroxide	0.173	0.079

Saturated solutions. A liquid will not dissolve an unlimited quantity of a solid. On adding the solid to the liquid in small portions at [Pg 98] a time, it will be found that a point is reached at which the liquid will not dissolve more of the solid at that temperature. The solid and the solution remain in contact with each other unchanged. This condition may be described by saying that they are in equilibrium with each other. A solution is said to be *saturated* when it remains unchanged in concentration in contact with some of the solid. The weight of the solid which will completely saturate a definite volume of a liquid at a given temperature is called the *solubility* of the substance at that temperature.

Supersaturated solutions. When a solution, saturated at a given temperature, is allowed to cool it sometimes happens that no solid crystallizes out. This is very likely to occur when the vessel used is perfectly smooth and the solution is not disturbed in any way. Such a solution is said to be *supersaturated.* That this condition is unstable can be shown by adding a crystal of the solid to the solution. All of the solid in excess of the quantity required to saturate the solution at this temperature will at once crystallize out, leaving the solution saturated. Supersaturation may also be overcome in many cases by vigorously shaking or stirring the solution.

General physical properties of solutions. A few general statements may be made in reference to the physical properties of solutions.

1. *Distribution of the solid in the liquid.* A solid, when dissolved, tends to distribute itself uniformly through the liquid, so that every part of the solution has the same concentration. The process goes on

very slowly unless hastened by stirring or shaking the solution. Thus, if a few crystals of a highly colored substance such as copper sulphate are placed in the bottom of a tall vessel full of water, it will take weeks for the solution to become uniformly colored.

2. *Boiling points of solutions.* The boiling point of a liquid is raised by the presence of a substance dissolved in it. In general the extent to which the boiling point of a solvent is raised by a given substance is proportional to the [Pg 99] concentration of the solution, that is, to the weight of the substance dissolved in a definite weight of the solvent.

3. *Freezing points of solutions.* A solution freezes at a lower temperature than the pure solvent. The lowering of the freezing point obeys the same law which holds for the raising of the boiling point: the extent of lowering is proportional to the weight of dissolved substance, that is, to the concentration of the solution.

Electrolysis of solutions. Pure water does not appreciably conduct the electric current. If, however, certain substances such as common salt are dissolved in the water, the resulting solutions are found to be conductors of electricity. Such solutions are called *electrolytes.* When the current passes through an electrolyte some chemical change always takes place. This change is called *electrolysis.*

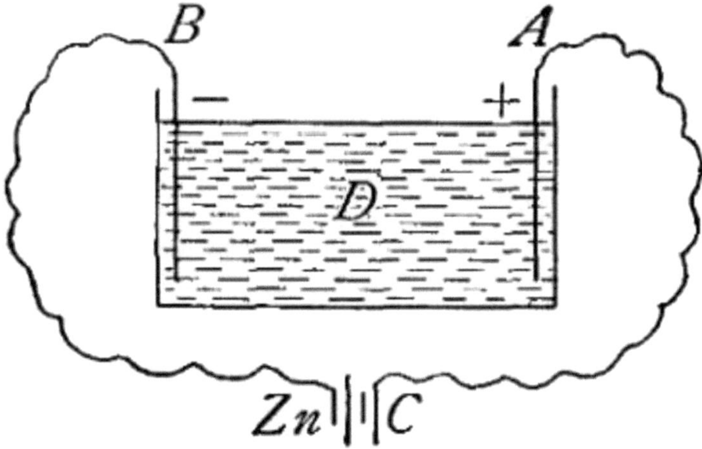

Fig. 31

The general method used in the electrolysis of a solution is illustrated in Fig. 31. The vessel D contains the electrolyte. Two plates or rods, A and B, made of suitable material, are connected with the wires from a battery (or dynamo) and dipped into the electrolyte, as shown in the figure. These plates or rods are called *electrodes*. The electrode connected with the zinc plate of the battery is the negative electrode or *cathode*, while that connected with the carbon plate is the positive electrode or *anode*.

Theory of electrolytic dissociation. The facts which have just been described in connection with solutions, together with many others, have led chemists to adopt a theory of solutions called *the theory of electrolytic dissociation*. The main assumptions in this theory are the following. [Pg 100]

1. *Formation of ions.* Many compounds when dissolved in water undergo an important change. A portion of their molecules fall apart, or *dissociate*, into two or more parts, called *ions*. Thus sodium nitrate ($NaNO_3$) dissociates into the ions Na and NO_3; sodium chloride, into the ions Na and Cl. These ions are free to move about in the solution independently of each other like independent molecules, and for this reason were given the name ion, which signifies a wanderer.

2. *The electrical charge of ions.* Each ion carries a heavy electrical charge, and in this respect differs from an atom or molecule. It is evident that the sodium in the form of an ion must differ in some important way from ordinary sodium, for sodium ions, formed from sodium nitrate, give no visible evidence of their presence in water, whereas metallic sodium at once decomposes the water. The electrical charge, therefore, greatly modifies the usual chemical properties of the element.

3. *The positive charges equal the negative charges.* The ions formed by the dissociation of any molecule are of two kinds. One kind is charged with positive electricity and the other with negative electricity; moreover the sum of all the positive charges is always equal to the sum of all the negative charges. The solution as a whole is therefore electrically neutral. If we represent dissociation by the usual chemical equations, with the electrical charges indicated by +

and - signs following the symbols, the dissociation of sodium chloride molecules is represented thus:

$$NaCl \longrightarrow Na^+, Cl^-.$$

The positive charge on each sodium ion exactly equals the negative charge on each chlorine ion. [Pg 101] Sodium sulphate dissociates, as shown in the equation

$$Na_2SO_4 \longrightarrow 2Na^+, SO_4^{--}.$$

Here the positive charge on the two sodium ions equals the double negative charge on the SO_4 ion.

4. *Not all compounds dissociate.* Only those compounds dissociate whose solutions form electrolytes. Thus salt dissociates when dissolved in water, the resulting solution being an electrolyte. Sugar, on the other hand, does not dissociate and its solution is not a conductor of the electric current.

5. *Extent of dissociation differs in different liquids.* While compounds most readily undergo dissociation in water, yet dissociation often occurs to a limited extent when solution takes place in liquids other than water. In the discussion of solutions it will be understood that the solvent is water unless otherwise noted.

The theory of electrolytic dissociation and the properties of solutions. In order to be of value, this theory must give a reasonable explanation of the properties of solutions. Let us now see if the theory is in harmony with certain of these properties.

The theory of electrolytic dissociation and the boiling and freezing points of solutions. We have seen that the boiling point of a solution of a substance is raised in proportion to the concentration of the dissolved substance. This is but another way of saying that the change in the boiling point of the solution is proportional to the number of molecules of the dissolved substance present in the solution.

It has been found, however, that in the case of electrolytes the boiling point is raised more than it should be to [Pg 102] conform to this law. If the solute dissociates into ions, the reason for this becomes clear. Each ion has the same effect on the boiling point as a molecule, and since their number is greater than the number of molecules from which they were formed, the effect on the boiling point is abnormally great.

In a similar way, the theory furnishes an explanation of the abnormal lowering of the freezing point of electrolytes.

The theory of electrolytic dissociation and electrolysis. The changes taking place during electrolysis harmonize very completely with the theory of dissociation. This will become clear from a study of the following examples.

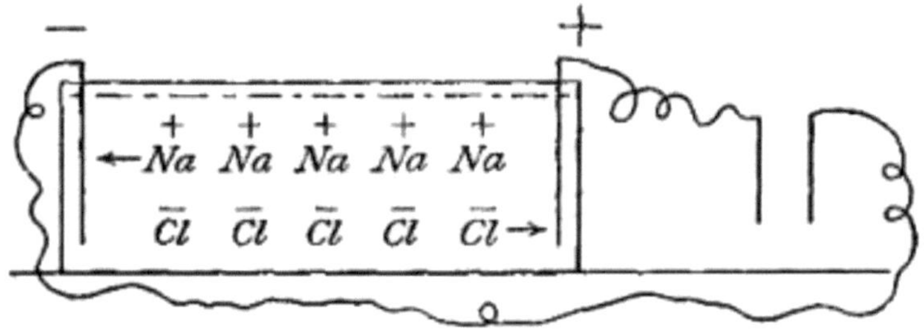

Fig. 32

1. *Electrolysis of sodium chloride.* Fig. 32 represents a vessel in which the electrolyte is a solution of sodium chloride (NaCl). According to the dissociation theory the molecules of sodium chloride dissociate into the ions Na^+ and Cl^-. The Na^+ ions are attracted to the cathode owing to its large negative charge. On coming into contact with the cathode, the Na^+ ions give up their positive charge and are then ordinary sodium atoms. They immediately decompose the water according to the equation

$$Na + H_2O = NaOH + H,$$

and hydrogen is evolved about the cathode.

The chlorine ions on being discharged at the anode in similar manner may either be given off as chlorine gas, or may attack the water, as represented in the equation

$$2Cl + H_2O = 2HCl + O.$$
[Pg 103]

2. *Electrolysis of water.* The reason for the addition of sulphuric acid to water in the preparation of oxygen and hydrogen by electrolysis will now be clear. Water itself is not an electrolyte to an appreciable extent; that is, it does not form enough ions to carry a current. Sulphuric acid dissolved in water is an electrolyte, and dissociates into the ions $2 H^+$ and $SO_4{}^-$. In the process of electrolysis of the solution, the hydrogen ions travel to the cathode, and on being discharged escape as hydrogen gas. The SO_4 ions, when discharged at the anode, act upon water, setting free oxygen and once more forming sulphuric acid:

$$SO_4 + H_2O = H_2SO_4 + O.$$

The sulphuric acid can again dissociate and the process repeat itself as long as any water is left. Hence the hydrogen and oxygen set free in the electrolysis of water really come directly from the acid but indirectly from the water.

3. *Electrolysis of sodium sulphate.* In a similar way, sodium sulphate (Na_2SO_4), when in solution, gives the ions $2 Na^+$ and $SO_4{}^-$. On being discharged, the sodium atoms decompose water about the cathode, as in the case of sodium chloride, while the SO_4 ions when discharged at the anode decompose the water, as represented in the equation

$$SO_4 + H_2O = H_2SO_4 + O$$

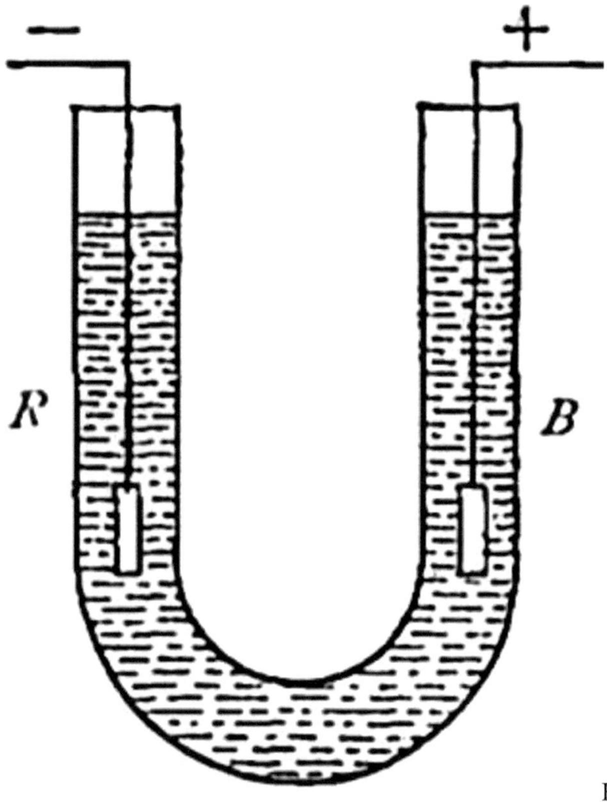

Fig. 33

That new substances are formed at the cathode and anode may be shown in the following way. A U-tube, such as is represented in Fig. 33, is partially filled with a solution of sodium sulphate, and the liquid in one arm is colored with red litmus, that in the other [Pg 104] with blue litmus. An electrode placed in the red solution is made to serve as cathode, while one in the blue solution is made the anode. On allowing the current to pass, the blue solution turns red, while the red solution turns blue. These are exactly the changes which would take place if sodium hydroxide and sulphuric acid were to be set free at the electrodes, as required by the theory.

The properties of electrolytes depend upon the ions present. When a substance capable of dissociating into ions is dissolved in water, the properties of the solution will depend upon two factors: (1) the ions formed from the substance; (2) the undissociated molecules. Since the ions are usually more active chemically than the molecules, most of the chemical properties of an electrolyte are due to the ions rather than to the molecules.

The solutions of any two substances which give the same ion will have certain properties in common. Thus all solutions containing the copper ion (Cu^{++}) are blue, unless the color is modified by the presence of ions or molecules having some other color.

EXERCISES

1. Distinguish clearly between the following terms: electrolysis, electrolyte, electrolytic dissociation, ions, solute, solvent, solution, saturated solution, and supersaturated solution.

2. Why does the water from some natural springs effervesce?

3. (*a*) Why does not the water of the ocean freeze? (*b*) Why will ice and salt produce a lower temperature than ice alone?

4. Why does shaking or stirring make a solid dissolve more rapidly in a liquid?

5. By experiment it was found that a certain volume of water was saturated at 100° with 114 g. of potassium nitrate. On cooling to 0° a portion of the substance crystallized. (*a*) How many grams of the substance remained in solution? (*b*) What was the strength [Pg 105] of the solution at 18°? (*c*) How much water had been used in the experiment?

6. (*a*) 10 g. of common salt were dissolved in water and the solution evaporated to dryness; what weight of solid was left? (*b*) 10 g. of zinc were dissolved in hydrochloric acid and the solution evaporated to dryness; what weight of solid was left?

7. Account for the fact that sugar sometimes deposits from molasses, even when no evaporation has taken place.

8. (*a*) From the standpoint of the theory of electrolytic dissociation, write the simple equation for a dilute solution of copper sulphate ($CuSO_4$); this solution is blue. (*b*) In the same manner, write one for sodium sulphate; this solution is colorless. (*c*) How would you account for the color of the copper sulphate solution?

9. (*a*) As in the preceding exercise, write a simple equation for a dilute solution of copper chloride (CuCl$_2$); this solution is blue. (*b*) In the same manner, write one for sodium chloride; this solution is colorless. To what is the blue color due?

10. What component is present in concentrated sulphuric acid that is almost wanting in very dilute sulphuric acid?

11. Why will vegetables cook faster when boiled in strong salt water than when boiled in pure water?

12. How do you explain the foaming of soda water?

[Pg 106]

CHAPTER X

ACIDS, BASES, AND SALTS; NEUTRALIZATION

Acids, bases, and salts. The three classes of compounds known respectively as acids, bases, and salts include the great majority of the compounds with which we shall have to deal. It is important, therefore, for us to consider each of these classes in a systematic way. The individual members belonging to each class will be discussed in detail in the appropriate places, but a few representatives of each class will be described in this chapter with special reference to the common properties in accordance with which they are classified.

The familiar acids. *Hydrochloric acid* is a gas composed of hydrogen and chlorine, and has the formula HCl. The substance is very soluble in water, and it is this solution which is usually called hydrochloric acid. *Nitric acid* is a liquid composed of hydrogen, nitrogen, and oxygen, having the formula HNO$_3$. As sold commercially it is mixed with about 32% of water. *Sulphuric acid*, whose composition is represented by the formula H$_2$SO$_4$, is an oily liquid nearly twice as heavy as water, and is commonly called *oil of vitriol*.

Characteristics of acids. (1) All acids contain hydrogen. (2) When dissolved in water the molecules of the acid dissociate into two

128

kinds of ions. One of these is always hydrogen and is the cation (+), while the other consists of the remainder of the molecule and is the anion (-). (3) The solution tastes sour. (4) It has the power to change the [Pg 107] color of certain substances called *indicators*. Thus blue litmus is changed to red, and yellow methyl orange is changed to red. Since all acids produce hydrogen cations, while the anions of each are different, the properties which all acids have in common when in solution, such as taste and action on indicators, must be attributed to the hydrogen ions.

DEFINITION: *An acid is a substance which produces hydrogen ions when dissolved in water or other dissociating liquids.*

Undissociated acids. When acids are perfectly free from water, or are dissolved in liquids like benzene which do not have the power of dissociating them into ions, they should have no real acid properties. This is found to be the case. Under these circumstances they do not affect the color of indicators or have any of the properties characteristic of acids.

The familiar bases. The bases most used in the laboratory are sodium hydroxide ($NaOH$), potassium hydroxide (KOH), and calcium hydroxide ($Ca(OH)_2$). These are white solids, soluble in water, the latter sparingly so. Some bases are very difficultly soluble in water. The very soluble ones with most pronounced basic properties are sometimes called the *alkalis*.

Characteristics of bases. (1) All bases contain hydrogen and oxygen. (2) When dissolved in water the molecules of the base dissociate into two kinds of ions. One of these is always composed of oxygen and hydrogen and is the anion. It has the formula OH and is called the *hydroxyl ion*. The remainder of the molecule, which usually consists of a single atom, is the cation. (3) The solution of a base has [Pg 108] a soapy feel and a brackish taste. (4) It reverses the color change produced in indicators by acids, turning red litmus blue, and red methyl orange yellow. Since all bases produce hydroxyl anions, while the cations of each are different, the properties which all bases have in common when in solution must be due to the hydroxyl ions.

DEFINITION: *A base is a substance which produces hydroxyl ions when dissolved in water or other dissociating liquids.*

Undissociated bases. Bases, in the absence of water or when dissolved in liquids which do not dissociate them, should have none of the properties characteristic of this class of substances. This has been found to be the case. For example, they have no effect upon indicators under these circumstances.

Neutralization. When an acid and a base are brought together in solution in proper proportion, the characteristic properties of each disappear. The solution tastes neither sour nor brackish; it has no effect upon indicators. There can therefore be neither hydrogen nor hydroxyl ions present in the solution. A study of reactions of this kind has shown that the hydrogen ions of the acid combine with the hydroxyl ions of the base to form molecules of water, water being a substance which is not appreciably dissociated into ions. This action of an acid on a base is called *neutralization*. The following equations express the neutralization of the three acids by three bases, water being formed in each case.

$$Na^+, OH^- + H^+, Cl^- = Na^+, Cl^- + H_2O.$$

$$K^+, OH^- + H^+, NO_3{}^- = K^+, NO_3{}^- + H_2O.$$

$$Ca^{++}, (OH)_2{}^- + H_2{}^{++}, SO_4{}^- = Ca^{++}, SO_4{}^- + 2H_2O.$$

[Pg 109]

DEFINITION: *Neutralization consists in the union of the hydrogen ion of an acid with the hydroxyl ion of a base to form water.*

Salts. It will be noticed that in neutralization the anion of the acid and the cation of the base are not changed. If, however, the water is expelled by evaporation, these two ions slowly unite, and when the water becomes saturated with the substance so produced, it separates in the form of a solid called a *salt*.

DEFINITION: *A salt is a substance formed by the union of the anion of an acid with the cation of a base.*

Characteristics of salts. (1) From the definition of a salt it will be seen that there is no element or group of elements which characterize salts. (2) Salts as a class have no peculiar taste. (3) In the absence of all other substances they are without action on indicators. (4) When dissolved in water they form two kinds of ions.

Heat of neutralization. If neutralization is due to the union of hydrogen ions with hydroxyl ions, and nothing more, it follows that when a given weight of water is formed in neutralization, the heat set free should always be the same, no matter from what acid and base the two kinds of ions have been supplied. Careful experiments have shown that this is the case, provided no other reactions take place at the same time. When 18g. of water are formed in neutralization, 13,700 cal. of heat are set free. This is represented in the equations

$$Na^+, OH^- + H^+, Cl^- = Na^+, Cl^- + H_2O + 13{,}700 \text{ cal.}$$

$$K^+, OH^- + H^+, NO_3{}^- = K^+, NO_3{}^- + H_2O + 13{,}700 \text{ cal.}$$

$$Ca^{++}, (OH)_2{}^- + H_2{}^{++}, SO_4{}^- = Ca^{++}, SO_4{}^- + 2H_2O + 2 \times 13{,}700 \text{ cal.}$$

Neutralization a quantitative act. Since neutralization is a definite chemical act, each acid will require a perfectly definite weight of each base for its neutralization. For [Pg 110] example, a given weight of sulphuric acid will always require a definite weight of sodium hydroxide, in accordance with the equation

$$H_2, SO_4 + 2Na, OH = Na_2, SO_4 + 2H_2O.$$

Determination of the ratio in neutralization. The quantities of acid and base required in neutralization may be determined in the following way. Dilute solutions of the two substances are prepared, the sulphuric acid being placed in one of the burettes (Fig. 34) and the sodium hydroxide in the other. The levels of the two liquids are then brought to the zero marks of the burettes by means of the stopcocks. A measured volume of the acid is drawn off into a beaker, a few drops of litmus solution added, and the sodium hydroxide is run in drop by drop until the red litmus just turns blue. The volume of the sodium hydroxide consumed is then noted. If the concentrations of the two solutions are known, it is easy to calculate what weight of sodium hydroxide is required to neutralize a given weight of sulphuric acid. By evaporating the neutralized solution to dryness, the weight of the sodium sulphate formed can be determined directly. Experiment shows that the weights are always in accordance with the equation in the preceding paragraph.

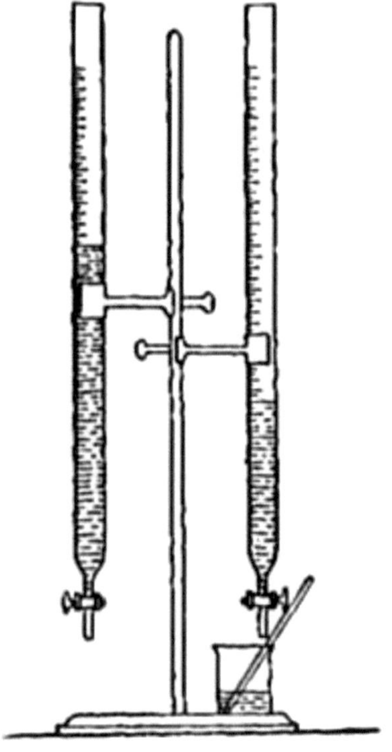

Fig. 34

Extent of dissociation. The question will naturally arise, When an acid, base, or salt dissolves in water, do all the molecules dissociate into ions, or only a part of them? The experiments by which this question can be answered cannot be described here. It has been found, however, that only a fraction of the molecules dissociate. The percentage which will dissociate in a given case depends upon several conditions, the chief of which are: (1) The concentration of the solution. In concentrated solutions only a very small [Pg 111] percentage of dissociation occurs. As the solution is diluted the percentage increases, and in very dilute solutions it may be very large, though it is never complete in any ordinary solution. (2) The nature of the dissolved compound. At equal concentrations substances differ much among themselves in the percentage of dissociation. The great majority of salts are about equally dissociated. Acids and bases, on the contrary, show great differences. Some are freely dissociated, while others are dissociated to but a slight extent.

Strength of acids and bases. Since acid and basic properties are due to hydrogen and hydroxyl ions respectively, the acid or base which will produce the greatest percentage of these ions at a given concentration must be regarded as the strongest representative of its class. The acids and bases described in the foregoing paragraphs are all quite strong. In 10% solutions they are dissociated to about 50%, and this is also approximately the extent to which most salts are dissociated at this same concentration.

Partial neutralization. 1. *Basic salts.* The chemical action between an acid and a base is not always as complete as has been represented in the foregoing paragraphs. For example, if the base magnesium hydroxide ($Mg(OH)_2$) and hydrochloric acid (HCl) are brought together in the ratio of an equal number of molecules of each, there will be only half enough hydrogen ions for the hydroxyl ions present.

$$Mg, (OH)_2 + H, Cl = Mg, OH, Cl + H_2O.$$

Magnesium, hydroxyl, and chlorine ions are left at the close of the reaction, and under the proper conditions unite to form molecules of the compound $Mg(OH)Cl$. This compound, when dissolved, can form hydroxyl ions and therefore possesses basic properties; it can also form the ions of a salt (Mg and Cl), and has properties characteristic of salts. Substances of this kind are called *basic salts.*

DEFINITION: *A basic salt is a substance which can give the ions both of a base and of a salt when dissolved in water.* [Pg 112]

2. *Acid salts.* In a similar way, when sulphuric acid and sodium hydroxide are brought together in the ratio of equal numbers of the molecules of each, it is possible to have a reaction expressed by the equation

$$Na, OH + H_2, SO_4 = Na, H, SO_4 + H_2O.$$

The ions remaining after all the hydroxyl ions have been used up are those of an acid (H) and those of a salt (Na and SO_4). These unite to form the substance $NaHSO_4$, and as the solution becomes saturated with this substance through evaporation, it separates in the form of crystals. In solution this substance can give hydrogen ions, and therefore possesses acid properties; it can also give the ions characteristic of a salt. It is therefore called an *acid salt.*

DEFINITION: *An acid salt is one which can give the ions of an acid and of a salt when in solution.*

3. *Normal salts.* Salts which are the products of complete neutralization, such as Na_2SO_4, and which in solution can give neither hydrogen nor hydroxyl ions, but only the ions of a salt, are called *normal salts* to distinguish them from acid and basic salts.

Methods of expressing reactions between compounds in solution. Chemical equations representing reactions between substances in solution may represent the details of the reaction, or they may simply indicate the final products formed. In the latter case the formation of ions is not indicated. Thus, if we wish to call attention to the details of the reaction between sodium hydroxide and hydrochloric acid in solution, the equation is written as follows:

$$Na^+, OH^- + H^+, Cl^- = Na^+, Cl^- + H_2O.$$

On the other hand, if we wish simply to represent the final products formed, the following is used.

$$NaOH + HCl = NaCl + H_2O.$$

Both of these methods will therefore be used:

Radicals. It has been emphasized that the hydroxyl group (OH) always [Pg 113] forms the anion of a base, while the group NO_3 forms the anion of nitric acid and sodium nitrate; the group SO_4, the anion of sulphuric acid and calcium sulphate. A group of elements which in this way constitutes a part of a molecule, acting as a unit in a chemical change, or forming ions in solution, is called a *radical*. Some of these radicals have been given special names, the names signifying the elements present in the radical. Thus we have the hydroxyl radical (OH) and the nitrate radical (NO_3).

DEFINITION: *A radical is a group of elements forming part of a molecule, and acting as a unit in chemical reactions.*

Names of acids, bases, and salts. Since acids, bases, and salts are so intimately related to each other, it is very advantageous to give names to the three classes in accordance with some fixed system. The system universally adopted is as follows:

Naming of bases. All bases are called *hydroxides*. They are distinguished from each other by prefixing the name of the element which is in combination with the hydroxyl group. Examples: sodium hydroxide (NaOH); calcium hydroxide ($Ca(OH)_2$); copper hydroxide ($Cu(OH)_2$).

Naming of acids. The method of naming acids depends upon whether the acid consists of two elements or three.

1. *Binary acids.* Acids containing only one element in addition to hydrogen are called *binary acids.* They are given names consisting of the prefix *hydro-*, the name of the second element present, and the termination *-ic.* Examples: hydrochloric acid (HCl); hydrosulphuric acid (H_2S).

2. *Ternary acids.* In addition to the two elements present in binary acids, the great majority of acids also contain oxygen. They therefore consist of three elements and [Pg 114] are called *ternary acids*. It usually happens that the same three elements can unite in different proportions to make several different acids. The most familiar one of these is given a name ending in the suffix *-ic,* while the one with less oxygen is given a similar name, but ending in the suffix *-ous.* Examples: nitric acid (HNO_3); nitrous acid (HNO_2). In cases where

more than two acids are known, use is made of prefixes in addition to the two suffixes -*ic* and -*ous*. Thus the prefix *per-* signifies an acid still richer in oxygen; the prefix *hypo-* signifies one with less oxygen.

Naming of salts. A salt derived from a binary acid is given a name consisting of the names of the two elements composing it, with the termination -*ide*. Example: sodium chloride (NaCl). All other binary compounds are named in the same way.

A salt of a ternary acid is named in accordance with the acid from which it is derived. A ternary acid with the termination -*ic* gives a salt with the name ending in -*ate*, while an acid with termination -*ous* gives a salt with the name ending in -*ite*. The following table will make the application of these principles clear:

ACIDS	SYMBOL	SALTS	SYMBOL
Hydrochloric	HCl	Sodium chloride	NaCl
Hypochlorous	HClO	Sodium hypochlorite	NaClO
Chlorous	HClO2	Sodium chlorite	NaClO2
Chloric	HClO3	Sodium chlorate	NaClO3
Perchloric	HClO4	Sodium perchlorate	NaClO4

[Pg 115]

EXERCISES

1. 25 cc. of a solution containing 40 g. of sodium hydroxide per liter was found to neutralize 25 cc. of a solution of hydrochloric acid. What was the strength of the acid solution?

2. After neutralizing a solution of sodium hydroxide with nitric acid, there remained after evaporation 100 g. of sodium nitrate. How much of each substance had been used?

3. A solution contains 18 g. of hydrochloric acid per 100 cc. It required 25 cc. of this solution to neutralize 30 cc. of a solution of sodium hydroxide. What was the strength of the sodium hydroxide solution in parts per hundred?

4. When perfectly dry sulphuric acid is treated with perfectly dry sodium hydroxide, no chemical change takes place. Explain.

5. When cold, concentrated sulphuric acid is added to zinc, no change takes place. Recall the action of dilute sulphuric acid on the same metal. How do you account for the difference?

6. A solution of hydrochloric acid in benzene does not conduct the electric current. When this solution is treated with zinc, will hydrogen be evolved? Explain.

7. (*a*) Write equation for preparation of hydrogen from zinc and dilute sulphuric acid. (*b*) Rewrite the same equation from the standpoint of the theory of electrolytic dissociation, (*c*) Subtract the common SO_4 ion from both members of the equation, (*d*) From the resulting equation, explain in what the preparation of hydrogen consists when examined from the standpoint of this theory.

8. In the same manner as in the preceding exercise, explain in what the action of sodium on water to give hydrogen consists.

[Pg 116]

CHAPTER XI

VALENCE

Definition of valence. A study of the formulas of various binary compounds shows that the elements differ between themselves in the number of atoms of other elements which they are able to hold in combination. This is illustrated in the formulas

HCl, H_2O, H_3N, H_4C.

(hydrochloric acid)   (water)   (ammonia)   (marsh gas)

It will be noticed that while one atom of chlorine combines with one atom of hydrogen, an atom of oxygen combines with two, an atom of nitrogen with three, one of carbon with four. The number which expresses this combining ratio between atoms is a definite property of each element and is called its *valence*.

DEFINITION: *The valence of an element is that property which determines the number of the atoms of another element which its atom can hold in combination.*

Valence a numerical property. Valence is therefore merely a numerical relation and does not convey any information in regard to the intensity of the affinity between atoms. Judging by the heat liberated in their union, oxygen has a far stronger affinity for hydrogen than does nitrogen, but an atom of oxygen can combine with two atoms only of hydrogen, while an atom of nitrogen can combine with three. [Pg 117]

Measure of valence. In expressing the valence of an element we must select some standard for comparison, just as in the measurement of any other numerical quantity. It has been found that an atom of hydrogen is never able to hold in combination more than one atom of any other element. Hydrogen is therefore taken as the standard, and other elements are compared with it in determining their valence. A number of other elements are like hydrogen in being able to combine with at most one atom of other elements, and such elements are called *univalent*. Among these are chlorine, iodine, and sodium. Elements such as oxygen, calcium, and zinc, which can combine with two atoms of hydrogen or other univalent elements, are said to be *divalent*. Similarly, we have *trivalent, tetravalent, pentavalent* elements. None have a valence of more than 8.

Indirect measure of valence. Many elements, especially among the metals, do not readily form compounds with hydrogen, and their valence is not easy to determine by direct comparison with the standard element. These elements, however, combine with other univalent elements, such as chlorine, and their valence can be determined from the compounds so formed.

Variable valence. Many elements are able to exert different valences under differing circumstances. Thus we have the compounds Cu_2O and CuO, CO and CO_2, $FeCl_2$ and $FeCl_3$. It is not always possible to assign a fixed valence to an element. Nevertheless each element tends to exert some normal valence, and the compounds in which it has a valence different from this are apt to be unstable and easily changed into compounds in which the valence of the element is normal. The valences of the various elements will become familiar as the elements are studied in detail. [Pg 118]

Valence and combining ratios. When elements combine to form compounds, the ratio in which they combine will be determined by their valences. In those compounds which consist of two elements directly combined, the union is between such numbers of the two atoms as have equal valences. Elements of the same valence will therefore combine atom for atom. Designating the valence of the atoms by Roman numerals placed above their symbols, we have the formulas

II II	II III	I II	IV IV
HCl,	ZnO,	BN,	CSi.

A divalent element, on the other hand, will combine with two atoms of a univalent element. Thus we have

II II $\qquad\qquad\qquad\qquad\qquad\qquad\qquad$ II II

$ZnCl_2$ $\qquad\qquad$ and H_2O

(the numerals above each symbol representing the sum of the valences of the atoms of the element present). A trivalent atom will combine with three atoms of a univalent element, as in the compound

III III
H_3N.

If a trivalent element combines with a divalent element, the union will be between two atoms of the trivalent element and three of the divalent element, since these numbers are the smallest which have equal valences. Thus the oxide of the trivalent metal aluminium has the formula Al_2O_3. Finally one atom of a tetravalent element such as carbon will combine with four atoms of a univalent element, as in the compound CH_4, or with two atoms of a divalent element, as in the compound CO_2.

We have no knowledge as to why elements differ in their combining power, and there is no way to determine their valences save by experiment.

Valence and the structure of compounds. Compounds will be met from time to time which are apparent exceptions to the general statements just made in regard to valence. Thus, from the formula for hydrogen dioxide (H_2O_2), it might be supposed that the oxygen is univalent; yet it is certainly [Pg 119] divalent in water (H_2O). That it may also be divalent in H_2O_2 may be made clear as follows: The unit valence of each element may be represented graphically by a line attached to its symbol. Univalent hydrogen and divalent oxygen will then have the symbols H- and -O-. When atoms combine, each unit valence of one atom combines with a unit valence of another atom. Thus the composition of water may be expressed by the formula H-O-H, which is meant to show that each of the unit valences of oxygen is satisfied with the unit valence of a single hydrogen atom.

The chemical conduct of hydrogen dioxide leads to the conclusion that the two oxygen atoms of its molecule are in direct combination with each other, and in addition each is in combination with a hydrogen atom. This may be expressed by the formula H-O-O-H. The oxygen in the compound is therefore divalent, just as it is in water. It will thus be seen that the structure of a compound must be known before the valences of the atoms making up the compound can be definitely decided upon.

Such formulas as H-O-H and H-O-O-H are known as *structural formulas,* because they are intended to show what is known in regard to the arrangement of the atoms in the molecules.

Valence and the replacing power of atoms. Just as elements having the same valence combine with each other atom for atom, so if they replace each other in a chemical reaction they will do so in the same ratio. This is seen in the following equations, in which a univalent hydrogen atom is replaced by a univalent sodium atom:

$$NaOH + HCl = NaCl + H_2O.$$

$$2NaOH + H_2SO_4 = Na_2SO_4 + 2H_2O.$$

$$Na + H_2O = NaOH + H.$$

Similarly, one atom of divalent calcium will replace two atoms of univalent hydrogen or one of divalent zinc:

$$Ca(OH)_2 + 2\,HCl = CaCl_2 + 2H_2O.$$

$$CaCl_2 + ZnSO_4 = CaSO_4 + ZnCl_2.$$

[Pg 120]

In like manner, one atom of a trivalent element will replace three of a univalent element, or two atoms will replace three atoms of a divalent element.

Valence and its applications to formulas of salts. While the true nature of valence is not understood and many questions connected with the subject remain unanswered, yet many of the main facts are of much help to the student. Thus the formula of a salt, differs from that of the acid from which it is derived in that the hydrogen of the acid has been replaced by a metal. If, then, it is known that a given metal forms a normal salt with a certain acid, the formula of the salt can at once be determined if the valence of the metal is known. Since sodium is univalent, the sodium salts of the acids HCl and H_2SO_4 will be respectively NaCl and Na_2SO_4. One atom of divalent zinc will replace 2 hydrogen atoms, so that the corresponding zinc salts will be $ZnCl_2$ and $ZnSO_4$.

The formula for aluminium sulphate is somewhat more difficult to determine. Aluminium is trivalent, and the simplest ratio in which the aluminium atom can replace the hydrogen in sulphuric acid is 2 atoms of aluminium (6 valences) to 3 molecules of sulphuric acid (6 hydrogen atoms). The formula of the sulphate will then be $Al_2(SO_4)_3$.

Valence and its application to equation writing. It will be readily seen that a knowledge of valence is also of very great assistance in writing the equations for reactions of double decomposition.

Thus, in the general reaction between an acid and a base, the essential action is between the univalent hydrogen ion and the univalent hydroxyl ion. The base and the acid must always be taken in such proportions as to secure an equal number of each of these ions. Thus, in the reaction between ferric hydroxide ($Fe(OH)_3$) and sulphuric acid (H_2SO_4), it will be necessary to take 2 molecules of the former and 3 of the latter in order to have an equal number of the two ions, namely, 6. The equation will then be

$$2Fe(OH)_3 + 3H_2SO_4 = Fe_2(SO_4)_3 + 6H_2O.$$

Under certain conditions the salts $Al_2(SO_4)_3$ and $CaCl_2$ undergo double decomposition, the two metals, aluminium and calcium, exchanging places. The simplest ratio of exchange in this case is 2 atoms of aluminium (6 valences) and 3 atoms of calcium (6 valences). [Pg 121] The reaction will therefore take place between 1 molecule of $Al_2(SO_4)_3$ and 3 of $CaCl_2$, and the equation is as follows:

$$Al_2(SO_4)_3 + 3\ CaCl_2 = 3CaSO_4 + 2AlCl_3.$$

EXERCISES

1. Sodium, calcium, and aluminium have valences of 1, 2, and 3 respectively; write the formulas of their chlorides, sulphates, and phosphates (phosphoric acid = H_3PO_4), on the supposition that they form salts having the normal composition.

2. Iron forms one series of salts in which it has a valence of 2, and another series in which it has a valence of 3; write the formulas for the two chlorides of iron, also for the two sulphates, on the supposition that these have the normal composition.

3. Write the equation representing the neutralization of each of the following bases by each of the acids whose formulas are given:

NaOH   HCl

$Ba(OH)_2$ H_2SO_4

$Al(OH)_3$ H_3PO_4

4. Silver acts as a univalent element and calcium as a divalent element in the formation of their respective nitrates and chlorides. (*a*) Write the formula for silver nitrate; for calcium chloride. (*b*) When solutions of these two salts are mixed, the two metals, silver and calcium, exchange places; write the equation for the reaction.

5. Antimony acts as a trivalent element in the formation of a chloride. (*a*) What is the formula for antimony chloride? (*b*) When hydrosulphuric acid (H_2S) is passed into a solution of this chloride the hydrogen and antimony exchange places; write the equation for the reaction.

6. Lead has a valence of 2 and iron of 3 in the compounds known respectively as lead nitrate and ferric sulphate. (*a*) Write the formulas for these two compounds. (*b*) When their solutions are mixed the two metals exchange places; write the equation for the reaction.

[Pg 122]

CHAPTER XII

COMPOUNDS OF NITROGEN

Occurrence. As has been stated in a former chapter, nitrogen constitutes a large fraction of the atmosphere. The compounds of nitrogen, however, cannot readily be obtained from this source, since at any ordinary temperature nitrogen is able to combine directly with very few of the elements.

In certain forms of combination nitrogen occurs in the soil from which it is taken up by plants and built into complex substances composed chiefly of carbon, hydrogen, oxygen, and nitrogen. Animals feeding on these plants assimilate the nitrogenous matter, so that this element is an essential constituent of both plants and animals.

Decomposition of organic matter by bacteria. When living matter dies and undergoes decay complicated chemical reactions take place, one result of which is that the nitrogen of the organic matter is set free either as the element nitrogen, or in the form of simple compounds, such as ammonia (NH_3) or oxides of nitrogen. Experiment has shown that all such processes of decay are due to the action of different kinds of bacteria, each particular kind effecting a different change.

Decomposition of organic matter by heat. When organic matter is strongly heated decomposition into simpler substances takes place in much the same way as in the case of bacterial decomposition. Coal is a complex substance of [Pg 123] vegetable origin, consisting largely of carbon, but also containing hydrogen, oxygen, and nitrogen. When this is heated in a closed vessel so that air is excluded, about one seventh of the nitrogen is converted into ammonia, and this is the chief source from which ammonia and its compounds are obtained.

COMPOUNDS OF NITROGEN WITH HYDROGEN

Ammonia (NH_3). Several compounds consisting exclusively of nitrogen and hydrogen are known, but only one, ammonia, need be considered here.

Preparation of ammonia. Ammonia is prepared in the laboratory by a different method from the one which is used commercially.

1. *Laboratory method.* In the laboratory ammonia is prepared from ammonium chloride, a compound having the formula NH_4Cl, and obtained in the manufacture of coal gas. As will be shown later in the chapter, the group NH_4 in this compound acts as a univalent radical and is known as *ammonium.* When ammonium chloride is warmed with sodium hydroxide, the ammonium and sodium change places, the reaction being expressed in the following equation.

$$NH_4Cl + NaOH = NaCl + NH_4OH.$$

The ammonium hydroxide (NH_4OH) so formed is unstable and breaks down into water and ammonia.

$$NH_4OH = NH_3 + H_2O.$$

Calcium hydroxide ($Ca(OH)_2$) is frequently used in place of the more expensive sodium hydroxide, the equations being

$$2NH_4Cl + Ca(OH)_2 = CaCl_2 + 2NH_4OH,$$

$$2NH_4OH = 2H_2O + 2NH_3.$$
[Pg 124]

In the preparation, the ammonium chloride and calcium hydroxide are mixed together and placed in a flask arranged as shown in Fig. 35. The mixture is gently warmed, when ammonia is evolved as a gas and is collected by displacement of air.

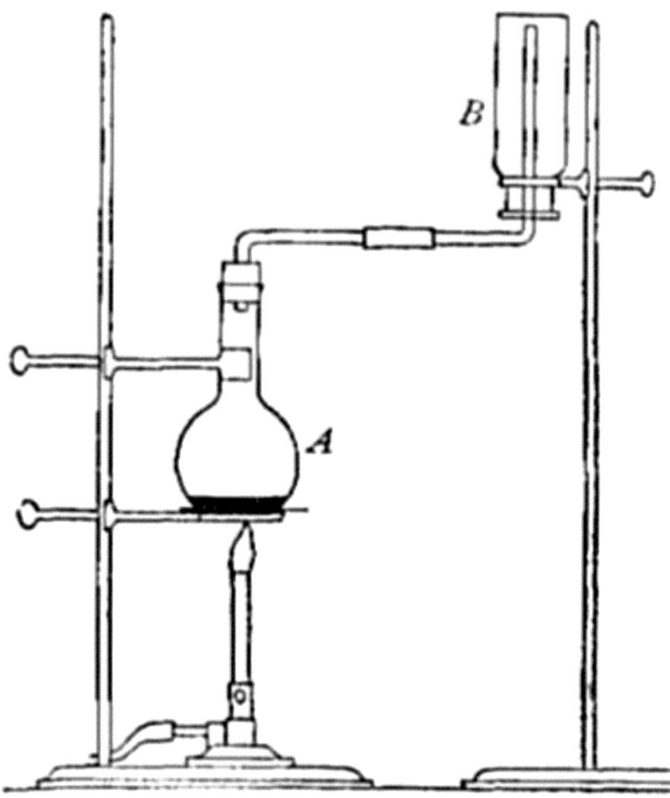

Fig. 35

2. *Commercial method.* Nearly all the ammonia of commerce comes from the gasworks. Ordinary illuminating gas is made by distilling coal, as will be explained later, and among the products of this distillation a solution of ammonia in water is obtained. This solution, known as *gas liquor,* contains not only ammonia but other soluble substances. Most of these combine chemically with lime, while ammonia does not; if then lime is added to the gas liquor and the liquor is heated, the ammonia is driven out from the mixture. It may be dissolved again in pure, cold water, forming *aqua ammonia,* or the ammonia water of commerce.

Preparation from hydrogen and nitrogen. When electric sparks are passed for some time through a mixture of hydrogen and nitrogen, a small percentage of the two elements in the mixture is changed into ammonia. The action soon ceases, however, for the reason that ammonia is decomposed by the electric discharge. The reaction expressed in the equation

$$N + 3H = NH_3$$

can therefore go in either direction depending upon the relative quantities of the substances present. This recalls the similar change from oxygen into ozone, which soon ceases because the ozone is in turn decomposed into oxygen.

[Pg 125]

Physical properties. Under ordinary conditions ammonia is a gas whose density is 0.59. It is therefore little more than half as heavy as air. It is easily condensed into a colorless liquid, and can now be purchased in liquid form in steel cylinders. The gas is colorless and has a strong, suffocating odor. It is extremely soluble in water, 1 l. of water at 0° and 760 mm. pressure dissolving 1148 l. of the gas. In dissolving this large volume of gas the water expands considerably, so that the density of the solution is less than that of water, the strongest solutions having a density of 0.88.

Chemical properties. Ammonia will not support combustion, nor will it burn under ordinary conditions. In an atmosphere of oxygen it burns with a feeble, yellowish flame. When quite dry it is not a very active substance, but when moist it combines with a great many substances, particularly with acids.

Uses. It has been stated that ammonia can be condensed to a liquid by the application of pressure. If the pressure is removed from the liquid so obtained, it rapidly passes again into the gaseous state and in so doing absorbs a large amount of heat. Advantage is taken of this fact in the preparation of artificial ice. Large quantities of ammonia are also used in the preparation of ammonium compounds.

The manufacture of artificial ice. Fig. 36 illustrates the method of preparing artificial ice. The ammonia gas is liquefied in the pipes X by means of the pump Y. The heat generated is absorbed by water flowing over the pipes. The pipes lead into a large brine tank, a cross section of which is shown in the figure. Into the brine (concentrated solution of common salt) contained in this tank are dipped the vessels A, B, C, filled with pure water. The pressure is removed from the liquid ammonia as it passes into the pipes immersed in the [Pg 126] brine, and the heat absorbed by the rapid evaporation of the liquid lowers the temperature of the brine below zero. The water in A, B, C is thereby frozen into cakes of ice. The gaseous ammonia resulting from the evaporation of the liquid ammonia is again condensed, so that the process is continuous.

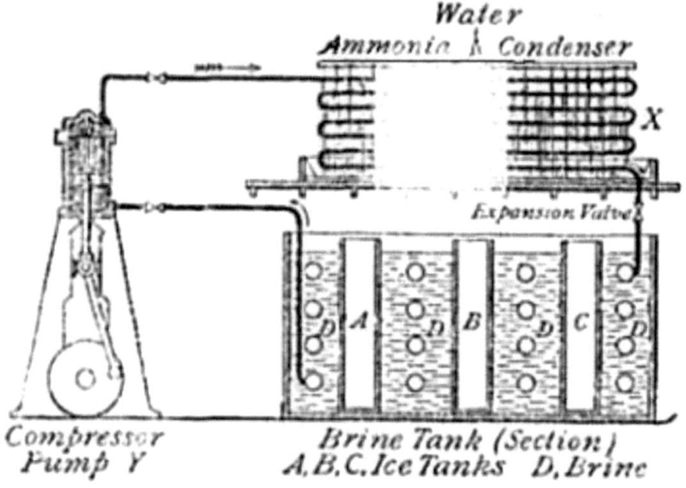

Fig. 36

Ammonium hydroxide (NH_4OH). The solution of ammonia in water is found to have strong basic properties and therefore contains hydroxyl ions. It turns red litmus blue; it has a soapy feel; it neutralizes acids, forming salts with them. It seems probable, therefore, that when ammonia dissolves in water it combines chemically with it according to the equation

$$NH_3 + H_2O = NH_4OH,$$

and that it is the substance NH_4OH, called ammonium hydroxide, which has the basic properties, dissociating into the ions NH_4 and OH. Ammonium hydroxide has never been obtained in a pure state. At every attempt to isolate it the substance breaks up into water and ammonia, —

$$NH_4OH = NH_3 + H_2O.$$

The ammonium radical. The radical NH_4 plays the part of a metal in many chemical reactions and is called ammonium. The ending *-ium* is given to the name to indicate the metallic properties of the substance, since the names [Pg 127] of the metals in general have that ending. The salts formed by the action of the base ammonium hydroxide on acids are called ammonium salts. Thus, with hydrochloric acid, ammonium chloride is formed in accordance with the equation

$$NH_4OH + HCl = NH_4Cl + H_2O.$$

Similarly, with nitric acid, ammonium nitrate (NH_4NO_3) is formed, and with sulphuric acid, ammonium sulphate (($NH_4)_2SO_4$).

It will be noticed that in the neutralization of ammonium hydroxide by acids the group NH_4 replaces one hydrogen atom of the acid, just as sodium does. The group therefore acts as a univalent metal.

Combination of nitrogen with hydrogen by volume. Under suitable conditions ammonia can be decomposed into nitrogen and hydrogen by passing electric sparks through the gas. Accurate measurement has shown that when ammonia is decomposed, two volumes of the gas yield one volume of nitrogen and three volumes of hydrogen. Consequently, if the two elements were to combine directly, one volume of nitrogen would combine with three volumes of hydrogen to form two volumes of ammonia. Here, as in the formation of steam from hydrogen and oxygen, small whole numbers serve to indicate the relation between the volumes of combining gases and that of the gaseous product.

COMPOUNDS OF NITROGEN WITH OXYGEN AND HYDRO-GEN

In addition to ammonium hydroxide, nitrogen forms several compounds with hydrogen and oxygen, of which nitric acid (HNO_3) and nitrous acid [Pg 128] (HNO_2) are the most familiar.

Nitric acid (HNO_3). Nitric acid is not found to any extent in nature, but some of its salts, especially sodium nitrate ($NaNO_3$) and potassium nitrate (KNO_3) are found in large quantities. From these salts nitric acid can be obtained.

Fig. 37

Preparation of nitric acid. When sodium nitrate is treated with concentrated cold sulphuric acid, no chemical action seems to take place. If, however, the mixture is heated in a retort, nitric acid is given off as a vapor and may be easily condensed to a liquid by passing the vapor into a tube surrounded by cold water, as shown in Fig. 37. An examination of the liquid left in the retort shows that it contains sodium acid sulphate ($NaHSO_4$), so that the reaction may be represented by the equation

$$NaNO_3 + H_2SO_4 = NaHSO_4 + HNO_3.$$

If a smaller quantity of sulphuric acid is taken and the mixture is heated to a high temperature, normal sodium sulphate is formed:

$$2NaNO_3 + H_2SO_4 = Na_2SO_4 + 2HNO_3.$$

In this case, however, the higher temperature required decomposes a part of the nitric acid.

The commercial preparation of nitric acid. Fig. 38 illustrates a form of apparatus used in the preparation of nitric acid on a large scale. Sodium nitrate and sulphuric acid are heated in the iron retort *A*. The resulting acid vapors pass in the direction indicated by the arrows, and are condensed in the glass tubes *B*, which are covered with cloth kept cool by streams of water. These tubes are inclined so that the liquid resulting from the condensation of the vapors runs back into *C* and is drawn off into large vessels (*D*).

[Pg 129]

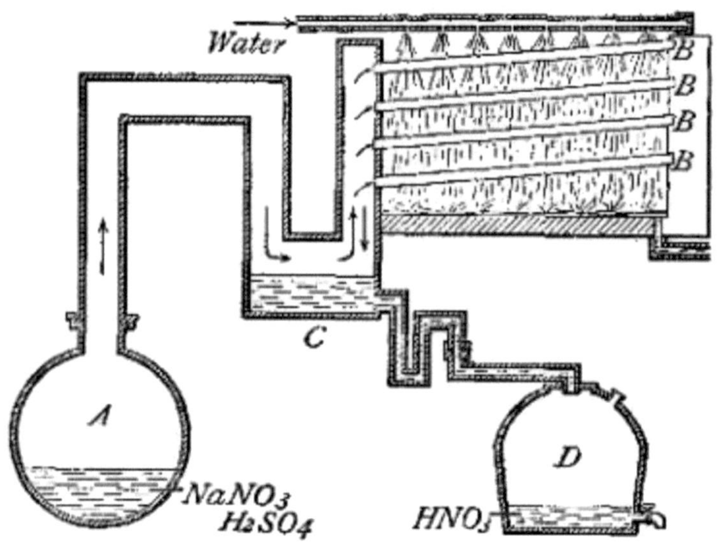

Fig. 38

Physical properties of nitric acid. Pure nitric acid is a colorless liquid, which boils at about 86° and has a density of 1.56. The concentrated acid of commerce contains about 68% of the acid, the remainder being water. Such a mixture has a density of 1.4. The concentrated acid fumes somewhat in moist air, and has a sharp choking odor.

Chemical properties. The most important chemical properties of nitric acid are the following.

1. *Acid properties.* As the name indicates, this substance is an acid, and has all the properties of that class of substances. It changes blue litmus red and has a sour taste in dilute solutions. It forms hydrogen ions in solution and neutralizes bases forming salts. It also acts upon the oxides of most metals, forming a salt and water. It is one of the strongest acids.

2. *Decomposition on heating.* When boiled, or exposed for some time to sunlight, it suffers a partial decomposition according to the equation

$$2HNO_3 = H_2O + 2NO_2 + O.$$

The substance NO_2, called nitrogen peroxide, is a brownish gas, which is readily soluble in water and in nitric acid. It therefore dissolves in the undecomposed acid, and imparts a yellowish or reddish color to it. Concentrated [Pg 130] nitric acid highly charged with this substance is called *fuming nitric acid*.

3. *Oxidizing action.* According to its formula, nitric acid contains a large percentage of oxygen, and the reaction just mentioned shows that the compound is not a very stable one, easily undergoing decomposition. These properties should make it a good oxidizing agent, and we find that this is the case. Under ordinary circumstances, when acting as an oxidizing agent, it is decomposed according to the equation

$$2HNO_3 = H_2O + 2NO + 3O.$$

The oxygen is taken up by the substance oxidized, and not set free, as is indicated in the equation. Thus, if carbon is oxidized by nitric acid, the oxygen combines with carbon, forming carbon dioxide (CO_2):

$$C + 2O = CO_2.$$

4. *Action on metals.* We have seen that when an acid acts upon a metal hydrogen is set free. Accordingly, when nitric acid acts upon a metal, such as copper, we should expect the reaction to take place which is expressed in the equation

$$Cu + 2HNO_3 = Cu(NO_3)_2 + 2H.$$

This reaction does take place, but the hydrogen set free is immediately oxidized to water by another portion of the nitric acid according to the equation

$$HNO_3 + 3H = 2H_2O + NO.$$

As these two equations are written, two atoms of hydrogen are given off in the first equation, while three are used up in the second. In order that the hydrogen may be equal in [Pg 131] the two equations, we must multiply the first by 3 and the second by 2. We shall then have

$$3Cu + 6HNO_3 = 3Cu(NO_3)_2 + 6H,$$

$$2HNO_3 + 6H = 4H_2O + 2NO.$$

The two equations may now be combined into one by adding the quantities on each side of the equality sign, canceling the hydrogen which is given off in the one reaction and used up in the other. We shall then have the equation

$$3Cu + 8HNO_3 = 3Cu(NO_3)_2 + 2NO + 4H_2O.$$

A number of other reactions may take place when nitric acid acts upon metals, resulting in the formation of other oxides of nitrogen, free nitrogen, or even ammonia. The reaction just given is, however, the usual one.

Importance of steps in a reaction. This complete equation has the advantage of making it possible to calculate very easily the proportions in which the various substances enter into the reaction or are formed in it. It is unsatisfactory in that it does not give full information about the way in which the reaction takes place. For example, it does not suggest that hydrogen is at first formed, and subsequently transformed into water. It is always much more important to remember the steps in a chemical reaction than to remember the equation expressing the complete action; for if these steps in the reaction are understood, the complete equation is easily obtained in the manner just described.

Salts of nitric acid, — nitrates. The salts of nitric acid are called nitrates. Many of these salts will be described in the study of the metals. They are all soluble in water, and when heated to a high temperature undergo decomposition. In a few cases a nitrate on being heated evolves oxygen, forming a nitrite:

$$NaNO_3 = NaNO_2 + O.$$

[Pg 132]

In other cases the decomposition goes further, and the metal is left as oxide:

$$Cu(NO_3)_2 = CuO + 2NO_2 + O.$$

Nitrous acid (HNO_2). It is an easy matter to obtain sodium nitrite ($NaNO_2$), as the reaction given on the previous page indicates. Instead of merely heating the nitrate, it is better to heat it together with a mild reducing agent, such as lead, when the reaction takes place which is expressed by the equation

$$NaNO_3 + Pb = PbO + NaNO_2.$$

When sodium nitrite is treated with an acid, such as sulphuric acid, it is decomposed and nitrous acid is set free:

$$NaNO_2 + H_2SO_4 = NaHSO_4 + HNO_2.$$

The acid is very unstable, however, and decomposes readily into water and nitrogen trioxide (N_2O_3):

$$2HNO_2 = H_2O + N_2O_3.$$

Dilute solutions of the acid, however, can be obtained.

COMPOUNDS OF NITROGEN WITH OXYGEN

Nitrogen combines with oxygen to form five different oxides. The formulas and names of these are as follows:

N_2O nitrous oxide.

NO nitric oxide.

NO_2 nitrogen peroxide.

N_2O_3 nitrogen trioxide, or nitrous anhydride.

N_2O_5 nitrogen pentoxide, or nitric anhydride.

These will now be briefly discussed.

Nitrous oxide (*laughing gas*) (N_2O). Ammonium nitrate, like all nitrates, undergoes decomposition when heated; and owing to the fact that it contains no metal, but does [Pg 133] contain both oxygen and hydrogen, the reaction is a peculiar one. It is represented by the equation

$$NH_4NO_3 = 2H_2O + N_2O.$$

The oxide of nitrogen so formed is called nitrous oxide or laughing gas. It is a colorless gas having a slight odor. It is somewhat soluble in water, and in solution has a slightly sweetish taste. It is easily converted into a liquid and can be purchased in this form. When inhaled it produces a kind of hysteria (hence the name "laughing gas"), and even unconsciousness and insensibility to pain if taken in large amounts. It has long been used as an anæsthetic for minor surgical operations, such as those of dentistry, but owing to its unpleasant after effects it is not so much in use now as formerly.

Chemically, nitrous oxide is remarkable for the fact that it is a very energetic oxidizing agent. Substances such as carbon, sulphur, iron, and phosphorus burn in it almost as brilliantly as in oxygen, forming oxides and setting free nitrogen. Evidently the oxygen in nitrous oxide cannot be held in very firm combination by the nitrogen.

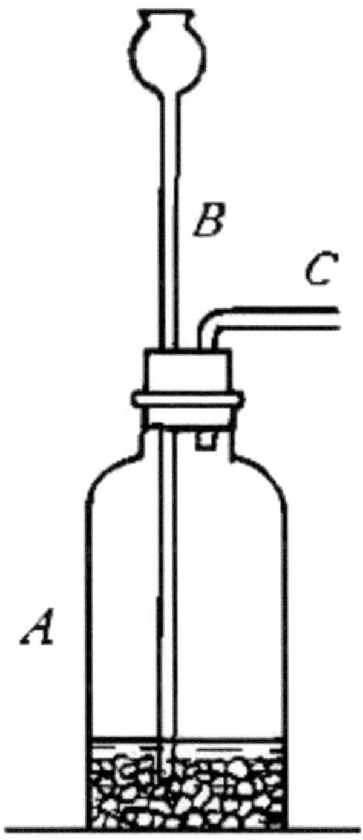

Fig. 39

Nitric oxide (NO). We have seen that when nitric acid acts upon metals, such as copper, the reaction represented by the following equation takes place:

$$3Cu + 8HNO_3 = 3Cu(NO_3)_3 + 2NO + 4H_2O.$$

Nitric oxide is most conveniently prepared in this way. The metal is placed in the flask *A* (Fig. 39) and the acid added slowly through the funnel tube *B*. The gas escapes through *C* and is collected over water. [Pg 134]

Pure nitric oxide is a colorless gas, slightly heavier than air, and is practically insoluble in water. It is a difficult gas to liquefy. Unlike nitrous oxide, nitric oxide does not part with its oxygen easily, and burning substances introduced into this gas are usually extinguished. A few substances like phosphorus, which have a very strong affinity for oxygen and which are burning energetically in the air, will continue to burn in an atmosphere of nitric oxide. In this case the nitric oxide loses all of its oxygen and the nitrogen is set free as gas.

Action of nitric oxide with oxygen. When nitric oxide comes into contact with oxygen or with the air, it at once combines with the oxygen even at ordinary temperatures, forming a reddish-yellow gas of the formula NO_2, which is called nitrogen peroxide. This action is not energetic enough to produce a flame, though considerable heat is set free.

Nitrogen peroxide (NO_2). This gas, as we have just seen, is formed by allowing nitric oxide to come into contact with oxygen. It can also be made by heating certain nitrates, such as lead nitrate:

$$Pb(NO_3)_2 = PbO + 2NO_2 + O.$$

It is a reddish-yellow gas of unpleasant odor, which is quite poisonous when inhaled. It is heavier than air and is easily condensed to a liquid. It dissolves in water, but this solution is not a mere physical solution; the nitrogen peroxide is decomposed, forming a mixture of nitric and nitrous acids:

$$2NO_2 + H_2O = HNO_2 + HNO_3.$$

Nitrogen peroxide will not combine with more oxygen; it will, however, give up a part of its oxygen to burning substances, acting as an oxidizing agent:

$$NO_2 = NO + O.$$

[Pg 135]

Acid anhydrides. The oxides N_2O_3 (nitrogen trioxide) and N_2O_5 (nitrogen pentoxide) are rarely prepared and need not be separately described. They bear a very interesting relation to the acids of nitrogen. When dissolved in water they combine with the water, forming acids:

$$N_2O_3 + H_2O = 2HNO_2,$$

$$N_2O_5 + H_2O = 2HNO_3.$$

On the other hand, nitrous acid very easily decomposes, yielding water and nitrogen trioxide, and by suitable means nitric acid likewise may be decomposed into water and nitrogen pentoxide:

$$2HNO_2 = H_2O + N_2O_3,$$

$$2HNO_3 = H_2O + N_2O_5.$$

In view of the close relation between these oxides and the corresponding acids, they are called *anhydrides* of the acids, N_2O_3 being nitrous anhydride and N_2O_5 nitric anhydride.

DEFINITION: *Any oxide which will combine with water to form an acid, or which together with water is formed by the decomposition of an acid, is called an anhydride of that acid.*

EXERCISES

1. Perfectly dry ammonia does not affect litmus paper. Explain.

2. Can ammonia be dried by passing the gas through concentrated sulphuric acid? Explain.

3. Ammonium hydroxide is a weak base, i.e. it is not highly dissociated. When it is neutralized by strong acids the heat of reaction is less than when strong bases are so neutralized. Suggest some possible cause for this.

4. Why is brine used in the manufacture of artificial ice?

5. Discuss the energy changes which take place in the manufacture of artificial ice. [Pg 136]

6. What weight of ammonium chloride is necessary to furnish enough ammonia to saturate 1 l. of water at 0° and 760 mm.?

7. What weight of sodium nitrate is necessary to prepare 100 cc. of commercial nitric acid? What weight of potassium nitrate is necessary to furnish the same weight of acid?

8. 100 l. of nitrogen peroxide were dissolved in water and neutralized with sodium hydroxide. What substances were formed and how much of each?(1 l. nitrogen peroxide weighs 2.05 grams.)

9. How many liters of nitrous oxide, measured under standard conditions, can be prepared from 10 g. of ammonium nitrate?

10. What weight of copper is necessary to prepare 50 l. of nitric oxide under standard conditions?

11. (*a*) Calculate the percentage composition of the oxides of nitrogen. (*b*) What important law does this series of substances illustrate?

12. Write the equations representing the reactions between ammonium hydroxide, and sulphuric acid and nitric acid respectively, in accordance with the theory of electrolytic dissociation.

13. In the same way, write the equations representing the reactions between nitric acid and each of the following bases: NaOH, KOH, NH_4OH, $Ca(OH)_2$.

[Pg 137]

CHAPTER XIII

REVERSIBLE REACTIONS AND CHEMICAL EQUILIBRIUM

Reversible reactions. The reactions so far considered have been represented as continuing, when once started, until one or the other substance taking part in the reaction has been used up. In some reactions this is not the case. For example, we have seen that when steam is passed over hot iron the reaction is represented by the equation

$$3Fe + 4H_2O = Fe_3O_4 + 8H.$$

On the other hand, when hydrogen is passed over hot iron oxide the reverse reaction takes place:

$$Fe_3O_4 + 8H = 3Fe + 4H_2O.$$

The reaction can therefore go in either direction, depending upon the conditions of the experiment. Such a reaction is called a *reversible reaction*. It is represented by an equation with double arrows in place of the equality sign, thus:

$$3Fe + 4H_2O <\text{--}> Fe_3O_4 + 8H.$$

In a similar way, the equation

$$N + 3H <\text{--}> NH_3$$

expresses the fact that under some conditions nitrogen may unite with hydrogen to form ammonia, while under other conditions ammonia decomposes into nitrogen and hydrogen.

The conversion of oxygen into ozone is also reversible and may be represented thus:

oxygen <--> ozone.

[Pg 138]

Chemical equilibrium. Reversible reactions do not usually go on to completion in one direction unless the conditions under which the reaction takes place are very carefully chosen. Thus, if iron and steam are confined in a heated tube, the steam acts upon the iron, producing iron oxide and hydrogen. But these substances in turn act upon each other to form iron and steam once more. When these two opposite reactions go on at such rates that the weight of the iron changed into iron oxide is just balanced by the weight of the iron oxide changed into iron, there will be no further change in the relative weights of the four substances present in the tube. The reaction is then said to have reached an equilibrium.

Factors which determine the point of equilibrium. There are two factors which have a great deal of influence in determining the point at which a given reaction will reach equilibrium.

1. *Influence of the chemical nature of the substances.* If two reversible reactions of the same general kind are selected, it has been found that the point of equilibrium is different in the two cases. For example, in the reactions represented by the equations

$$3Fe + 4H_2O <--> Fe_3O_4 + 8H,$$

$$Zn + H_2O <--> ZnO + 2H,$$

the equilibrium will be reached when very different quantities of the iron and zinc have been changed into oxides. The individual chemical properties of the iron and zinc have therefore marked influence upon the point at which equilibrium will be reached.

2. *Influence of relative mass.* If the tube in which the reaction

$$3Fe + 4H_2O \longleftrightarrow Fe_3O_4 + 8H$$

[Pg 139]

has come to an equilibrium is opened and more steam is admitted, an additional quantity of the iron will be changed into iron oxide. If more hydrogen is admitted, some of the oxide will be reduced to metal. The point of equilibrium is therefore dependent upon the relative masses of the substances taking part in the reaction. When one of the substances is a solid, however, its mass has little influence, since it is only the extent of its surface which can affect the reaction.

Conditions under which reversible reactions are complete. If, when the equilibrium between iron and steam has been reached, the tube is opened and a current of steam is passed in, the hydrogen is swept away as fast as it is formed. The opposing reaction of hydrogen upon iron oxide must therefore cease, and the action of steam on the iron will go on until all of the iron has been transformed into iron oxide.

On the other hand, if a current of hydrogen is admitted into the tube, the steam will be swept away by the hydrogen, and all of the iron oxide will be reduced to iron. *A reversible reaction can therefore be completed in either direction when one of the products of the reaction is removed as fast as it is formed.*

Equilibrium in solution. When reactions take place in solution in water the same general principles hold good. The matter is not so simple, however, as in the case just described, owing to the fact that many of the reactions in solution are due to the presence of ions. The substances most commonly employed in solution are acids, bases, or salts, and all of these undergo dissociation. Any equilibrium which may be reached in solutions of these substances must take place between the various ions formed, on the [Pg 140] one hand, and the undissociated molecules, on the other. Thus, when nitric acid is dissolved in water, equilibrium is reached in accordance with the equation

$$H^+ + NO_3^- \longleftrightarrow HNO_3.$$

Conditions under which reversible reactions in solution are complete. The equilibrium between substances in solution may be disturbed and the reaction caused to go on in one direction to completion in either of three ways.

1. *A gas may be formed which escapes from the solution.* When sodium nitrate and sulphuric acid are brought together in solution all four ions, Na^+, NO_3^-, H^+, SO_4^-, are formed. These ions are free to rearrange themselves in various combinations. For example, the H^+ and the NO_3^- ions will reach the equilibrium

$$H^+ + NO_3^- \longleftrightarrow HNO_3.$$

If the experiment is performed with very little water present, as is the case in the preparation of nitric acid, the equilibrium will be reached when most of the H^+ and the NO_3^- ions have combined to form undissociated HNO_3.

Finally, if the mixture is now heated above the boiling point of nitric acid, the acid distills away as fast as it is formed. More and more H^+ and NO_3^- ions will then combine, and the process will continue until one or the other of them has all been removed from the solution. The substance remaining is sodium acid sulphate ($NaHSO_4$), and the reaction can therefore be expressed by the equation

$$NaNO_3 + H_2SO_4 = NaHSO_4 + HNO_3.$$

2. *An insoluble solid may be formed.* When hydrochloric acid (HCl) and [Pg 141] silver nitrate ($AgNO_3$) are brought together in solution the following ions will be present: H^+, Cl^-, Ag^+, NO_3^-. The ions Ag^+ and Cl^- will then set up the equilibrium

$$Ag^+ + Cl^- \longleftrightarrow AgCl.$$

But silver chloride (AgCl) is almost completely insoluble in water, and as soon as a very little of it has formed the solution becomes supersaturated, and the excess of the salt precipitates. More silver and chlorine ions then unite, and this continues until practically all of the silver or the chlorine ions have been removed from the solution. We then say that the following reaction is complete:

$$AgNO_3 + HCl = AgCl + HNO_3.$$

3. *Two different ions may form undissociated molecules.* In the neutralization of sodium hydroxide by hydrochloric acid the ions H^+ and OH^- come to the equilibrium

$$H^+ + OH^- \longleftrightarrow H_2O.$$

But since water is almost entirely undissociated, equilibrium can only be reached when there are very few hydroxyl or hydrogen ions present. Consequently the two ions keep uniting until one or the other of them is practically removed from the solution. When this occurs the neutralization expressed in the following equation is complete:

$$NaOH + HCl = H_2O + NaCl.$$

Preparation of acids. The principle of reversible reactions finds practical application in the preparation of most of the common acids. An acid is usually prepared by treating the most common of its salts with some other acid of high boiling point. The mixture is then heated until the lower boiling acid desired distills out. Owing to [Pg

142] its high boiling point (338°), sulphuric acid is usually employed for this purpose, most other acids boiling below that temperature.

EXERCISES

1. What would take place when solutions of silver nitrate and sodium chloride are brought together? What other chlorides would act in the same way?

2. Is the reaction expressed by the equation $NH_3 + H_2O = NH_4OH$ reversible? If so, state the conditions under which it will go in each direction.

3. Is the reaction expressed by the equation $2H + O = H_2O$ reversible? If so, state the conditions under which it will go in each direction.

4. Suggest a method for the preparation of hydrochloric acid.

[Pg 143]

CHAPTER XIV

SULPHUR AND ITS COMPOUNDS

Occurrence. The element sulphur has been known from the earliest times, since it is widely distributed in nature and occurs in large quantities in the uncombined form, especially in the neighborhood of volcanoes. Sicily has long been famous for its sulphur mines, and smaller deposits are found in Italy, Iceland, Mexico, and especially in Louisiana, where it is mined extensively. In combination, sulphur occurs abundantly in the form of sulphides and sulphates. In smaller amounts it is found in a great variety of minerals, and it is a constituent of many animal and vegetable substances.

Extraction of sulphur. Sulphur is prepared from the native substance, the separation of crude sulphur from the rock and earthy materials with which it is mixed being a very simple process. The ore from the mines is merely heated until the sulphur melts and drains away from the earthy impurities. The crude sulphur obtained in this way is distilled in a retort-shaped vessel made of iron, the exit tube of which opens into a cooling chamber of brickwork. When the sulphur vapor first enters the cooling chamber it condenses as a fine crystalline powder called *flowers of sulphur*. As the condensing chamber becomes warm, the sulphur collects as a liquid in it, and is drawn off into cylindrical molds, the product being called *roll sulphur* or *brimstone*. [Pg 144]

Physical properties. Roll sulphur is a pale yellow, crystalline solid, without marked taste and with but a faint odor. It is insoluble in water, but is freely soluble in a few liquids, notably in carbon disulphide. Roll sulphur melts at 114.8°. Just above the melting point it forms a rather thin, straw-colored liquid. As the temperature is raised, this liquid turns darker in color and becomes thicker, until at about 235° it is almost black and is so thick that the vessel containing it can be inverted without danger of the liquid running out. At higher temperatures it becomes thin once more, and boils at 448°, forming a yellowish vapor. On cooling the same changes take place in reverse order.

Varieties of sulphur. Sulphur is known in two general forms, crystalline and amorphous. Each of these forms exists in definite modifications.

Crystalline sulphur. Sulphur occurs in two crystalline forms, namely, rhombic sulphur and monoclinic sulphur.

1. *Rhombic sulphur.* When sulphur crystallizes from its solution in carbon disulphide it separates in crystals which have the same color and melting point as roll sulphur, and are rhombic in shape. Roll sulphur is made up of minute rhombic crystals.

2. *Monoclinic sulphur.* When melted sulphur is allowed to cool until a part of the liquid has solidified, and the remaining liquid is then poured off, it is found that the solid sulphur remaining in the vessel has assumed the form of fine needle-shaped crystals. These differ much in appearance from the rhombic crystals obtained by crystallizing sulphur from its solution in carbon disulphide. The needle-shaped form is called *monoclinic sulphur.* The two varieties differ also in density and in melting point, the monoclinic sulphur melting at 120°. [Pg 145]

Monoclinic and rhombic sulphur remain unchanged in contact with each other at 96°. Above this temperature the rhombic changes into monoclinic; at lower temperatures the monoclinic changes into rhombic. The temperature 96° is therefore called the transition point of sulphur. Heat is set free when monoclinic sulphur changes into rhombic.

Amorphous sulphur. Two varieties of amorphous sulphur can be readily obtained. These are white sulphur and plastic sulphur.

1. *White sulphur.* Flowers of sulphur, the preparation of which has been described, consists of a mixture of rhombic crystals and amorphous particles. When treated with carbon disulphide, the crystals dissolve, leaving the amorphous particles as a white residue.

2. *Plastic sulphur.* When boiling sulphur is poured into cold water it assumes a gummy, doughlike form, which is quite elastic. This can be seen in a very striking manner by distilling sulphur from a small, short-necked retort, such as is represented in Fig. 40, and allowing the liquid to run directly into water. In a few days it becomes quite brittle and passes over into ordinary rhombic sulphur.

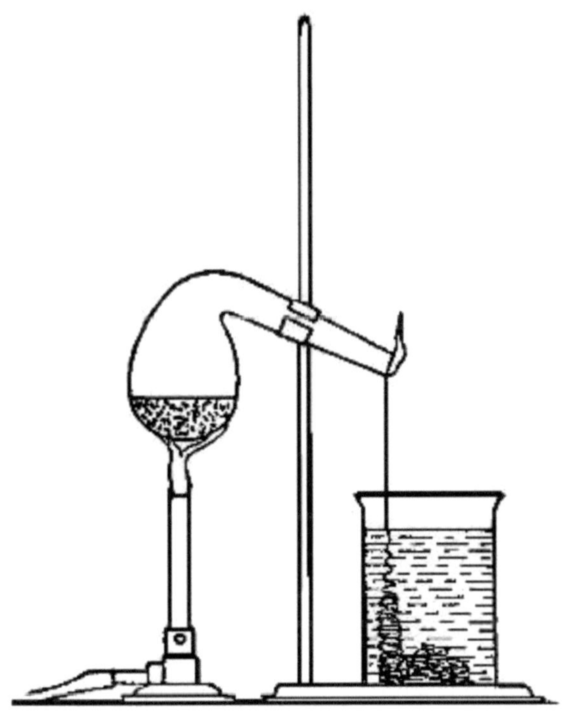

Fig. 40

Chemical properties of sulphur. When sulphur is heated to its kindling temperature in oxygen or in the air it burns with a pale blue flame, forming sulphur dioxide (SO_2). Small quantities of sulphur trioxide [Pg 146] (SO_3) may also be formed in the combustion of sulphur. Most metals when heated with sulphur combine directly with it, forming metallic sulphides. In some cases the action is so energetic that the mass becomes incandescent, as has been seen in the case of iron uniting with sulphur. This property recalls the action of oxygen upon metals, and in general the metals which combine readily with oxygen are apt to combine quite readily with sulphur.

Uses of sulphur. Large quantities of sulphur are used as a germicide in vineyards, also in the manufacture of gunpowder, matches, vulcanized rubber, and sulphuric acid.

COMPOUNDS OF SULPHUR WITH HYDROGEN

Hydrosulphuric acid (H_2S). This substance is a gas having the composition expressed by the formula H_2S and is commonly called hydrogen sulphide. It is found in the vapors issuing from volcanoes, and in solution in the so-called sulphur waters of many springs. It is formed when organic matter containing sulphur undergoes decay, just as ammonia is formed under similar circumstances from nitrogenous matter.

Preparation. Hydrosulphuric acid is prepared in the laboratory by treating a sulphide with an acid. Iron sulphide (FeS) is usually employed:

$$FeS + 2HCl = FeCl_2 + H_2S.$$

A convenient apparatus is shown in Fig. 41. A few lumps of iron sulphide are placed in the bottle A, and dilute acid is added in small quantities at a time through the funnel tube B, the gas escaping through the tube C.

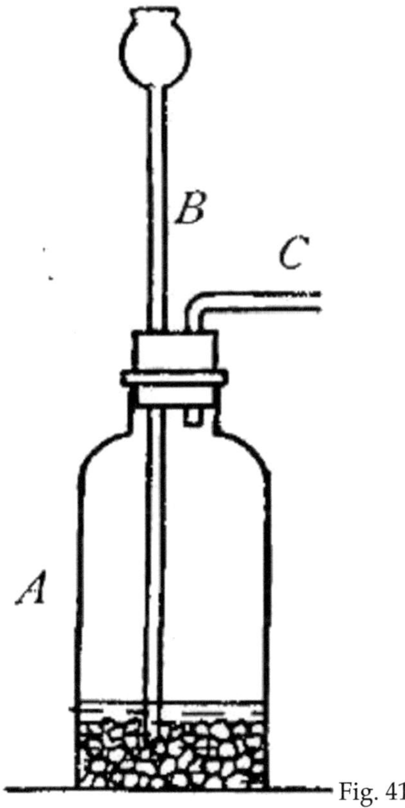

Fig. 41

[Pg 147]

Explanation of the reaction. Iron sulphide is a salt of hydrosulphuric acid, and this reaction is therefore similar to the one which takes place when sulphuric acid acts upon a nitrate. In both cases a salt and an acid are brought together, and there is a tendency for the reaction to go on until a state of equilibrium is reached. This equilibrium is constantly disturbed by the escape of the gaseous acid set free, so that the reaction goes on until all of the original salt has been decomposed. The two reactions differ in that the first one is complete at ordinary temperatures, while in the case of sulphuric acid acting upon sodium nitrate, the reacting substances must be heated so as to secure a temperature at which nitric acid is a gas.

Physical properties. Hydrosulphuric acid is a colorless gas, having a weak, disagreeable taste and an exceedingly offensive odor. It is rather sparingly soluble in water at ordinary temperatures, about three volumes dissolving in one of water. In boiling water it is not soluble at all. In pure form it acts as a violent poison, and even when diluted largely with air produces headache, dizziness, and nausea. It is a little heavier than air, having a density of 1.18.

Chemical properties. The most important chemical properties of hydrosulphuric acid are the following:

1. *Acid properties.* Hydrosulphuric acid is a weak acid. In solution in water it turns blue litmus red and neutralizes bases, forming salts called *sulphides.*

2. *Action on oxygen.* The elements composing hydrosulphuric acid have each a strong affinity for oxygen, and are not held together very firmly. Consequently the gas burns readily in oxygen or the air, according to the equation

$$H_2S + 3O = H_2O + SO_2.$$

When there is not enough oxygen for both the sulphur and the hydrogen, the latter element combines with the oxygen and the sulphur is set free:

$$H_2S + O = H_2O + S.$$

[Pg 148]

3. *Reducing action.* Owing to the ease with which hydrosulphuric acid decomposes and the strong affinity of both sulphur and hydrogen for oxygen, the substance is a strong reducing agent, taking oxygen away from many substances which contain it.

4. *Action on metals.* Hydrosulphuric acid acts towards metals in a way very similar to water. Thus, when it is passed over heated iron in a tube, the reaction is represented by the equation

$$3Fe + 4H_2S = Fe_3S_4 + 8H.$$

Water in the form of steam, under similar circumstances, acts according to the equation

$$3Fe + 4H_2O = Fe_3O_4 + 8H.$$

Salts of hydrosulphuric acid,—sulphides. The salts of hydrosulphuric acid, called sulphides, form an important class of salts. Many of them are found abundantly in nature, and some of them are important ores. They will be frequently mentioned in connection with the metals.

Most of the sulphides are insoluble in water, and some of them are insoluble in acids. Consequently, when hydrosulphuric acid is passed into a solution of a salt, it often happens that a sulphide is precipitated. With copper chloride the equation is

$$CuCl_2 + H_2S = CuS + 2HCl.$$

Because of the fact that some metals are precipitated in this way as sulphides while others are not, hydrosulphuric acid is extensively used in the separation of the metals in the laboratory.

[Pg 149]

Explanation of the reaction. When hydrosulphuric acid and copper chloride are brought together in solution, both copper and sulphur ions are present, and these will come to an equilibrium, as represented in the equation

$$Cu^+ + S^- \longleftrightarrow CuS.$$

Since copper sulphide is almost insoluble in water, as soon as a very small quantity has formed the solution becomes supersaturated, and the excess keeps precipitating until nearly all the copper or sulphur ions have been removed from the solution. With some other ions, such as iron, the sulphide formed does not saturate the solution, and no precipitate results.

OXIDES OF SULPHUR

Sulphur forms two well-known compounds with oxygen: sulphur dioxide (SO_2), sometimes called sulphurous anhydride; and sulphur trioxide (SO_3), frequently called sulphuric anhydride.

Sulphur dioxide (SO_2). Sulphur dioxide occurs in nature in the gases issuing from volcanoes, and in solution in the water of many springs. It is likely to be found wherever sulphur compounds are undergoing oxidation.

Preparation. Three general ways may be mentioned for the preparation of sulphur dioxide:

1. *By the combustion of sulphur.* Sulphur dioxide is readily formed by the combustion of sulphur in oxygen or the air:

$$S + 2O = SO_2.$$

It is also formed when substances containing sulphur are burned:

$$ZnS + 3O = ZnO + SO_2.$$

2. *By the reduction of sulphuric acid.* When concentrated sulphuric acid is heated with certain metals, such as copper, part of the acid is changed into copper sulphate, and part is reduced to sulphurous acid. The latter then decomposes into sulphur dioxide and water, the complete equation being

$$Cu + 2H_2SO_4 = CuSO_4 + SO_2 + 2H_2O.$$

[Pg 150]

3. *By the action of an acid on a sulphite.* Sulphites are salts of sulphurous acid (H_2SO_3). When a sulphite is treated with an acid, sulphurous acid is set free, and being very unstable, decomposes into water and sulphur dioxide. These reactions are expressed in the equations

$$Na_2SO_3 + 2HCl = 2NaCl + H_2SO_3,$$

$$H_2SO_3 = H_2O + SO_2.$$

Explanation of the reaction. In this case we have two reversible reactions depending on each other. In the first reaction,

$$(1)\ Na_2SO_3 + 2HCl <\!\!-\!\!> 2NaCl + H_2SO_3,$$

we should expect an equilibrium to result, for none of the four substances in the equation are insoluble or volatile when water is present to hold them in solution. But the quantity of the H_2SO_3 is constantly diminishing, owing to the fact that it decomposes, as represented in the equation

$$(2)\ H_2SO_3 <\!\!-\!\!> H_2O + SO_2,$$

and the sulphur dioxide, being a gas, escapes. No equilibrium can therefore result, since the quantity of the sulphurous acid is constantly being diminished because of the escape of sulphur dioxide.

Physical properties. Sulphur dioxide is a colorless gas, which at ordinary temperatures is 2.2 times as heavy as air. It has a peculiar, irritating odor. The gas is very soluble in water, one volume of water dissolving eighty of the gas under standard conditions. It is easily condensed to a colorless liquid, and can be purchased in this condition stored in strong bottles, such as the one represented in Fig. 42.

Fig. 42

Chemical properties. Sulphur dioxide has a marked tendency to combine with other substances, and is therefore an [Pg 151] active substance chemically. It combines with oxygen gas, but not very easily. It can, however, take oxygen away from some other substances, and is therefore a good reducing agent. Its most marked chemical property is its ability to combine with water to form sulphurous acid (H_2SO_3).

Sulphurous acid (H_2SO_3). When sulphur dioxide dissolves in water it combines chemically with it to form sulphurous acid, an unstable substance having the formula H_3SO_3. It is impossible to prepare this acid in pure form, as it breaks down very easily into water and sulphur dioxide. The reaction is therefore reversible, and is expressed by the equation

$$H_2O + SO_2 \longleftrightarrow H_2SO_3.$$

Solutions of the acid in water have a number of interesting properties.

1. *Acid properties.* The solution has all the properties typical of an acid. When neutralized by bases, sulphurous acid yields a series of salts called *sulphites.*

2. *Reducing properties.* Solutions of sulphurous acid act as good reducing agents. This is due to the fact that sulphurous acid has the power of taking up oxygen from the air, or from substances rich in oxygen, and is changed by this reaction into sulphuric acid:

$$H_2SO_3 + O = H_2SO_4,$$

$$H_2SO_3 + H_2O_2 = H_2SO_4 + H_2O.$$

3. *Bleaching properties.* Sulphurous acid has strong bleaching properties, acting upon many colored substances in such a way as to destroy their color. It is on this account used to bleach paper, straw goods, and even such foods as canned corn.

4. *Antiseptic properties.* Sulphurous acid has marked antiseptic properties, and on this account has the power [Pg 152] of arresting fermentation. It is therefore used as a preservative.

Salts of sulphurous acid, — sulphites. The sulphites, like sulphurous acid, have the power of taking up oxygen very readily, and are good reducing agents. On account of this tendency, commercial sulphites are often contaminated with sulphates. A great deal of sodium sulphite is used in the bleaching industry, and as a reagent for softening paper pulp.

Sulphur trioxide (SO_3). When sulphur dioxide and oxygen are heated together at a rather high temperature, a small amount of sulphur trioxide (SO$_3$) is formed, but the reaction is slow and incomplete. If, however, the heating takes place in the presence of very fine platinum dust, the reaction is rapid and nearly complete.

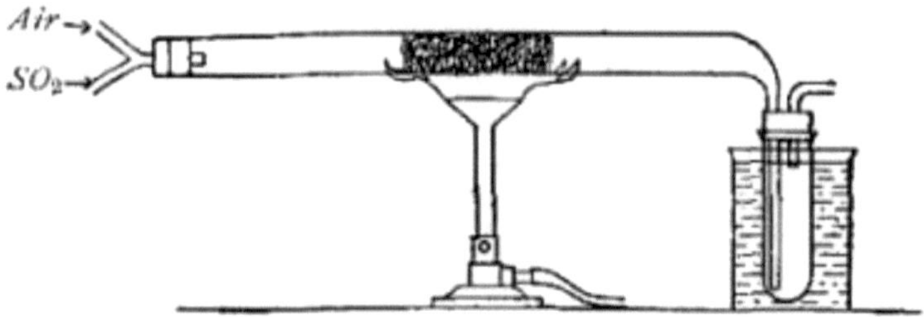

Fig. 43

Experimental preparation of sulphur trioxide. The experiment can be performed by the use of the apparatus shown in Fig. 43, the fine platinum being secured by moistening asbestos fiber with a solution of platinum chloride and igniting it in a flame. The fiber, covered with fine platinum, is placed in a tube of hard glass, which is then heated with a burner to about 350°, while sulphur dioxide and air are passed into the tube. Union takes place at once, and the strongly fuming sulphur trioxide escapes from the jet at the end of the tube, and may be condensed by surrounding the receiving tube with a freezing mixture.

Properties of sulphur trioxide. Sulphur trioxide is a colorless liquid, which solidifies at about 15° and boils at 46°. [Pg 153] A trace of moisture causes it to solidify into a mass of silky white crystals, somewhat resembling asbestos fiber in appearance. In contact with the air it fumes strongly, and when thrown upon water it dissolves with a hissing sound and the liberation of a great deal of heat. The product of this reaction is sulphuric acid, so that sulphur trioxide is the anhydride of that acid:

$$SO_3 + H_2O = H_2SO_4.$$

Catalysis. It has been found that many chemical reactions, such as the union of sulphur dioxide with oxygen, are much influenced by the presence of substances which do not themselves seem to take a part in the reaction, and are left apparently unchanged after it has ceased. These reactions go on very slowly under ordinary circumstances, but are greatly hastened by the presence of the foreign substance. Substances which hasten very slow reactions in this way are said to act as catalytic agents or *catalyzers*, and the action is called *catalysis*. Just how the action is brought about is not well understood.

DEFINITION: *A catalyzer is a substance which changes the velocity of a reaction, but does not change its products.*

Examples of Catalysis. We have already had several instances of such action. Oxygen and hydrogen combine with each other at ordinary temperatures in the presence of platinum powder, while if no catalytic agent is present they do not combine in appreciable quantities until a rather high temperature is reached. Potassium chlorate, when heated with manganese dioxide, gives up its oxygen at a much lower temperature than when heated alone. Hydrogen dioxide decomposes very rapidly when powdered manganese dioxide is sifted into its concentrated solution. [Pg 154]

On the other hand, the catalytic agent sometimes retards chemical action. For example, a solution of hydrogen dioxide decomposes more slowly when it contains a little phosphoric acid than when perfectly pure. For this reason commercial hydrogen dioxide always contains phosphoric acid.

Many reactions are brought about by the catalytic action of traces of water. For example, phosphorus will not burn in oxygen in the absence of all moisture. Hydrochloric acid will not unite with ammonia if the reagents are perfectly dry. It is probable that many of the chemical transformations in physiological processes, such as digestion, are assisted by certain substances acting as catalytic agents. The principle of catalysis is therefore very important.

Sulphuric acid (*oil of vitriol*) (H_2SO_4). Sulphuric acid is one of the most important of all manufactured chemicals. Not only is it one of the most common reagents in the laboratory, but enormous quantities of it are used in many of the industries, especially in the refining of petroleum, the manufacture of nitroglycerin, sodium carbonate, and fertilizers.

Manufacture of sulphuric acid. 1. *Contact process.* The reactions taking place in this process are represented by the following equations:

$$SO_2 + O = SO_3,$$

$$SO_3 + H_2O = H_2SO_4.$$

To bring about the first of these reactions rapidly, a catalyzer is employed, and the process is carried out in the following way: Large iron tubes are packed with some porous material, such as calcium and magnesium sulphates, which contains a suitable catalytic substance scattered through it. The catalyzers most used are platinum powder, [Pg 155] vanadium oxide, and iron oxide. Purified sulphur dioxide and air are passed through the tubes, which are kept at a temperature of about 350°. Sulphur trioxide is formed, and as it issues from the tube it is absorbed in water or dilute sulphuric acid. The process is continued until all the water in the absorbing vessel has been changed into sulphuric acid, so that a very concentrated acid is made in this way. An excess of the trioxide may dissolve in the strong sulphuric acid, forming what is known as *fuming sulphuric acid.*

2. *Chamber process.* The method of manufacture exclusively employed until recent years, and still in very extensive use, is much more complicated. The reactions are quite involved, but the conversion of water, sulphur dioxide, and oxygen into sulphuric acid is accomplished by the catalytic action of oxides of nitrogen. The reactions are brought about in large lead-lined chambers, into which oxides of nitrogen, sulphur dioxide, steam, and air are introduced in suitable proportions.

Reactions of the chamber process. In a very general way, the various reactions which take place in the lead chambers may be expressed in two equations. In the first reaction sulphur dioxide, nitrogen peroxide, steam, and oxygen unite, as shown in the equation

(1) $2SO_2 + 2NO_2 + H_2O + O = 2SO_2 (OH) (NO_2).$

The product formed in this reaction is called nitrosulphuric acid or "chamber crystals." It actually separates on the walls of the chambers when the process is not working properly. Under normal conditions, it is decomposed as fast as it is formed by the action of excess of steam, as shown in the equation

(2) $2SO_2 (OH) (NO_2) + H_2O + O = 2H_2SO_4 + 2NO_2$.

The nitrogen dioxide formed in this reaction can now enter into combination with a new quantity of sulphur dioxide, steam, and oxygen, and the series of reactions go on indefinitely. Many other reactions occur, but these two illustrate the principle of the process.

[Pg 156]

The relation between sulphuric acid and nitrosulphuric acid can be seen by comparing their structural formulas:

O= -OH   O= -OH

       S   S

O= -OH   O= -NO2

The latter may be regarded as derived from the former by the substitution of the nitro group (NO₂) for the hydroxyl group (OH).

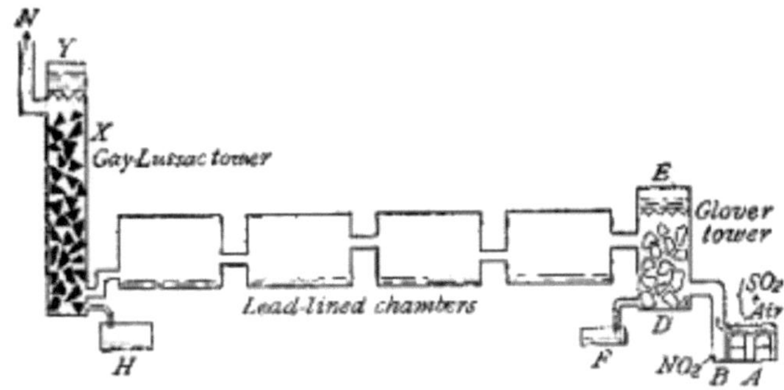

Fig. 44.

The sulphuric acid plant. Fig. 44 illustrates the simpler parts of a plant used in the manufacture of sulphuric acid by the chamber process. Sulphur or some sulphide, as FeS_2, is burned in furnace A. The resulting sulphur dioxide, together with air and some nitrogen peroxide, are conducted into the large chambers, the capacity of each chamber being about 75,000 cu. ft. Steam is also admitted into these chambers at different points. These compounds react to form sulphuric acid, according to the equations given above. The nitrogen left after the withdrawal of the oxygen from the admitted air escapes through the Gay-Lussac tower X. In order to prevent the escape of the oxides of nitrogen regenerated in the reaction, the tower is filled with lumps of coke, over which trickles concentrated sulphuric acid admitted from Y. The nitrogen peroxide dissolves in the acid and the resulting solution collects in H. This is pumped into E, where it is mixed with dilute acid and allowed to trickle down through the chamber D (Glover tower), which is filled with some acid-resisting rock. Here the nitrogen peroxide is expelled from the solution by the action of the hot gases entering from A, and together with them enters the first chamber again. The acid from which the nitrogen peroxide is expelled collects in F. Theoretically, a small amount of nitrogen peroxide would suffice to prepare an unlimited amount of sulphuric acid; practically, some of it escapes, and this is replaced by small amounts admitted at B. [Pg 157]

The sulphuric acid so formed, together with the excess of condensed steam, collect upon the floor of the chambers in the form of a liquid containing from 62% to 70% of sulphuric acid. The product is called *chamber acid* and is quite impure; but for many purposes, such as the manufacture of fertilizers, it needs no further treatment. It can be concentrated by boiling it in vessels made of iron or platinum, which resist the action of the acid, nearly all the water boiling off. Pure concentrated acid can be made best by the contact process, while the chamber process is cheaper for the dilute impure acid.

Physical properties. Sulphuric acid is a colorless, oily liquid, nearly twice as heavy as water. The ordinary concentrated acid contains about 2% of water, has a density of 1.84, and boils at 338°. It is sometimes called *oil of vitriol*, since it was formerly made by distilling a substance called *green vitriol*.

Chemical properties. Sulphuric acid possesses chemical properties which make it one of the most important of chemical substances.

1. *Action as an acid.* In dilute solution sulphuric acid acts as any other acid, forming salts with oxides and hydroxides.

2. *Action as an oxidizing agent.* Sulphuric acid contains a large percentage of oxygen and is, like nitric acid, a very good oxidizing agent. When the concentrated acid is heated with sulphur, carbon, and many other substances, oxidation takes place, the sulphuric acid decomposing according to the equation

$$H_2SO_4 = H_2SO_3 + O.$$

3. *Action on metals.* In dilute solution sulphuric acid acts upon many metals, such as zinc, forming a sulphate and liberating hydrogen. When the concentrated acid is employed the hydrogen set free is oxidized by a new portion [Pg 158] of the acid, with the liberation of sulphur dioxide. With copper the reactions are expressed by the equations

$$(1)\ Cu + H_2SO_4 = CuSO_4 + 2H,$$

$$(2)\ H_2SO_4 + 2H = H_2SO_3 + H_2O,$$

$$(3)\ H_2SO_3 = H_2O + SO_2.$$

By combining these equations the following one is obtained:

$$Cu + 2H_2SO_4 = CuSO_4 + SO_2 + 2H_2O.$$

4. *Action on salts.* We have repeatedly seen that an acid of high boiling point heated with the salt of some acid of lower boiling point will drive out the low boiling acid. The boiling point of sul-

phuric acid (338°) is higher than that of almost any common acid; hence it is used largely in the preparation of other acids.

5. *Action on water.* Concentrated sulphuric acid has a very great affinity for water, and is therefore an effective dehydrating agent. Gases which have no chemical action upon sulphuric acid can be freed from water vapor by bubbling them through the strong acid. When the acid is diluted with water much heat is set free, and care must be taken to keep the liquid thoroughly stirred during the mixing, and to pour the acid into the water,—never the reverse.

Not only can sulphuric acid absorb water, but it will often withdraw the elements hydrogen and oxygen from a compound containing them, decomposing the compound, and combining with the water so formed. For this reason most organic substances, such as sugar, wood, cotton, and woolen fiber, and even flesh, all of which contain much oxygen and hydrogen in addition to carbon, are charred or burned by the action of the concentrated acid. [Pg 159]

Salts of sulphuric acid,—sulphates. The sulphates form a very important class of salts, and many of them have commercial uses. Copperas (iron sulphate), blue vitriol (copper sulphate), and Epsom salt (magnesium sulphate) serve as examples. Many sulphates are important minerals, prominent among these being gypsum (calcium sulphate) and barytes (barium sulphate).

Thiosulphuric acid ($H_2S_2O_3$); **Thiosulphates.** Many other acids of sulphur containing oxygen are known, but none of them are of great importance. Most of them cannot be prepared in a pure state, and are known only through their salts. The most important of these is thiosulphuric acid.

When sodium sulphite is boiled with sulphur the two substances combine, forming a salt which has the composition represented in the formula $Na_2S_2O_3$:

$$Na_2SO_3 + S = Na_2S_2O_3.$$

The substance is called sodium thiosulphate, and is a salt of the easily decomposed acid $H_2S_2O_3$, called thiosulphuric acid. This reaction is quite similar to the action of oxygen upon sulphites:

$$Na_2SO_3 + O = Na_2SO_4.$$

More commonly the salt is called sodium hyposulphite, or merely "hypo." It is a white solid and is extensively used in photography, in the bleaching industry, and as a disinfectant.

Monobasic and dibasic acids. Such acids as hydrochloric and nitric acids, which have only one replaceable hydrogen atom in the molecule, or in other words yield one hydrogen ion in solution, are called monobasic acids. Acids yielding two hydrogen ions in solution are called dibasic acids. Similarly, we may have tribasic and tetrabasic acids. The three acids of sulphur are dibasic acids. It is therefore possible for each of them to form both normal and acid salts. The acid salts can be made in two ways: the acid may be treated with only half enough base to neutralize it, —

$$NaOH + H_2SO_4 = NaHSO_4 + H_2O;$$

[Pg 160]

or a normal salt may be treated with the free acid, —

$$Na_2SO_4 + H_2SO_4 = 2NaHSO_4.$$

Acid sulphites and sulphides may be made in the same ways.

Carbon disulphide (CS_2). When sulphur vapor is passed over highly heated carbon the two elements combine, forming carbon disulphide (CS_2), just as oxygen and carbon unite to form carbon dioxide (CO_2). The substance is a heavy, colorless liquid, possessing, when pure, a pleasant ethereal odor. On standing for some time, especially when exposed to sunlight, it undergoes a slight decomposition and acquires a most disagreeable, rancid odor. It has the property of dissolving many substances, such as gums, resins, and waxes, which are insoluble in most liquids, and it is extensively used as a solvent for such substances. It is also used as an insecticide. It boils at a low temperature (46°), and its vapor is very inflammable, burning in the air to form carbon dioxide and sulphur dioxide, according to the equation

$$CS_2 + 6O = CO_2 + 2SO_2.$$

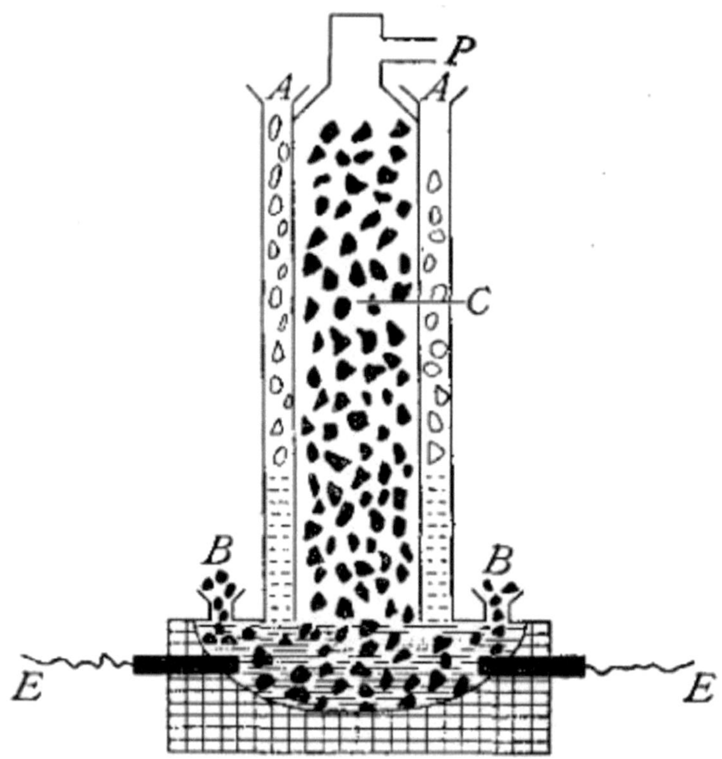

Fig. 45

Commercial preparation of carbon disulphide. In the preparation of carbon disulphide an electrical furnace is employed, such as is represented in Fig. 45. The furnace is packed with carbon C, and this is fed in through the hoppers B, as fast as that which is present in the hearth of the furnace is used up. Sulphur is introduced at A, and at the lower ends of the tubes it is melted by the heat of the furnace and flows into the hearth as a liquid. An electrical current is passed through the carbon and melted sulphur from the electrodes E, heating the charge. The vapors of carbon disulphide pass up

186

through the furnace and escape at *D*, from which they pass to a suitable condensing apparatus.

[Pg 161]

Comparison of sulphur and oxygen. A comparison of the formulas and the chemical properties of corresponding compounds of oxygen and sulphur brings to light many striking similarities. The conduct of hydrosulphuric acid and water toward many substances has been seen to be very similar; the oxides and sulphides of the metals have analogous formulas and undergo many parallel reactions. Carbon dioxide and disulphide are prepared in similar ways and undergo many analogous reactions. It is clear, therefore, that these two elements are far more closely related to each other than to any of the other elements so far studied.

Selenium and tellurium. These two very uncommon elements are still more closely related to sulphur than is oxygen. They occur in comparatively small quantities and are usually found associated with sulphur and sulphides, either as the free elements or more commonly in combination with metals. They form compounds with hydrogen of the formulas H_2Se and H_2Te; these bodies are gases with properties very similar to those of H_2S. They also form oxides and oxygen acids which resemble the corresponding sulphur compounds. The elements even have allotropic forms corresponding very closely to those of sulphur. Tellurium is sometimes found in combination with gold and copper, and occasions some difficulties in the refining of these metals. The elements have very few practical applications.

Crystallography. In order to understand the difference between the two kinds of sulphur crystals, it is necessary to know something about crystals in general and the forms which they may assume. An examination of a large number of crystals has shown that although they may differ much in geometric form, they can all be considered as modifications of a few simple plans. The best way to understand the relation of one crystal to another is to look upon every crystal as having its faces and angles arranged in definite fashion about [Pg 162] certain imaginary lines drawn through the crystal. These lines are called axes, and bear much the same relation to a crystal as do the axis and parallels of latitude and longitude to the earth and a geographical study of it. All crystals can be referred to one of six simple plans or systems, which have their axes as shown in the following drawings.

The names and characteristics of these systems are as follows:

1. Isometric or regular system (Fig. 46). Three equal axes, all at right angles.

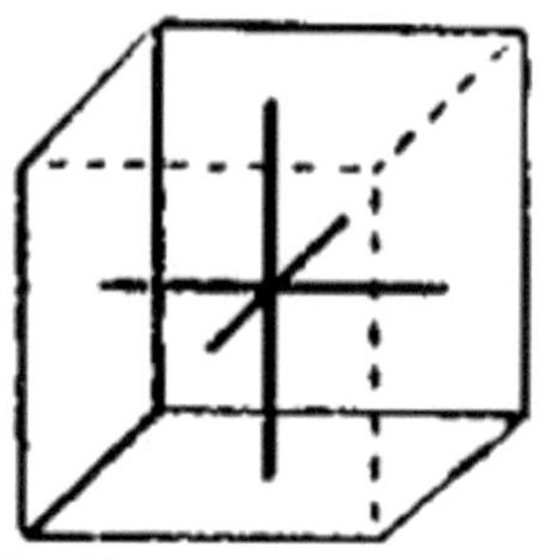
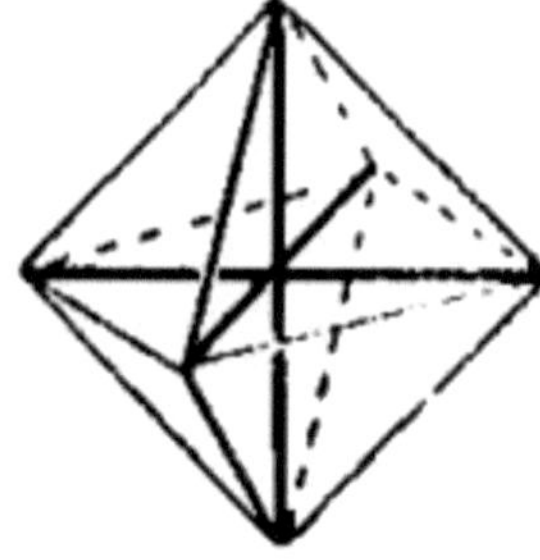

Fig. 46

2. Tetragonal system (Fig. 47). Two equal axes and one of different length, all at right angles to each other.

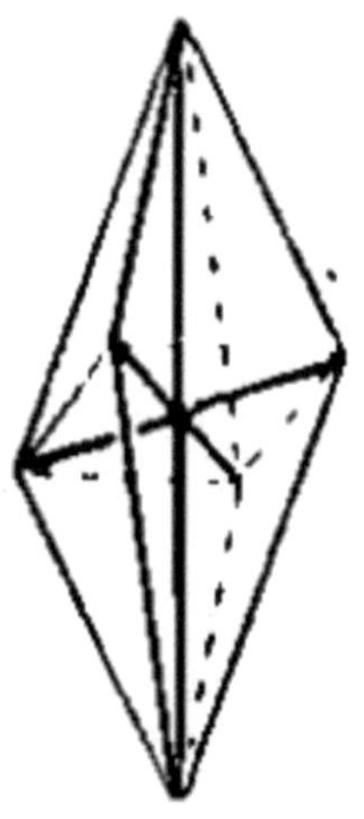
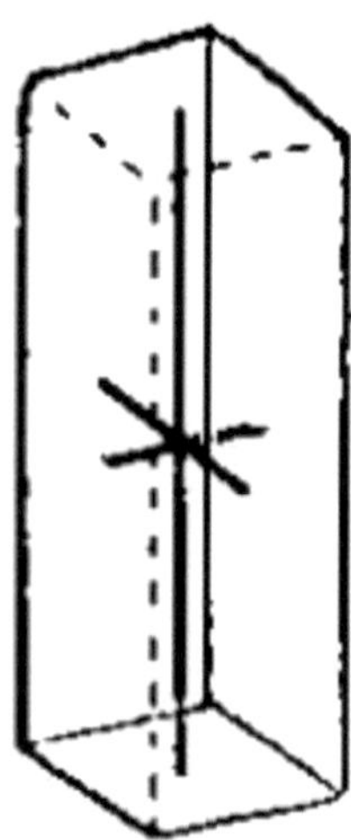

Fig. 47

3. Orthorhombic system (Fig. 48). Three unequal axes, all at right angles to each other.

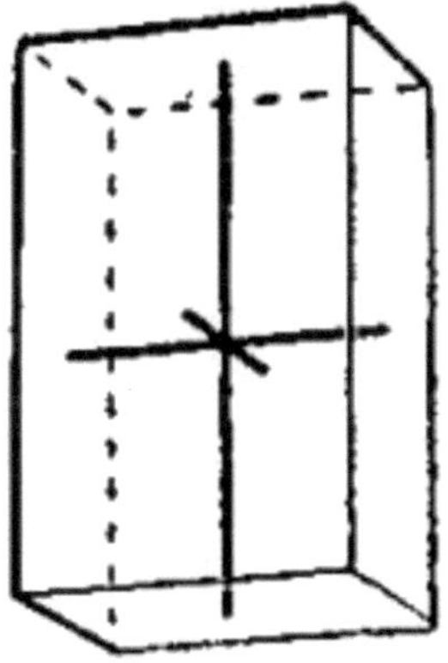 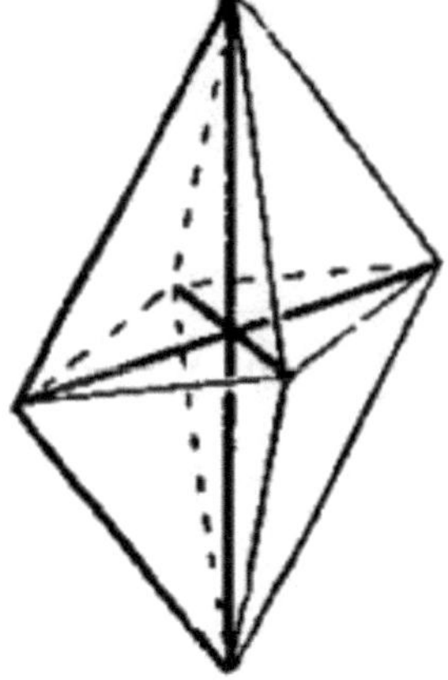

Fig. 48

4. Monoclinic system (Fig. 49). Two axes at right angles, and a third at right angles to one of these, but inclined to the other.

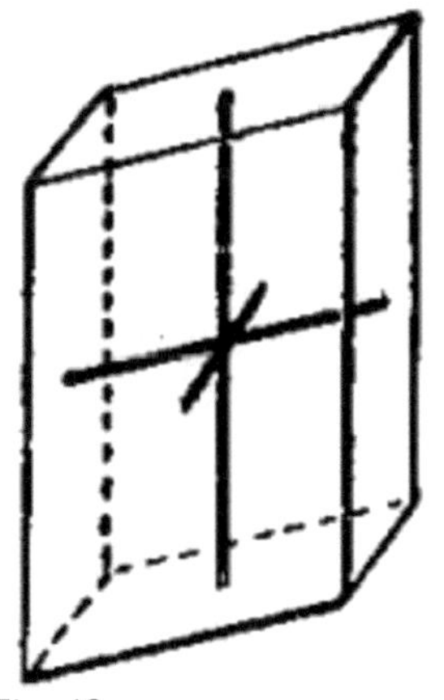 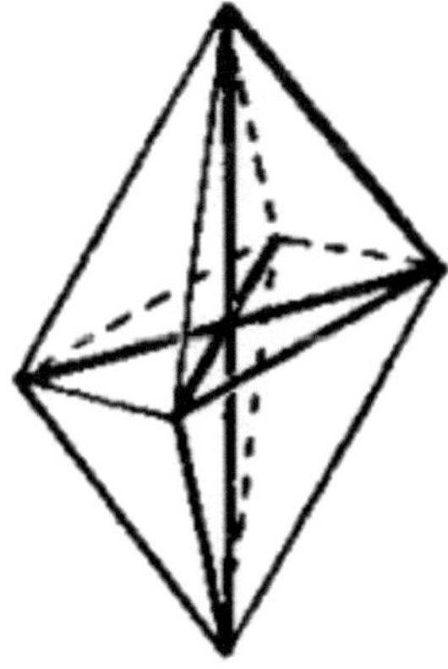

Fig. 49

5. Triclinic system (Fig. 50). Three axes, all inclined to each other.

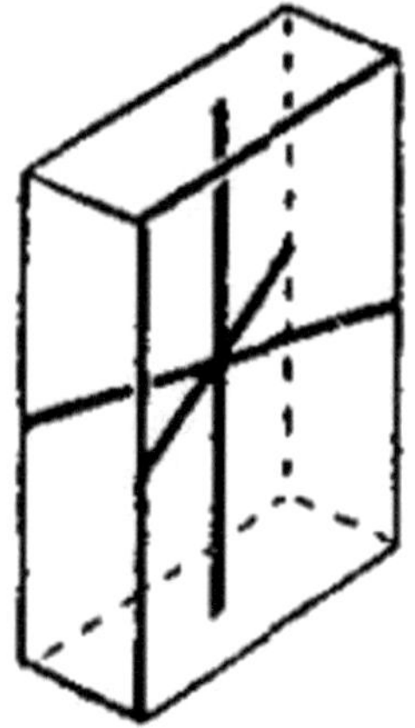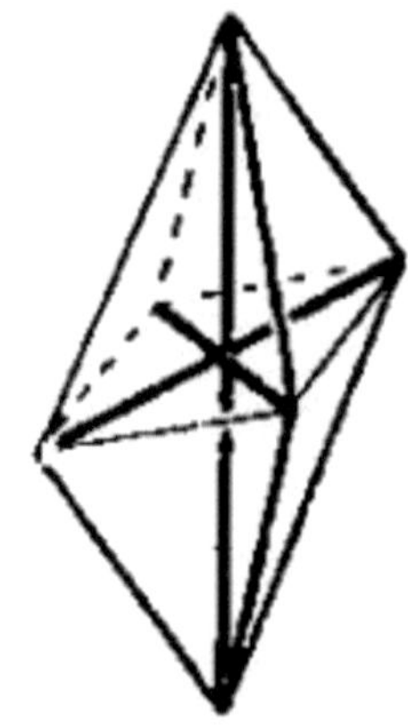

Fig. 50

6. Hexagonal system (Fig. 51). Three equal axes in the same plane intersecting at angles of 60°, and a fourth at right angles to all of these.

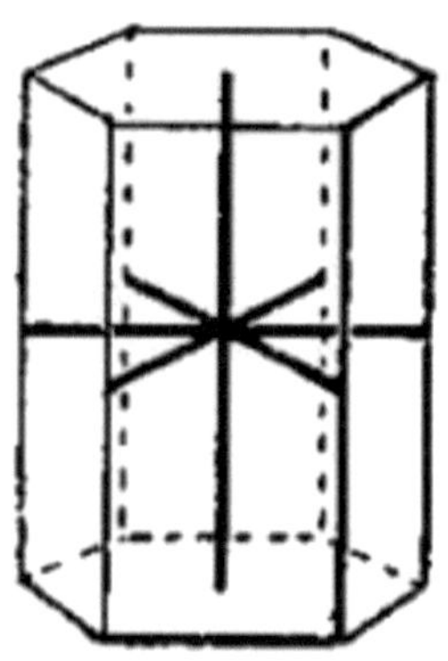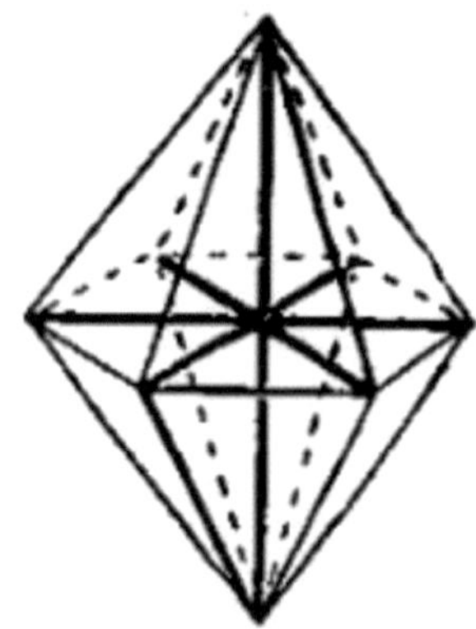

Fig. 51

Every crystal can be imagined to have its faces and angles arranged in a definite way around one of these systems of axes. A cube, for instance, is referred to Plan 1, an axis ending in the center of each face; while in a regular octohedron an axis ends in each solid angle. These forms are shown in Fig. 46. It will be seen that both of these figures belong to the same system, though they are very different in appearance. In the same way, many geometric [Pg 163] forms may be derived from each of the systems, and the light lines about the axes in the drawings show two of the simplest forms of each of the systems.

In general a given substance always crystallizes in the same system, and two corresponding faces of each crystal of it always make the same angle with each other. A few substances, of which sulphur is an example, crystallize in two different systems, and the crystals differ in such physical properties as melting point and density. Such substances are said to be *dimorphous*.

EXERCISES

1. (*a*) Would the same amount of heat be generated by the combustion of 1 g. of each of the allotropic modifications of sulphur? (*b*) Would the same amount of sulphur dioxide be formed in each case?

2. Is the equation for the preparation of hydrosulphuric acid a reversible one? As ordinarily carried out, does the reaction complete itself?

3. Suppose that hydrosulphuric acid were a liquid, would it be necessary to modify the method of preparation?

4. Can sulphuric acid be used to dry hydrosulphuric acid? Give reason for answer.

5. Does dry hydrosulphuric acid react with litmus paper? State reason for answer.

6. How many grams of iron sulphide are necessary to prepare 100 l. of hydrosulphuric acid when the laboratory conditions are 17° and 740 mm. pressure?

7. Suppose that the hydrogen in 1 l. of hydrosulphuric acid were liberated; what volume would it occupy, the gases being measured under the same conditions?

8. Write the equations representing the reaction between hydrosulphuric acid and sodium hydroxide and ammonium hydroxide respectively.

9. Show that the preparation of sulphur dioxide from a sulphite is similar in principle to the preparation of hydrogen sulphide.

10. (*a*) Does dry sulphur dioxide react with litmus paper? (*b*) How can it be shown that a solution of sulphur dioxide in water acts like an acid? [Pg 164]

11. (*a*) Calculate the percentage composition of sulphurous anhydride and sulphuric anhydride. (*b*) Show how these two substances are in harmony with the law of multiple proportion.

12. How many pounds of sulphur would be necessary in the preparation of 100 lb. of 98% sulphuric acid?

13. What weight of sulphur dioxide is necessary in the preparation of 1 kg. of sodium sulphite?

14. What weight of copper sulphate crystals can be obtained by dissolving 1 kg. of copper in sulphuric acid and crystallizing the product from water?

15. Write the names and formulas of the oxides and oxygen acids of selenium and tellurium.

16. In the commercial preparation of carbon disulphide, what is the function of the electric current?

17. If the Gay-Lussac tower were omitted from the sulphuric acid factory, what effect would this have on the cost of production of sulphuric acid?

[Pg 165]

CHAPTER XV

PERIODIC LAW

A number of the elements have now been studied somewhat closely. The first three of these, oxygen, hydrogen, and nitrogen, while having some physical properties in common with each other, have almost no point of similarity as regards their chemical conduct. On the other hand, oxygen and sulphur, while quite different physically, have much in common in their chemical properties.

About eighty elements are now known. If all of these should have properties as diverse as do oxygen, hydrogen, and nitrogen, the study of chemistry would plainly be a very difficult and complicated one. If, however, the elements can be classified in groups, the members of which have very similar properties, the study will be very much simplified.

Earlier classification of the elements. Even at an early period efforts were made to discover some natural principle in accordance with which the elements could be classified. Two of these classifications may be mentioned here.

1. *Classification into metals and non-metals.* The classification into metals and non-metals most naturally suggested itself. This grouping was based largely on physical properties, the metals being heavy, lustrous, malleable, ductile, and good conductors of heat and electricity. Elements possessing these properties are usually base-forming in character, and the ability to form bases came to be regarded as a characteristic property of the metals. The [Pg 166] non-metals possessed physical properties which were the reverse of those of the metals, and were acid-forming in character.

Not much was gained by this classification, and it was very imperfect. Some metals, such as potassium, are very light; some non-metals, such as iodine, have a high luster; some elements can form either an acid or a base.

2. *Classification into triad families.* In 1825 Döbereiner observed that an interesting relation exists between the atomic weights of chemically similar elements. To illustrate, lithium, sodium, and potassium resemble each other very closely, and the atomic weight of sodium is almost exactly an arithmetical mean between those of the other two: $(7.03 + 39.15)/2 = 23.09$. In many chemical and physical properties sodium is midway between the other two.

A number of triad families were found, but among eighty elements, whose atomic weights range all the way from 1 to 240, such agreements might be mere chance. Moreover many elements did not appear to belong to such families.

Periodic division. In 1869 the Russian chemist Mendeléeff devised an arrangement of the elements based on their atomic weights, which has proved to be of great service in the comparative study of the elements. A few months later the German, Lothar Meyer, independently suggested the same ideas. This arrangement brought to light a great generalization, now known as the *periodic law*. An exact statement of the law will be given after the method of arranging the elements has been described.

DMITRI IVANOVITCH MENDELÉEFF (Russian) (1834-1907)

Author of the periodic law; made many investigations on the physical constants of elements and compounds; wrote an important book entitled "Principles of Chemistry"; university professor and government official

[Pg 167]

Arrangement of the periodic table. The arrangement suggested by Mendeléeff, modified somewhat by more recent investigations, is as follows: Beginning with lithium, which has an atomic weight of 7, the elements are arranged in a horizontal row in the order of their atomic weights, thus:

Li (7.03), Be (9.1), B (11), C (12), N (14.04), O (16), F (19).

These seven elements all differ markedly from each other. The eighth element, sodium, is very similar to lithium. It is placed just under lithium, and a new row follows:

Na(23.05), Mg (24.36), Al (27.1), Si (28.4), P (31), S (32.06), Cl(35.45).

When the fifteenth element, potassium, is reached, it is placed under sodium, to which it is very similar, and serves to begin a third row:

K (39.15), Ca (40.1), Sc (44.1,) Ti (48.1), V (51.2), Cr (52.1), Mn(55).

Not only is there a strong similarity between lithium, sodium, and potassium, which have been placed in a vertical row because of this resemblance, but the elements in the other vertical rows exhibit much of the same kind of similarity among themselves, and evidently form little natural groups.

The three elements following manganese, namely, iron, nickel, and cobalt, have atomic weights near together, and are very similar chemically. They do not strongly resemble any of the elements so far considered, and are accordingly placed in a group by themselves, following manganese. A new row is begun with copper, which somewhat resembles the elements of the first vertical column. Following the fifth and seventh rows are groups of three closely related elements, so that the completed arrangement has the appearance represented in the table on page 168.

[Pg 168]

Period	Group 0	Group I A	Group I B	Group II A	Group II B	Group III A	Group III B	Group IV A	Group IV B	Group V A	Group V B	Group VI A	Group VI B	Group VII A	Group VII B	Group VIII
1, 2	H = 1.008, He = 4	Li = 7.03		Be = 9.1		B = 11		C = 12		N = 14.04		O = 16		F = 19		
3	Ne = 20	Na = 23.05		Mg = 24.36		Al = 27.1		Si = 28.4		P = 31		S = 32.06		Cl = 35.45		
4	A = 39.9	K = 39.15		Ca = 40.1		Sc = 44.1		Ti = 48.1		V = 51.2		Cr = 52.1		Mn = 55		Fe = 55.9, Ni = 58.7, Co = 59
5			Cu = 63.6		Zn = 65.4		Ga = 70		Ge = 72.5		As = 75		Se = 79.2		Br = 79.96	
6	Kr = 81.8	Rb = 85.5		Sr = 87.6		Y = 89		Zr = 90.6		Cb = 94		Mo = 96				Ru = 101.7, Rh = 103, Pd = 106.5
7			Ag = 107.93		Cd = 112.4		In = 115		Sn = 119		Sb = 120.2		Te = 127.6		I = 126.97	
8	X = 128	Cs = 132.9		Ba = 137.4		La = 138.9		Ce = 140*, 140-?-175		Ta = 183		W = 184				Os = 191, Ir = 193, Pt = 194.8
9			Au = 197.2		Hg = 200		Tl = 204.1		Pb = 206.9		Bi = 208.5					
10				Ra = 225				Th = 232.5		U = 238.5						
		R_2O, RH		RO, RH_2		R_2O_3, RH_3		RO_2, RH_4		R_2O_5, RH_3		RO_3, RH_2		R_2O_7, RH		RO_4

* This includes a number of elements whose atomic weights lie between 140 and 175, but which have not been accurately studied, and so their proper arrangement is uncertain.

THE PERIODIC ARRANGEMENT OF THE ELEMENTS

[Pg 169]

Place of the atmospheric elements. When argon was discovered it was seen at once that there was no place in the table for an element of atomic weight approximately 40. When the other inactive elements were found, however, it became apparent that they form a group just preceding Group 1. They are accordingly arranged in this way in Group 0 (see table on opposite page). A study of this table brings to light certain very striking facts.

Properties of elements vary with atomic weights. There is evidently a close relation between the properties of an element and its atomic weight. Lithium, at the beginning of the first group, is a very strong base-forming element, with pronounced metallic properties. Beryllium, following lithium, is less strongly base-forming, while boron has some base-forming and some acid-forming properties. In carbon all base-forming properties have disappeared, and the acid-forming properties are more marked than in boron. These become still more emphasized as we pass through nitrogen and oxygen, until on reaching fluorine we have one of the strongest acid-forming elements. The properties of these seven elements therefore vary regularly with their atomic weights, or, in mathematical language, are regular functions of them.

Periodic law. The properties of the first seven elements vary *continuously*—that is steadily—away from base-forming and toward acid-forming properties. If lithium had the smallest atomic weight of any of the elements, and fluorine the greatest, so that in passing from one to the other we had included all the elements, we could say that the properties of elements are continuous functions of their atomic weights. But fluorine is an element of small atomic weight, and the one following it, sodium, breaks the regular order, for in it reappear all the characteristic properties of lithium. Magnesium, following sodium, bears much the same relation to [Pg 170] beryllium that sodium does to lithium, and the properties of the elements in the second row vary much as they do in the first row until potassium is reached, when another repetition begins. The properties of the elements do not vary continuously, therefore, with atomic weights, but at regular intervals there is a repetition, or *period*. This generalization is known as the *periodic law*, and may be stated thus: *The properties of elements are periodic functions of their atomic weights.*

The two families in a group. While all the elements in a given vertical column bear a general resemblance to each other, it has been noticed that those belonging to periods having even numbers are very strikingly similar to each other. They are placed at the left side of the group columns. In like manner, the elements belonging to the odd periods are very similar and are arranged at the right side of the group columns. Thus calcium, strontium, and barium are very much alike; so, too, are magnesium, zinc, and cadmium. The resemblance between calcium and magnesium, or strontium and zinc, is much less marked. This method of arrangement therefore divides each group into two families, each containing four or five members, between which there is a great similarity.

Family resemblances. Let us now inquire more closely in what respects the elements of a family resemble each other.

1. *Valence.* In general the valence of the elements in a family is the same, and the formulas of their compounds are therefore similar. If we know that the formula of sodium chloride is NaCl, it is pretty certain that the formula of potassium chloride will be KCl—not

KCl_2 or [Pg 171] KCl_3. The general formulas R_2O, RO, etc., placed below the columns show the formulas of the oxides of the elements in the column provided they form oxides. In like manner the formulas RH, RH_2, etc., show the composition of the compounds formed with hydrogen or chlorine.

2. *Chemical properties.* The chemical properties of the members of a family are quite similar. If one member is a metal, the others usually are; if one is a non-metal, so, too, are the others. The families in the first two columns consist of metals, while the elements found in the last two columns form acids. There is in addition a certain regularity in properties of the elements in each family. If the element at the head of the family is a strong acid-forming element, this property is likely to diminish gradually, as we pass to the members of the family with higher atomic weights. Thus phosphorus is strongly acid-forming, arsenic less so, antimony still less so, while bismuth has almost no acid-forming properties. We shall meet with many illustrations of this fact.

3. *Physical properties.* In the same way, the physical properties of the members of a family are in general somewhat similar, and show a regular gradation as we pass from element to element in the family. Thus the densities of the members of the magnesium family are

Mg = 1.75, Zn = 7.00, Cd = 8.67, Hg = 13.6.

Their melting points are

Mg = 750°, Zn = 420°, Cd = 320°, Hg = -39.5°.

Value of the periodic law. The periodic law has proved of much value in the development of the science of chemistry.

1. *It simplifies study.* It is at once evident that such regularities very much simplify the study of chemistry. [Pg 172] A thorough study of one element of a family makes the study of the other members a much easier task, since so many of the properties and chemical reactions of the elements are similar. Thus, having studied the element

sulphur in some detail, it is not necessary to study selenium and tellurium so closely, for most of their properties can be predicted from the relation which they sustain to sulphur.

2. *It predicts new elements.* When the periodic law was first formulated there were a number of vacant places in the table which evidently belonged to elements at that time unknown. From their position in the table, Mendeléeff predicted with great precision the properties of the elements which he felt sure would one day be discovered to fill these places. Three of them, scandium, germanium, and gallium, were found within fifteen years, and their properties agreed in a remarkable way with the predictions of Mendeléeff. There are still some vacant places in the table, especially among the heavier elements.

3. *It corrects errors.* The physical constants of many of the elements did not at first agree with those demanded by the periodic law, and a further study of many such cases showed that errors had been made. The law has therefore done much service in indicating probable error.

Imperfections of the law. There still remain a good many features which must be regarded as imperfections in the law. Most conspicuous is the fact that the element hydrogen has no place in the table. In some of the groups elements appear in one of the families, while all of their properties show that they belong in the other. Thus sodium belongs with lithium and not with copper; fluorine belongs with chlorine and not with manganese. There are [Pg 173] two instances where the elements must be transposed in order to make them fit into their proper group. According to their atomic weights, tellurium should follow iodine, and argon should follow potassium. Their properties show in each case that this order must be reversed. The table separates some elements altogether which, in many respects have closely agreeing properties. Iron, chromium, and manganese are all in different groups, although they are similar in many respects.

The system is therefore to be regarded as but a partial and imperfect expression of some very important and fundamental relation between the substances which we know as elements, the exact nature of this relation being as yet not completely clear to us.

EXERCISES

1. Suppose that an element were discovered that filled the blank in Group O, Period 5; what properties would it probably have?

2. Suppose that an element were discovered that filled the blank in Group VI, Period 9, family *B*; what properties would it have?

3. Sulphur and oxygen both belong in Group VI, although in different families; in what respects are the two similar?

[Pg 174]

CHAPTER XVI

THE CHLORINE FAMILY

	ATOMIC WEIGHT	MELTING POINT	BOILING POINT	COLOR AND STATE
Fluorine (F)	19.00	-223°	-187°	Pale yellowish gas.
Chlorine (Cl)	35.45	-102°	-33.6°	Greenish-yellow gas.
Bromine (Br)	79.96	-7°	59°	Red liquid.
Iodine (I)	126.97	107°	175°	Purplish-black solid.

The family. The four elements named in the above table form a strongly marked family of elements and illustrate very clearly the way in which the members of a family in a periodic group resemble each other, as well as the character of the differences which we may expect to find between the individual members.

1. *Occurrence.* These elements do not occur in nature in the free state. The compounds of the last three elements of the family are found extensively in sea water, and on this account the name *halogens,* signifying "producers of sea salt," is sometimes applied to the family.

2. *Properties.* As will be seen by reference to the table, the melting points and boiling points of the elements of the family increase with their atomic weights. A somewhat similar gradation is noted in their color and state. One atom of each of the elements combines with one atom of hydrogen to form acids, which are gases very soluble in water. The affinity of the elements for hydrogen is in [Pg 175] the inverse order of their atomic weights, fluorine having the strongest affinity and iodine the weakest. Only chlorine and iodine form oxides, and those of the former element are very unstable. The elements of the group are univalent in their compounds with hydrogen and the metals.

FLUORINE

Occurrence. The element fluorine occurs in nature most abundantly as the mineral fluorspar (CaF_2), as cryolite (Na_3AlF_6), and in the complex mineral apatite ($3\ Ca_3(PO_4)_2 \cdot CaF_2$).

Preparation. All attempts to isolate the element resulted in failure until recent years. Methods similar to those which succeed in the preparation of the other elements of the family cannot be used; for as soon as the fluorine is liberated it combines with the materials of which the apparatus is made or with the hydrogen of the water which is always present. The preparation of fluorine was finally accomplished by the French chemist Moissan by the electrolysis of hydrofluoric acid. Perfectly dry hydrofluoric acid (HF) was condensed to a liquid and placed in a U-shaped tube made of platinum (or copper), which was furnished with electrodes and delivery tubes, as shown in Fig. 52. This liquid is not an electrolyte, but becomes such when potassium fluoride is dissolved in it. When this solution was electrolyzed hydrogen was set free at the cathode and fluorine at the anode.

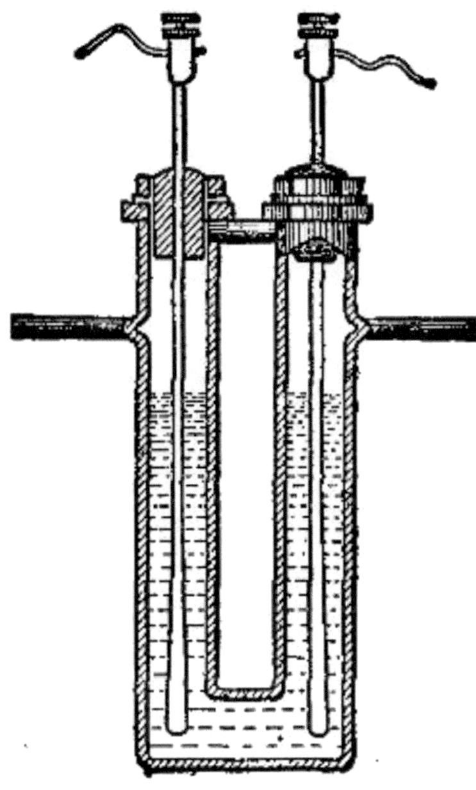

Fig. 52

[Pg 176]

Properties. Fluorine is a gas of slightly yellowish color, and can be condensed to a liquid boiling at -187° under atmospheric pressure. It solidifies at -223°. It is extremely active chemically, being the most active of all the elements at ordinary temperatures.

It combines with all the common elements save oxygen, very often with incandescence and the liberation of much heat. It has a strong affinity for hydrogen and is able to withdraw it from its compounds with other elements. Because of its great activity it is extremely poisonous. Fluorine does not form any oxides, neither does it form any oxygen acids, in which respects it differs from the other members of the family.

Hydrofluoric acid (HF). Hydrofluoric acid is readily obtained from fluorspar by the action of concentrated sulphuric acid. The equation is

$$CaF_2 + H_2SO_4 = CaSO_4 + 2HF.$$

In its physical properties it resembles the binary acids of the other elements of this family, being, however, more easily condensed to a liquid. The anhydrous acid boils at 19° and can therefore be prepared at ordinary pressures. It is soluble in all proportions in water, and a concentrated solution — about 50% — is prepared for the market. Its fumes are exceedingly irritating to the respiratory organs, and several chemists have lost their lives by accidentally breathing them.

HENRI MOISSAN (French) (1853-1907)

Famous for his work with the electric furnace at high temperatures; prepared artificial diamonds, together with many new binary compounds such as carbides, silicides, borides, and nitrides; isolated fluorine and studied its properties and its compounds very thoroughly

[Pg 177]

Chemical properties. Hydrofluoric acid, like other strong acids, readily acts on bases and metallic oxides and forms the corresponding fluorides. It also dissolves certain metals such as silver and copper. It acts very vigorously upon organic matter, a single drop of the concentrated acid making a sore on the skin which is very painful and slow in healing. Its most characteristic property is its action upon silicon dioxide (SiO_2), with which it forms water and the gas silicon tetrafluoride (SiF_4), as shown in the equation

$$SiO_2 + 4HF = SiF_4 + 2H_2O.$$

Glass consists of certain compounds of silicon, which are likewise acted on by the acid so that it cannot be kept in glass bottles. It is preserved in flasks made of wax or gutta-percha.

Etching. Advantage is taken of this reaction in etching designs upon glass. The glass vessel is painted over with a protective paint upon which the acid will not act, the parts which it is desired to make opaque being left unprotected. A mixture of fluorspar and sulphuric acid is then painted over the vessel and after a few minutes the vessel is washed clean. Wherever the hydrofluoric acid comes in contact with the glass it acts upon it, destroying its luster and making it opaque, so that the exposed design will be etched upon the clear glass. Frosted glass globes are often made in this way.

The etching may also be effected by covering the glass with a thin layer of paraffin, cutting the design through the wax and then exposing the glass to the fumes of the acid.

Salts of hydrofluoric acid, — fluorides. A number of the fluorides are known, but only one of them, calcium fluoride (CaF_2), is of importance. This is the well-known mineral fluorspar.

CHLORINE

Historical. While studying the action of hydrochloric acid upon the mineral pyrolusite, in 1774, Scheele obtained a yellowish, gaseous substance to which he gave a name in keeping with the phlogiston theory then current. Later it was supposed to be a compound containing oxygen. In [Pg 178] 1810, however, the English chemist Sir Humphry Davy proved it to be an element and named it chlorine.

Occurrence. Chlorine does not occur free in nature, but its compounds are widely distributed. For the most part it occurs in combination with the metals in the form of chlorides, those of sodium, potassium, and magnesium being most abundant. Nearly all salt water contains these substances, particularly sodium chloride, and very large salt beds consisting of chlorides are found in many parts of the world.

Preparation. Two general methods of preparing chlorine may be mentioned, namely, the laboratory method and the electrolytic method.

1. *Laboratory method.* In the laboratory chlorine is made by warming the mineral pyrolusite (manganese dioxide, MnO_2) with concentrated hydrochloric acid. The first reaction, which seems to be similar to the action of acids upon oxides in general, is expressed in the equation

$$MnO_2 + 4HCl = MnCl_4 + 2H_2O.$$

The manganese compound so formed is very unstable, however, and breaks clown according to the equation

$$MnCl_4 = MnCl_2 + 2Cl.$$

Instead of using hydrochloric acid in the preparation of chlorine it will serve just as well to use a mixture of sodium chloride and sulphuric acid, since these two react to form hydrochloric acid. The following equations will then express the changes:

$$(1)\ 2NaCl + H_2SO_4 = Na_2SO_4 + 2HCl.$$

$$(2)\ MnO_2 + 4\ HCl = MnCl_2 + 2Cl + 2H_2O.$$

(3) $MnCl_2 + H_2SO_4 = MnSO_4 + 2HCl.$

[Pg 179]

Combining these equations, the following equation expressing the complete reaction is obtained:

$$2NaCl + MnO_2 + 2H_2SO_4 = MnSO_4 + Na_2SO_4 + 2H_2O + 2Cl.$$

Since the hydrochloric acid liberated in the third equation is free to act upon manganese dioxide, it will be seen that all of the chlorine originally present in the sodium chloride is set free.

The manganese dioxide and the hydrochloric acid are brought together in a flask, as represented in Fig. 53, and a gentle heat is applied. The rate of evolution of the gas is regulated by the amount of heat applied, and the gas is collected by displacement of air. As the equations show, only half of the chlorine present in the hydrochloric acid is liberated.

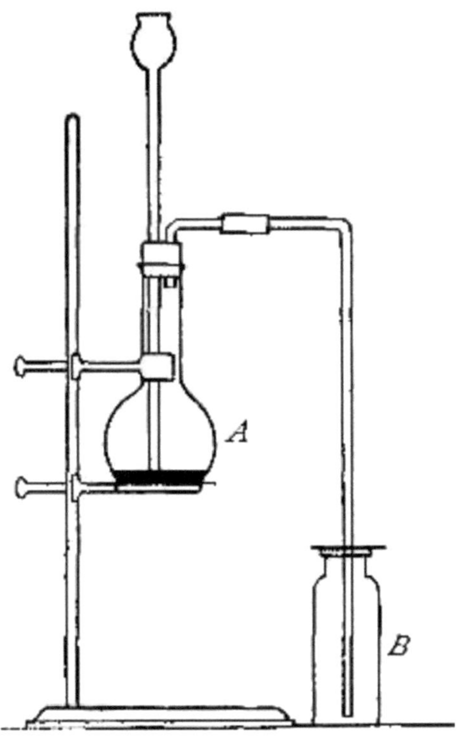

Fig. 53

2. *Electrolytic method.* Under the discussion of electrolysis (p. 102) it was shown that when a solution of sodium chloride is electrolyzed chlorine is evolved at the anode, while the sodium set free at the cathode reacts with the water to form hydrogen, which is evolved, and sodium hydroxide, which remains in solution. A great deal of the chlorine required in the chemical industries is now made in this way in connection with the manufacture of sodium hydroxide.

Physical properties. Chlorine is a greenish-yellow gas, which has a peculiar suffocating odor and produces a very violent effect upon the throat and lungs. Even when inhaled in small quantities it often produces all the symptoms of a [Pg 180] hard cold, and in larger quantities may have serious and even fatal action. It is quite heavy (density = 2.45) and can therefore be collected by displacement of air. One volume of water under ordinary conditions dissolves about three volumes of chlorine. The gas is readily liquefied, a pressure of six atmospheres serving to liquefy it at 0°. It forms a yellowish liquid which solidifies at -102°.

Chemical properties. At ordinary temperatures chlorine is far more active chemically than any of the elements we have so far considered, with the exception of fluorine; indeed, it is one of the most active of all elements.

1. *Action on metals.* A great many metals combine directly with chlorine, especially when hot. A strip of copper foil heated in a burner flame and then dropped into chlorine burns with incandescence. Sodium burns brilliantly when heated strongly in slightly moist chlorine. Gold and silver are quickly tarnished by the gas.

2. *Action on non-metals.* Chlorine has likewise a strong affinity for many of the non-metals. Thus phosphorus burns in a current of the gas, while antimony and arsenic in the form of a fine powder at once burst into flame when dropped into jars of the gas. The products formed in all cases where chlorine combines with another element are called *chlorides*.

3. *Action on hydrogen.* Chlorine has a strong affinity for hydrogen, uniting with it to form hydrochloric acid. A jet of hydrogen burning in the air continues to burn when introduced into a jar of chlorine, giving a somewhat luminous flame. A mixture of the two gases explodes violently when a spark is passed through it or when it is exposed to bright sunlight. In the latter case it is the light and not the heat which starts the action. [Pg 181]

4. *Action on substances containing hydrogen.* Not only will chlorine combine directly with free hydrogen but it will often abstract the element from its compounds. Thus, when chlorine is passed into a solution containing hydrosulphuric acid, sulphur is precipitated and Hydrochloric acid formed. The reaction is shown by the following equation:

$$H_2S + 2Cl = 2HCl + S.$$

With ammonia the action is similar:

$$NH_3 + 3Cl = 3HCl + N.$$

The same tendency is very strikingly seen in the action of chlorine upon turpentine. The latter substance is largely made up of compounds having the composition represented by the formula $C_{10}H_{16}$. When a strip of paper moistened with warm turpentine is placed in a jar of chlorine dense fumes of hydrochloric acid appear and a black deposit of carbon is formed. Even water, which is a very stable compound, can be decomposed by chlorine, the oxygen being liberated. This may be shown in the following way:

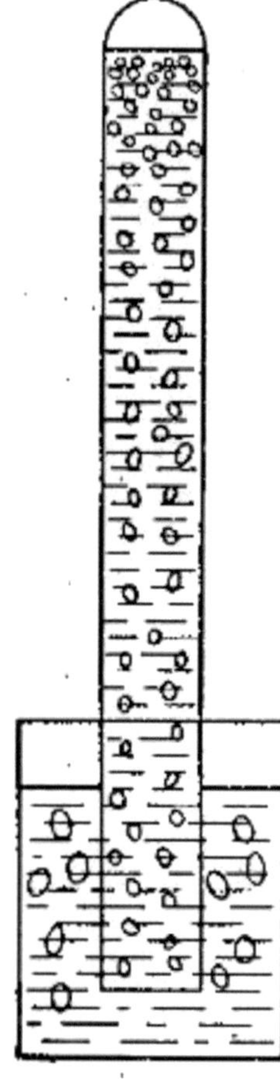

Fig. 54

If a long tube of rather large diameter is filled with a strong solution of chlorine in water and inverted in a vessel of the same solution, as shown in Fig. 54, and the apparatus is placed in bright sunlight, very soon bubbles of a gas will be observed to rise through the solution and collect in the tube. An examination of this gas will show that it is oxygen. It is liberated from water in accordance with the following equation:

$$H_2O + 2Cl = 2HCl + O.$$

5. *Action on color substances, – bleaching action.* If strips of brightly colored cloth or some highly colored flowers are placed in quite dry chlorine, no marked change [Pg 182] in color is noticed as a rule. If, however, the cloth and flowers are first moistened, the color rapidly disappears, that is, the objects are bleached. Evidently the moisture as well as the chlorine is concerned in the action, and a study of the case shows that the chlorine has combined with the hydrogen of the water. The oxygen set free oxidizes the color substance, converting it into a colorless compound. It is evident from this explanation that chlorine will only bleach those substances which are changed into colorless compounds by oxidation.

6. *Action as a disinfectant.* Chlorine has also marked germicidal properties, and the free element, as well as compounds from which it is easily liberated, are used as disinfectants.

Nascent state. It will be noticed that oxygen when set free from water by chlorine is able to do what ordinary oxygen cannot do, for both the cloth and the flowers are unchanged in the air which contains oxygen. It is generally true that the activity of an element is greatest at the instant of liberation from its compounds. To express this fact elements at the instant of liberation are said to be in the *nascent state.* It is nascent oxygen which does the bleaching.

Hydrochloric acid (*muriatic acid*) (HCl). The preparation of hydrochloric acid may be discussed under two general heads:

1. *Laboratory preparation.* The product formed by the burning of hydrogen in chlorine is the gas hydrochloric acid. This substance is much more easily obtained, however, by treating common salt (sodium chloride) with sulphuric acid. The following equation shows the reaction:

$$2NaCl + H_2SO_4 = Na_2SO_4 + 2HCl.$$

[Pg 183]

The dry salt is placed in a flask furnished with a funnel tube and an exit tube, the sulphuric acid is added, and the flask gently warmed. The hydrochloric acid gas is rapidly given off and can be collected by displacement of air. The same apparatus can be used as was employed in the preparation of chlorine (Fig. 53).

When a *solution* of salt is treated with sulphuric acid there is no very marked action. The hydrochloric acid formed is very soluble in water, and so does not escape from the solution; hence a state of equilibrium is soon reached between the four substances represented in the equation. When *concentrated* sulphuric acid, in which hydrochloric acid is not soluble, is poured upon dry salt the reaction is complete.

2. *Commercial preparation.* Commercially, hydrochloric acid is prepared in connection with the manufacture of sodium sulphate, the reaction being the same as that just given. The reaction is carried out in a furnace, and the hydrochloric acid as it escapes in the form of gas is passed into water in which it dissolves, the solution forming the hydrochloric acid of commerce. When the materials are pure a colorless solution is obtained. The most concentrated solution has a density of 1.2 and contains 40% HCl. The commercial acid, often called *muriatic acid*, is usually colored yellow by impurities.

Composition of hydrochloric acid. When a solution of hydrochloric acid is electrolyzed in an apparatus similar to the one in which water was electrolyzed (Fig. 18), chlorine collects at the anode and hydrogen at the cathode. At first the chlorine dissolves in the water, but soon the water in the one tube becomes saturated with it, and if the stopcocks are left open until this is the case, and are then closed, it will be seen that the two gases are set free in equal volumes. [Pg 184]

When measured volumes of the two gases are caused to unite it is found that one volume of hydrogen combines with one of chlorine. Other experiments show that the volume of hydrochloric acid formed is just equal to the sum of the volumes of hydrogen and chlorine. Therefore one volume of hydrogen combines with one volume of chlorine to form two volumes of hydrochloric acid gas. Since chlorine is 35.18 times as heavy as hydrogen, it follows that one part of hydrogen by weight combines with 35.18 parts of chlorine to form 36.18 parts of hydrochloric acid.

Physical properties. Hydrochloric acid is a colorless gas which has an irritating effect when inhaled, and possesses a sour, biting taste, but no marked odor. It is heavier than air (density = 1.26) and is very soluble in water. Under standard conditions 1 volume of water dissolves about 500 volumes of the gas. On warming such a solution the gas escapes, until at the boiling point the solution contains about 20% by weight of HCl. Further boiling will not drive out any more acid, but the solution will distill with unchanged concentration. A more dilute solution than this will lose water on boiling until it has reached the same concentration, 20%, and will then distill unchanged. Under high pressure the gas can be liquefied, 28 atmospheres being required at 0°. Under these conditions it forms a colorless liquid which is not very active chemically. It boils at -80° and solidifies at -113°. The solution of the gas in water is used almost entirely in the place of the gas itself, since it is not only far more convenient but also more active.

Chemical properties. The most important chemical properties of hydrochloric acid are the following:

1. *Action as an acid.* In aqueous solution hydrochloric acid has very strong acid properties; indeed, it is one of [Pg 185] the strongest acids. It acts upon oxides and hydroxides, converting them into salts:

$$NaOH + HCl = NaCl + H_2O,$$
$$CuO + 2HCl = CuCl_2 + H_2O.$$

It acts upon many metals, forming chlorides and liberating hydrogen:

$$Zn + 2HCl = ZnCl_2 + 2H,$$
$$Al + 3HCl = AlCl_3 + 3H.$$

Unlike nitric and sulphuric acids it has no oxidizing action, so that when it acts on metals hydrogen is always given off.

2. *Relation to combustion.* Hydrochloric acid gas is not readily decomposed, and is therefore neither combustible nor a supporter of combustion.

3. *Action on oxidizing agents.* Although hydrochloric acid is incombustible, it can be oxidized under some circumstances, in which case the hydrogen combines with oxygen, while the chlorine is set free. Thus, when a solution of hydrochloric acid acts upon manganese dioxide part of the chlorine is set free:

$$MnO_2 + 4HCl = MnCl_2 + 2H_2O + 2Cl.$$

Aqua regia. It has been seen that when nitric acid acts as an oxidizing agent it usually decomposes, as represented in the equation

$$2HNO_3 = H_2O + 2NO + 3O.$$

The oxygen so set free may act on hydrochloric acid:

$$6HCl + 3O = 3H_2O + 6Cl.$$

The complete equation therefore is

$$2HNO_3 + 6HCl = 4H_2O + 2NO + 6Cl.$$

[Pg 186]

When concentrated nitric and hydrochloric acids are mixed this reaction goes on slowly, chlorine and some other substances not represented in the equation being formed. The mixture is known as *aqua regia* and is commonly prepared by adding one volume of nitric acid to three volumes of hydrochloric acid. It acts more powerfully upon metals and other substances than either of the acids separately, and owes its strength not to acid properties but to the action of the nascent chlorine which it liberates. Consequently, when it acts upon metals such as gold it converts them into chlorides, and the reaction can be represented by such equations as

$$Au + 3Cl = AuCl_3.$$

Salts of hydrochloric acid,—chlorides. The chlorides of all the metals are known and many of them are very important compounds. Some of them are found in nature, and all can be prepared by the general method of preparing salts. Silver chloride, lead chloride, and mercurous chloride are insoluble in water and acids, and can be prepared by adding hydrochloric acid to solutions of compounds of the respective elements. While the chlorides have formulas similar to the fluorides, their properties are often quite different. This is seen in the solubility of the salts. Those metals whose chlorides are insoluble form soluble fluorides, while many of the metals which form soluble chlorides form insoluble fluorides.

Compounds of chlorine with oxygen and hydrogen. Chlorine combines with oxygen and hydrogen to form four different acids. They are all quite unstable, and most of them cannot be prepared in pure form; their salts can easily be made, however, and some of them will be met with in the [Pg 187] study of the metals. The formulas and names of these acids are as follows:

$HClO$	hypochlorous acid.
$HClO_2$	chlorous acid.
$HClO_3$	chloric acid.
$HClO_4$	perchloric acid.

Oxides of chlorine. Two oxides are known, having the formulas Cl_2O and ClO_2. They decompose very easily and are good oxidizing agents.

BROMINE

Historical. Bromine was discovered in 1826 by the French chemist Ballard, who isolated it from sea salt. He named it bromine (stench) because of its unbearable fumes.

Occurrence. Bromine occurs almost entirely in the form of bromides, especially as sodium bromide and magnesium bromide,

which are found in many salt springs and salt deposits. The Stassfurt deposits in Germany and the salt waters of Ohio and Michigan
are especially rich in bromides.

Preparation of bromine. The laboratory method of preparing
bromine is essentially different from the commercial method.

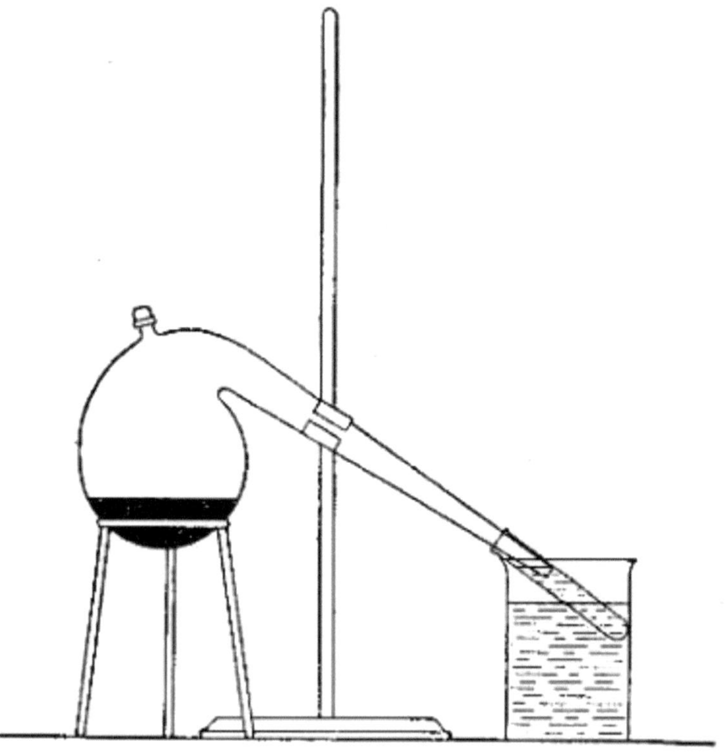

Fig. 55

1. *Laboratory method.* As in the case of chlorine, bromine can be
prepared by the action of hydrobromic acid (HBr) on manganese
dioxide. Since hydrobromic acid is not an article of commerce, a
mixture of sulphuric acid [Pg 188] and a bromide is commonly substituted for it. The materials are placed in a retort arranged as
shown in Fig. 55. The end of the retort just touches the surface of the

water in the test tube. On heating, the bromine distills over and is collected in the cold receiver. The equation is

$$2NaBr + 2H_2SO_4 + MnO_2 = Na_2SO_4 + MnSO_4 + 2H_2O + 2Br.$$

2. *Commercial method.* Bromine is prepared commercially from the waters of salt wells which are especially rich in bromides. On passing a current of electricity through such waters the bromine is first liberated. Any chlorine liberated, however, will assist in the reaction, since free chlorine decomposes bromides, as shown in the equation

$$NaBr + Cl = NaCl + Br.$$

When the water containing the bromine is heated, the liberated bromine distills over into the receiver.

Physical properties. Bromine is a dark red liquid about three times as heavy as water. Its vapor has a very offensive odor and is most irritating to the eyes and throat. The liquid boils at 59° and solidifies at -7°; but even at ordinary temperatures it evaporates rapidly, forming a reddish-brown gas very similar to nitrogen peroxide in appearance. Bromine is somewhat soluble in water, 100 volumes of water under ordinary conditions dissolving 1 volume of the liquid. It is readily soluble in carbon disulphide, forming a yellow solution.

Chemical properties and uses. In chemical action bromine is very similar to chlorine. It combines directly with many of the same elements with which chlorine unites, but with less energy. It combines with hydrogen and takes away [Pg 189] the latter element from some of its compounds, but not so readily as does chlorine. Its bleaching properties are also less marked.

Bromine finds many uses in the manufacture of organic drugs and dyestuffs and in the preparation of bromides.

Hydrobromic acid (HBr). When sulphuric acid acts upon a bromide hydrobromic acid is set free:

$$2NaBr + H_2SO_4 = Na_2SO_4 + 2HBr.$$

At the same time some bromine is set free, as may be seen from the red fumes which appear, and from the odor. The explanation of this is found in the fact that hydrobromic acid is much less stable than hydrochloric acid, and is therefore more easily oxidized. Concentrated sulphuric acid is a good oxidizing agent, and oxidizes a part of the hydrobromic acid, liberating bromine:

$$H_2SO_4 + 2HBr = 2H_2O + SO_2 + 2Br.$$

Preparation of pure hydrobromic acid. A convenient way to make pure hydrobromic acid is by the action of bromine upon moist red phosphorus. This can be done with the apparatus shown in Fig. 56. Bromine is put into the dropping funnel A, and red phosphorus, together with enough water to cover it, is placed in the flask B. By means of the stopcock the bromine is allowed to flow drop by drop into the flask, the reaction taking place without the application of heat. The equations are

(1) $P + 3Br = PBr_3,$

(2) $PBr_3 + 3H_2O = P(OH)_3 + 3HBr.$

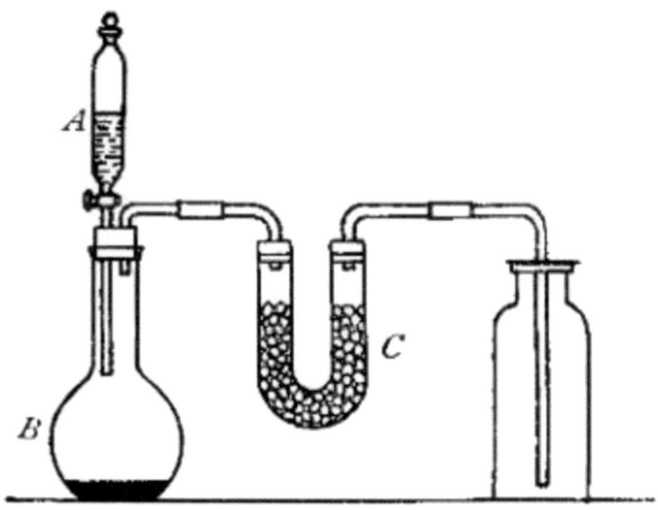

Fig. 56

[Pg 190]

The U-tube C contains glass beads which have been moistened with water and rubbed in red phosphorus. Any bromine escaping action in the flask acts upon the phosphorus in the U-tube. The hydrobromic acid is collected in the same way as hydrochloric acid.

Properties. Hydrobromic acid very strikingly resembles hydrochloric acid in physical and chemical properties. It is a colorless, strongly fuming gas, heavier than hydrochloric acid and, like it, is very soluble in water. Under standard conditions 1 volume of water dissolves 610 volumes of the gas. Chemically, the chief point in which it differs from hydrochloric acid is in the fact that it is much more easily oxidized, so that bromine is more readily set free from it than chlorine is from hydrochloric acid.

Salts of hydrobromic acid,—bromides. The bromides are very similar to the chlorides in their properties. Chlorine acts upon both bromides and free hydrobromic acid, liberating bromine from them:

KBr + Cl = KCl + Br,

HBr + Cl = HCl + Br.

Silver bromide is extensively used in photography, and the bromides of sodium and potassium are used as drugs.

Oxygen compounds. No oxides of bromine are surely known, and bromine does not form so many oxygen acids as chlorine does. Salts of hypobromous acid (HBrO) and bromic acid ($HBrO_3$) are known.

IODINE

Historical. Iodine was discovered in 1812 by Courtois in the ashes of certain sea plants. Its presence was revealed by its beautiful violet vapor, and this suggested the name iodine (from the Greek for violet appearance).

Occurrence. In the combined state iodine occurs in very small quantities in sea water, from which it is absorbed by [Pg 191] certain sea plants, so that it is found in their ashes. It occurs along with bromine in salt springs and beds, and is also found in Chili saltpeter.

Preparation. Iodine may be prepared in a number of ways, the principal methods being the following:

1. *Laboratory method.* Iodine can readily be prepared in the laboratory from an iodide by the method used in preparing bromine, except that sodium iodide is substituted for sodium bromide. It can also be made by passing chlorine into a solution of an iodide.

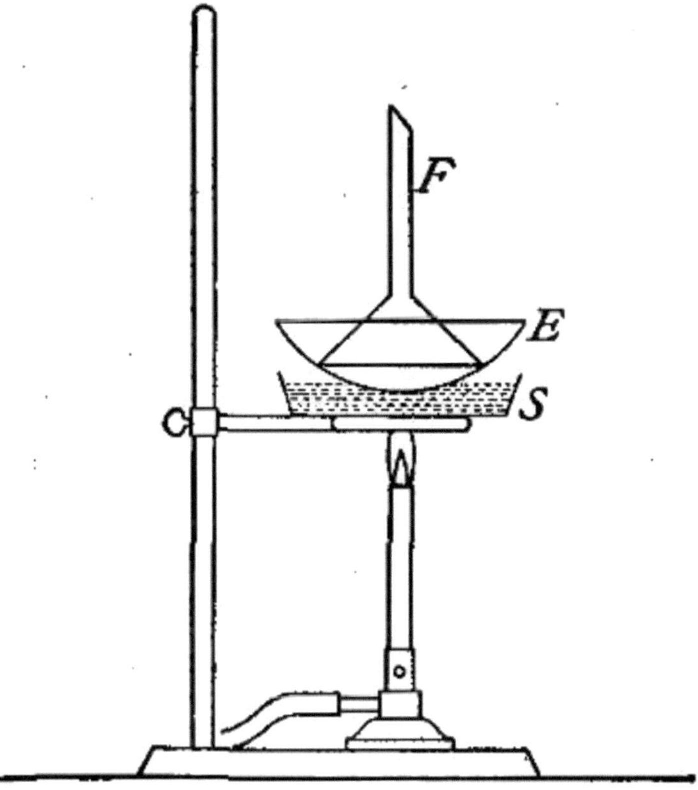

Fig. 57

2. *Commercial method.* Commercially iodine was formerly prepared from seaweed (kelp), but is now obtained almost entirely
from the deposits of Chili saltpeter. The crude saltpeter is dissolved
in water and the solution evaporated until the saltpeter crystallizes.
The remaining liquors, known as the "mother liquors," contain sodium iodate ($NaIO_3$), in which form the iodine is present in the
saltpeter. The chemical reaction by which the iodine is liberated
from this compound is a complicated one, depending on the fact
that sulphurous acid acts upon iodic acid, setting iodine free. This
reaction is shown as follows:

$$2HIO_3 + 5H_2SO_3 = 5H_2SO_4 + H_2O + 2I.$$

Purification of iodine. Iodine can be purified very conveniently in the following way. The crude iodine is placed in an evaporating dish E (Fig. 57), and the dish is set upon the sand bath S. The iodine is covered with the inverted funnel F, and the sand bath is [Pg 192] gently heated with a Bunsen burner. As the dish becomes warm the iodine rapidly evaporates and condenses again on the cold surface of the funnel in shining crystals.

This process, in which a solid is converted into a vapor and is again condensed into a solid without passing through the liquid state, is called *sublimation*.

Physical properties. Iodine is a purplish-black, shining, heavy solid which crystallizes in brilliant plates. Even at ordinary temperatures it gives off a beautiful violet vapor, which increases in amount as heat is applied. It melts at 107° and boils at 175°. It is slightly soluble in water, but readily dissolves in alcohol, forming a brown solution (tincture of iodine), and in carbon disulphide, forming a violet solution. The element has a strong, unpleasant odor, though by no means as irritating as that of chlorine and bromine.

Chemical properties. Chemically iodine is quite similar to chlorine and bromine, but is still less active than bromine. It combines directly with many elements at ordinary temperatures. At elevated temperatures it combines with hydrogen, but the reaction is reversible and the compound formed is quite easily decomposed. Both chlorine and bromine displace it from its salts:

$$KI + Br = KBr + I,$$

$$KI + Cl = KCl + I.$$

When even minute traces of iodine are added to thin starch paste a very intense blue color develops, and this reaction forms a delicate test for iodine. Iodine is extensively used in medicine, especially in the form of a tincture. It is also largely used in the preparation of dyes and organic drugs, iodoform, a substance used as an antiseptic, has the [Pg 193] formula CHI_3.

Hydriodic acid (HI). This acid cannot be prepared in pure condition by the action of sulphuric acid upon an iodide, since the hydriodic acid set free is oxidized by the sulphuric acid just as in the case of hydrobromic acid, but to a much greater extent. It can be prepared in exactly the same way as hydrobromic acid, iodine being substituted for bromine. It can also be prepared by passing hydrosulphuric acid into water in which iodine is suspended. The equation is

$$H_2S + 2I = 2HI + S.$$

The hydriodic acid formed in this way dissolves in the water.

Properties and uses. Hydriodic acid resembles the corresponding acids of chlorine and bromine in physical properties, being a strongly fuming, colorless gas, readily soluble in water. Under standard conditions 1 volume of water dissolves about 460 volumes of the gas. It is, however, more unstable than either hydrochloric or hydrobromic acids, and on exposure to the air it gradually decomposes in accordance with the equation

$$2HI + O = H_2O + 2I.$$

Owing to the slight affinity between iodine and hydrogen the acid easily gives up its hydrogen and is therefore a strong reducing agent. This is seen in its action on sulphuric acid.

The salts of hydriodic acid, the iodides, are, in general, similar to the chlorides and bromides. Potassium iodide (KI) is the most familiar of the iodides and is largely used in medicine.

Oxygen compounds. Iodine has a much greater affinity for oxygen than has either chlorine or bromine. When heated with nitric [Pg 194] acid it forms a stable oxide (I_2O_5). Salts of iodic acid (HIO_3)

and periodic acid (HIO_4) are easily prepared, and the free acids are much more stable than the corresponding acids of the other members of this family.

GAY-LUSSAC'S LAW OF VOLUMES

In the discussion of the composition of hydrochloric acid it was stated that one volume of hydrogen combines with one volume of chlorine to form two volumes of hydrochloric acid. With bromine and iodine similar combining ratios hold good. These facts recall the simple volume relations already noted in the study of the composition of steam and ammonia. These relations may be represented graphically in the following way:

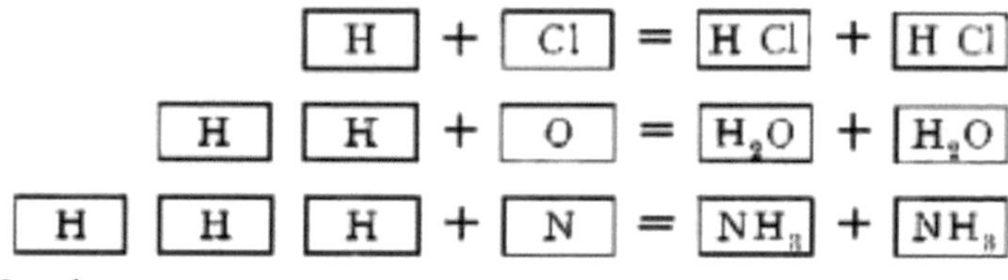

Graph

In the early part of the past century Gay-Lussac, a distinguished French chemist, studied the volume relations of many combining gases, and concluded that similar relations always hold. His observations are summed up in the following law: *When two gases combine chemically there is always a simple ratio between their volumes, and between the volume of either one of them and that of the product, provided it is a gas.* By a simple ratio is meant of course the ratio of small whole numbers, as 1 : 2, 2 : 3.

EXERCISES

1. How do we account for the fact that liquid hydrofluoric acid is not an electrolyte?

2. Why does sulphuric acid liberate hydrofluoric acid from its salts? [Pg 195]

3. In the preparation of chlorine, what advantages are there in treating manganese dioxide with a mixture of sodium chloride and sulphuric acid rather than with hydrochloric acid?

4. Why must chlorine water be kept in the dark?

5. What is the derivation of the word nascent?

6. What substances studied are used as bleaching agents? To what is the bleaching action due in each case?

7. What substances studied are used as disinfecting agents?

8. What is meant by the statement that hydrochloric acid is one of the strongest acids?

9. What is the meaning of the phrase *aqua regia*?

10. Cl_2O is the anhydride of what acid?

11. A solution of hydriodic acid on standing turns brown. How is this accounted for?

12. How can bromine vapor and nitrogen peroxide be distinguished from each other?

13. Write the equations for the reaction taking place when hydriodic acid is prepared from iodine, phosphorus, and water.

14. From their behavior toward sulphuric acid, to what class of agents do hydrobromic and hydriodic acids belong?

15. Give the derivation of the names of the elements of the chlorine family.

16. Write the names and formulas for the binary acids of the group in the order of the stability of the acids.

17. What is formed when a metal dissolves in each of the following? nitric acid; dilute sulphuric acid; concentrated sulphuric acid; hydrochloric acid; aqua regia.

18. How could you distinguish between a chloride, a bromide, and an iodide?

19. What weight of sodium chloride is necessary to prepare sufficient hydrochloric acid to saturate 1 l. of water under standard conditions?

20. On decomposition 100 l. of hydrochloric acid would yield how many liters of hydrogen and chlorine respectively, the gases being measured under the same conditions? Are your results in accord with the experimental facts?

[Pg 196]

CHAPTER XVII

CARBON AND SOME OF ITS SIMPLER COMPOUNDS

The family. Carbon stands at the head of a family of elements in the fourth group in the periodic table. The resemblances between the elements of this family, while quite marked, are not so striking as in the case of the elements of the chlorine family. With the exception of carbon, these elements are comparatively rare, and need not be taken up in detail in this chapter. Titanium will be referred to again in connection with silicon which it very closely resembles.

Occurrence. Carbon is found in nature in the uncombined state in several forms. The diamond is practically pure carbon, while graphite and coal are largely carbon, but contain small amounts of other substances. Its natural compounds are exceedingly numerous and occur as gases, liquids, and solids. Carbon dioxide is its most familiar gaseous compound. Natural gas and petroleum are largely compounds of carbon with hydrogen. The carbonates, especially calcium carbonate, constitute great strata of rocks, and are found in almost every locality. All living organisms, both plant and animal, contain a large percentage of this element, and the number of its compounds which go to make up all the vast variety of animate nature is almost limitless. Over one hundred thousand definite compounds containing carbon have been prepared. In the free state carbon occurs in three allotropic forms, two of which are crystalline and one amorphous. [Pg 197]

Crystalline carbon. Crystalline carbon occurs in two forms,— diamond and graphite.

1. *Diamond.* Diamonds are found in considerable quantities in several localities, especially in South Africa, the East Indies, and Brazil. The crystals belong to the regular system, but the natural stones do not show this very clearly. When found they are usually

covered with a rough coating which is removed in the process of cutting. Diamond cutting is carried on most extensively in Holland.

The density of the diamond is 3.5, and, though brittle, it is one of the hardest of substances. Black diamonds, as well as broken and imperfect stones which are valueless as gems, are used for grinding hard substances. Few chemical reagents have any action on the diamond, but when heated in oxygen or the air it blackens and burns, forming carbon dioxide.

Lavoisier first showed that carbon dioxide is formed by the combustion of the diamond, and Sir Humphry Davy in 1814 showed that this is the only product of combustion, and that the diamond is pure carbon.

The diamond as a gem. The pure diamond is perfectly transparent and colorless, but many are tinted a variety of colors by traces of foreign substances. Usually the colorless ones are the most highly prized, although in some instances the color adds to the value; thus the famous Hope diamond is a beautiful blue. Light passing through a diamond is very much refracted, and to this fact the stone owes its brilliancy and sparkle.

Artificial preparation of diamonds. Many attempts have been made to produce diamonds artificially, but for a long time these always ended in failure, graphite and not diamonds being the product obtained. The French chemist Moissan, in his extended study of chemistry at high temperatures, finally succeeded (1893) in making some small ones. He accomplished this by dissolving carbon in boiling iron and plunging the crucible containing the mixture into water, [Pg 198] as shown in Fig. 58. Under these conditions the carbon crystallized in the iron in the form of the diamond. The diamonds were then obtained by dissolving away the iron in hydrochloric acid.

Fig. 58

2. *Graphite.* This form of carbon is found in large quantities, especially in Ceylon, Siberia, and in some localities of the United States and Canada. It is a shining black substance, very soft and greasy to the touch. Its density is about 2.15. It varies somewhat in properties according to the locality in which it is found, and is more easily attacked by reagents than is the diamond. It is also manufactured by heating carbon with a small amount of iron (3%) in an electric furnace. It is used in the manufacture of lead pencils and crucibles, as a lubricant, and as a protective covering for iron in the form of a polish or a paint.

Amorphous carbon. Although there are many varieties of amorphous carbon known, they are not true allotropic modifications. They differ merely in their degree of purity, their fineness of division, and in their mode of preparation. These substances are of the greatest importance, owing to their many uses in the arts and industries. As they occur in nature, or are made artificially, they are nearly all impure carbon, the impurity depending on the particular substance in question.

1. *Pure carbon.* Pure amorphous carbon is best prepared by charring sugar. This is a substance consisting of carbon, hydrogen, and oxygen, the latter two elements being present in the ratio of one oxygen atom to two of hydrogen. [Pg 199] When sugar is strongly heated the oxygen and hydrogen are driven off in the form of water and pure carbon is left behind. Prepared in this way it is a soft, lustrous, very bulky, black powder.

2. *Coal and coke.* Coals of various kinds were probably formed from vast accumulations of vegetable matter in former ages, which became covered over with earthy material and were thus protected from rapid decay. Under various natural agencies the organic matter was slowly changed into coal. In anthracite these changes have gone the farthest, and this variety of coal is nearly pure carbon. Soft or bituminous coals contain considerable organic matter besides carbon and mineral substances. When heated strongly out of contact with air the organic matter is decomposed and the resulting volatile matter is driven off in the form of gases and vapors, and only the mineral matter and carbon remain behind. The gaseous product is chiefly illuminating gas and the solid residue is *coke.* Some of the coke is found as a dense cake on the sides and roof of the retort. This is called retort carbon and is quite pure.

3. *Charcoal.* This is prepared from wood in the same way that coke is made from coal. When the process is carried on in retorts the products expelled by the heat are saved. Among these are many valuable substances such as wood alcohol and acetic acid. Where timber is abundant the process is carried out in a wasteful way, by merely covering piles of wood with sod and setting the wood on fire. Some wood burns and the heat from this decomposes the wood

not burned, forming charcoal from it. The charcoal, of course, contains the mineral part of the wood from which it is formed. [Pg 200]

4. *Bone black.* This is sometimes called animal charcoal, and is made by charring bones and animal refuse. The organic part of the materials is thus decomposed and carbon is left in a very finely divided state, scattered through the mineral part which consists largely of calcium phosphate. For some uses this mineral part is removed by treatment with hydrochloric acid and prolonged washing.

5. *Lampblack.* Lampblack and soot are products of imperfect combustion of oil and coal, and are deposited from a smoky flame on a cold surface. The carbon in this form is very finely divided and usually contains various oily materials.

Properties. While the various forms of carbon differ in many properties, especially in color and hardness, yet they are all odorless, tasteless solids, insoluble in water and characterized by their stability towards heat. Only in the intense heat of the electric arc does carbon volatilize, passing directly from the solid state into a vapor. Owing to this fact the inside surface of an incandescent light bulb after being used for some time becomes coated with a dark film of carbon. It is not acted on at ordinary temperatures by most reagents, but at a higher temperature it combines directly with many of the elements, forming compounds called *carbides.* When heated in the presence of sufficient oxygen it burns, forming carbon dioxide.

Uses of carbon. The chief use of amorphous carbon is for fuel to furnish heat and power for all the uses of civilization. An enormous quantity of carbon in the form of the purer coals, coke, and charcoal is used as a reducing agent in the manufacture of the various metals, especially in the metallurgy of iron. Most of the metals are found in nature as oxides, or in forms which can readily be [Pg 201] converted into oxides. When these oxides are heated with carbon the oxygen is abstracted, leaving the metal. Retort carbon and coke are used to make electric light carbons and battery plates, while lampblack is used for indelible inks, printer's ink, and black varnishes. Bone black and charcoal have the property of absorbing large volumes of certain gases, as well as smaller amounts of organic matter; hence they are used in filters to remove noxious gases and objectionable colors and odors from water. Bone black is used extensively in the sugar refineries to remove coloring matter from the impure sugars.

Chemistry of carbon compounds. Carbon is remarkable for the very large number of compounds which it forms with the other elements, especially with oxygen and hydrogen. Compounds containing carbon are more numerous than all others put together, and the chemistry of these substances presents peculiarities not met with in the study of other substances. For these reasons the systematic study of carbon compounds, or of *organic chemistry* as it is usually called, must be deferred until the student has gained some knowledge of the chemistry of other elements. An acquaintance with a few of the most familiar carbon compounds is, however, essential for the understanding of the general principles of chemistry.

Compounds of carbon with hydrogen,—the hydrocarbons. Carbon unites with hydrogen to form a very large number of compounds called *hydrocarbons*. Petroleum and natural gas are essentially mixtures of a great variety of these hydrocarbons. Many others are found in living plants, and still others are produced by the decay of organic matter in the absence of air. Only two of them, methane and acetylene, will be discussed here. [Pg 202]

Methane (*marsh gas*) (CH_4). This is one of the most important of these hydrocarbons, and constitutes about nine tenths of natural gas. As its name suggests, it is formed in marshes by the decay of vegetable matter under water, and bubbles of the gas are often seen to rise when the dead leaves on the bottom of pools are stirred. It also collects in mines, and, when mixed with air, is called *fire damp* by the miners because of its great inflammability, damp being an old name for a gas. It is formed when organic matter, such as coal or wood, is heated in closed vessels, and is therefore a principal constituent of coal gas.

Preparation. Methane is prepared in the laboratory by heating sodium or calcium acetate with soda-lime. Equal weights of fused sodium acetate and soda-lime are thoroughly dried, then mixed and placed in a good-sized, hard-glass test tube fitted with a one-holed stopper and delivery tube. The mixture is gradually heated, and when the air has been displaced from the tube the gas is collected in bottles by displacement of water. Soda-lime is a mixture of sodium and calcium hydroxides. Regarding it as sodium hydroxide alone, the equation is

$$NaC_2H_3O_2 + NaOH = Na_2CO_3 + CH_4.$$

Properties. Methane is a colorless, odorless gas whose density is 0.55. It is difficult to liquefy, boiling at -155° under standard pressure, and is almost insoluble in water. It burns with a pale blue flame, liberating much heat, and when mixed with oxygen is very explosive.

Davy's safety lamp. In 1815 Sir Humphry Davy invented a lamp for the use of miners, to prevent the dreadful mine explosions then common, due to methane mixed with air. The invention consisted in surrounding the upper part of the common miner's lamp with a mantle of wire gauze and the lower part with glass (Fig. 59). It has been seen that two gases will not combine until raised to their [Pg 203] kindling temperature, and if while combining they are cooled below this point, the combination ceases. A flame will not pass through a wire gauze because the metal, being a good conductor of heat, takes away so much heat from the flame that the gases are cooled below the kindling temperature. When a lamp so protected is brought into an explosive mixture the gases inside the wire mantle burn in a series of little explosions, giving warning to the miner that the air is unsafe.

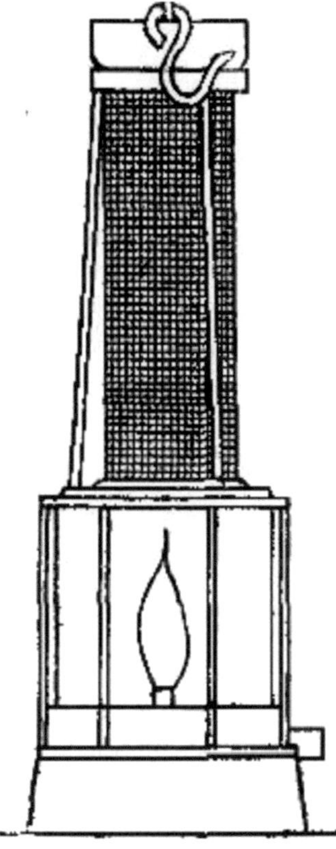

Fig. 59

Acetylene (C_2H_2). This is a colorless gas usually having a disagreeable odor due to impurities. It is now made in large quantities from calcium carbide (CaC_2). This substance is formed when coal and lime are heated together in an electric furnace. When treated with water the carbide is decomposed, yielding acetylene:

$$CaC_2 + 2H_2O = C_2H_2 + Ca(OH)_2.$$

Under ordinary conditions the gas burns with a very smoky flame; in burners constructed so as to secure a large amount of oxygen it burns with a very brilliant white light, and hence is used as an illuminant.

Laboratory preparation. The gas can be prepared readily in a generator such as is shown in Fig. 60. The inner tube contains fragments of calcium carbide, while the outer one is filled with water. As long as the stopcock is closed the water cannot rise in the inner tube. When the stopcock is open the water rises, and, coming into contact with the carbide in the inner tube, generates acetylene. This escapes through the stopcock, and after the air has been expelled may be lighted as it issues from the burner.

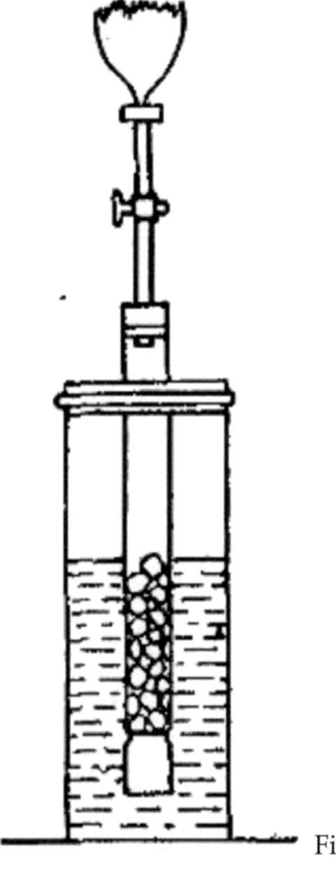

Fig. 60

Carbon forms two oxides, namely, carbon dioxide (CO_2) and carbon monoxide (CO). [Pg 204]

236

Carbon dioxide (CO_2). Carbon dioxide is present in the air to the extent of about 3 parts in 10,000, and this apparently small amount is of fundamental importance in nature. In some localities it escapes from the earth in great quantities, and many spring waters carry large amounts of it in solution. When these highly charged spring waters reach the surface of the earth, and the pressure on them is removed, the carbon dioxide escapes with effervescence. It is a product of the oxidation of all organic matter, and is therefore formed in fires as well as in the process of decay. It is thrown off from the lungs of all animals in respiration, and is a product of many fermentation processes such as vinegar making and brewing. Combined with metallic oxides it forms vast deposits of carbonates in nature.

Preparation. In the laboratory carbon dioxide is always prepared by the action of an acid upon a carbonate, usually calcium carbonate, the apparatus shown in Fig. 39 serving the purpose very well. This reaction might be expected to produce carbonic acid, thus:

$$CaCO_3 + 2HCl = CaCl_2 + H_2CO_3.$$

Carbonic acid is very unstable, however, and decomposes into its anhydride, CO_2, and water, thus:

$$H_2CO_3 = H_2O + CO_2.$$

The complete reaction is represented by the equation

$$CaCO_3 + 2HCl = CaCl_2 + CO_2 + H_2O.$$

Physical properties. Carbon dioxide is a colorless, practically odorless gas whose density is 1.5. Its weight may be inferred from the fact that it can be siphoned, or poured like water, from one vessel downward into another. At 15° [Pg 205] and under ordinary pressure it dissolves in its own volume of water and imparts a somewhat biting, pungent taste to it. It is easily condensed, and is now prepared commercially in this form by pumping the gas into steel cylinders (see Fig. 6) which are kept cold during the process. When the liquid is permitted to escape into the air part of it instantly evaporates, and in so doing absorbs so much heat that another portion is solidified, the solid form strikingly resembling snow in appearance. This snow is very cold and mercury can easily be frozen with it.

Solid carbon dioxide. Cylinders of liquid carbon dioxide are inexpensive, and should be available in every school. To demonstrate the properties of solid carbon dioxide, the cylinder should be placed across the table and supported in such a way that the stopcock end is several inches lower than the other end. A loose bag is made by holding the corners of a handkerchief around the neck of the stopcock, and the cock is then turned on so that the gas rushes out in large quantities. Very quickly a considerable quantity of the snow collects in the handkerchief. To freeze mercury, press a piece of filter paper into a small evaporating dish and pour the mercury upon it. Coil a flat spiral upon the end of a wire, and dip the spiral into the mercury. Place a quantity of solid carbon dioxide upon the mercury and pour 10 cc.-15 cc. of ether over it. In a minute or two the mercury will solidify and may be removed from the dish by the wire serving as a handle. The filter paper is to prevent the mercury from sticking to the dish; the ether dissolves the solid carbon dioxide and promotes its rapid conversion into gas.

Chemical properties. Carbon dioxide is incombustible, since it is, like water, a product of combustion. It does not support combustion, as does nitrogen peroxide, because the oxygen in it is held in very firm chemical union with the carbon. Very strong reducing agents, such as highly heated carbon, can take away half of its oxygen:

$$CO_2 + C = 2CO.$$

[Pg 206]

238

Uses. The relation of carbon dioxide to plant life has been discussed in a previous chapter. Water highly charged with carbon dioxide is used for making soda water and similar beverages. Since it is a non-supporter of combustion and can be generated readily, carbon dioxide is also used as a fire extinguisher. Some of the portable fire extinguishers are simply devices for generating large amounts of the gas. It is not necessary that all the oxygen should be kept away from the fire in order to smother it. A burning candle is extinguished in air which contains only 2.5% of carbon dioxide.

Carbonic acid (H_2CO_3). Like most of the oxides of the non-metallic elements, carbon dioxide is an acid anhydride. It combines with water to form an acid of the formula H_2CO_3, called carbonic acid:

$$H_2O + CO_2 = H_2CO_3.$$

The acid is, however, very unstable and cannot be isolated. Only a very small amount of it is actually formed when carbon dioxide is passed into water, as is evident from the small solubility of the gas. If, however, a base is present in the water, salts of carbonic acid are formed, and these are quite stable:

$$2NaOH + H_2O + CO_2 = Na_2CO_3 + 2H_2O.$$

Action of carbon dioxide on bases. This conduct is explained by the principles of reversible reactions. The equation

$$H_2O + CO_2 \longleftrightarrow H_2CO_3$$

is a reversible equation, and the extent to which the reaction progresses depends upon the relative concentrations of each of the three factors in it. Equilibrium is ordinarily reached when very little H_2CO_3 is formed. If a base is present in the water to combine with the H_2CO_3 as fast as it is formed, all of the CO_2 is converted [Pg 207] into H_2CO_3, and thence into a carbonate.

Salts of carbonic acid,—carbonates. The carbonates form a very important class of salts. They are found in large quantities in nature, and are often used in chemical processes. Only the carbonates of sodium, potassium, and ammonium are soluble, and these can be made by the action of carbon dioxide on solutions of the bases, as has just been explained.

The insoluble carbonates are formed as precipitates when soluble salts are treated with a solution of a soluble carbonate. Thus the insoluble calcium carbonate can be made by bringing together solutions of calcium chloride and sodium carbonate:

$$CaCl_2 + Na_2CO_3 = CaCO_3 + 2NaCl.$$

Most of the carbonates are decomposed by heat, yielding an oxide of the metal and carbon dioxide. Thus lime (calcium oxide) is made by strongly heating calcium carbonate:

$$CaCO_3 = CaO + CO_2.$$

Acid carbonates. Like all acids containing two acid hydrogen atoms, carbonic acid can form both normal and acid salts. The acid carbonates are made by treating a normal carbonate with an excess of carbonic acid. With few exceptions they are very unstable, heat decomposing them even when in solution.

Action of carbon dioxide on calcium hydroxide. If carbon dioxide is passed into clear lime water, calcium carbonate is at first precipitated:

$$H_2O + CO_2 = H_2CO_3,$$

$$Ca(OH)_2 + H_2CO_3 = CaCO_3 + 2H_2O.$$

Advantage is taken of this reaction in testing for the presence of carbon dioxide, as already explained in the chapter on the atmosphere. If the current of carbon dioxide is continued, the precipitate [Pg 208] soon dissolves, because the excess of carbonic acid forms calcium acid carbonate which is soluble:

$$CaCO_3 + H_2CO_3 = Ca(HCO_3)_2.$$

If now the solution is heated, the acid carbonate is decomposed and calcium carbonate once more precipitated:

$$Ca(HCO_3)_2 = CaCO_3 + H_2CO_3.$$

Carbon monoxide (CO). Carbon monoxide can be made in a number of ways, the most important of which are the three following:

1. *By the partial oxidation of carbon.* If a slow current of air is conducted over highly heated carbon, the monoxide is formed, thus:

$$C + O = CO$$

It is therefore often formed in stoves when the air draught is insufficient. Water gas, which contains large amounts of carbon monoxide, is made by partially oxidizing carbon with steam:

$$C + H_2O = CO + 2H.$$

2. *By the partial reduction of carbon dioxide.* When carbon dioxide is conducted over highly heated carbon it is reduced to carbon monoxide by the excess of carbon:

$$CO_2 + C = 2CO.$$

When coal is burning in a stove or grate carbon dioxide is at first formed in the free supply of air, but as the hot gas rises through the glowing coal it is reduced to carbon monoxide. When the carbon monoxide reaches the free air above the coal it takes up oxygen to form carbon dioxide, burning with the blue flame so familiar above a bed of coals, especially in the case of hard coals.

3. *By the decomposition of oxalic acid.* In the laboratory carbon monoxide is usually prepared by the action of [Pg 209] concentrated sulphuric acid upon oxalic acid. The latter substance has the formula $C_2H_2O_4$. The sulphuric acid, owing to its affinity for water, decomposes the oxalic acid, as represented in the equation

$$C_2H_2O_4 + (H_2SO_4) = (H_2SO_4) + H_2O + CO_2 + CO.$$

Properties. Carbon monoxide is a light, colorless, almost odorless gas, very difficult to liquefy. Chemically it is very active, combining directly with a great many substances. It has a great affinity for oxygen and is therefore combustible and a good reducing agent. Thus, if carbon monoxide is passed over hot copper oxide, the copper is reduced to the metallic state:

$$CuO + CO = Cu + CO_2.$$

When inhaled it combines with the red coloring matter of the blood and in this way prevents the absorption of oxygen, so that even a small quantity of the gas may prove fatal.

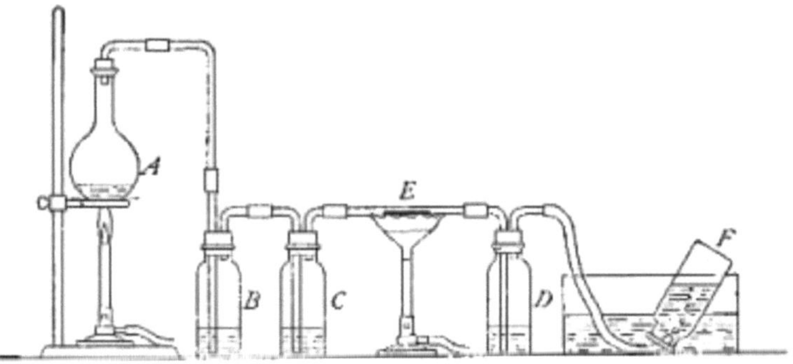

Fig. 61

The reducing power of carbon monoxide. Fig. 61 illustrates a method of showing the reducing power of carbon monoxide. The gas is generated by gently heating 7 or 8 g. of oxalic acid with 25 cc. of concentrated sulphuric acid in a 200 cc. flask A. The bottle B contains a solution of sodium hydroxide, which removes the carbon dioxide formed along with the monoxide. C contains a solution of calcium hydroxide to show that the carbon dioxide is completely removed. E is a hard-glass tube containing 1 or 2 g. of copper oxide, which is heated by a burner. The black copper oxide is reduced to reddish metallic copper by the carbon monoxide, which is thereby changed to carbon dioxide. The presence of the carbon dioxide is shown by the precipitate in the calcium hydroxide solution in D. Any unchanged carbon monoxide is collected over water in F.

[Pg 210]

Carbon disulphide (CS_2). Just as carbon combines with oxygen to form carbon dioxide, so it combines with sulphur to form carbon disulphide (CS_2). This compound has been described in the chapter on sulphur.

Hydrocyanic acid (*prussic acid*)(HCN). Under the proper conditions carbon unites with nitrogen and hydrogen to form the acid HCN, called hydrocyanic acid. It is a weak, volatile acid, and is therefore easily prepared by treating its salts with sulphuric acid:

$$KCN + H_2SO_4 = KHSO_4 + HCN.$$

It is most familiar as a gas, though it condenses to a colorless liquid boiling at 26°. It has a peculiar odor, suggesting bitter almonds, and is extremely poisonous either when inhaled or when taken into the stomach. A single drop may cause death. It dissolves readily in water, its solution being commonly called prussic acid.

The salts of hydrocyanic acid are called *cyanides*, the cyanides of sodium and potassium being the best known. These are white solids and are extremely poisonous.

Solutions of potassium cyanide are alkaline. A solution of potassium cyanide turns red litmus blue, and must therefore contain hydroxyl ions. The presence of these ions is accounted for in the following way.

Although water is so little dissociated into its ions H^+ and OH^- that for most purposes we may neglect the dissociation, it is nevertheless measurably dissociated. Hydrocyanic acid is one of the weakest of acids, and dissociates [Pg 211] to an extremely slight extent. When a cyanide such as potassium cyanide dissolves it freely dissociates, and the CN^- ions must come to an equilibrium with the H^+ ions derived from the water:

$$H^+ + CN^- \longleftrightarrow HCN.$$

The result of this equilibrium is that quite a number of H^+ ions from the water are converted into undissociated HCN molecules. But for every H^+ ion so removed an OH^- ion remains free, and this will give the solution alkaline properties.

EXERCISES

1. How can you prove that the composition of the different allotropic forms of carbon is the same?

2. Are lampblack and bone black allotropic forms of carbon? Will equal amounts of heat be liberated in the combustion of 1 g. of each?

3. How could you judge of the relative purity of different forms of carbon?

4. Apart from its color, why should carbon be useful in the preparation of inks and paints?

5. Could asbestos fibers be used to replace the wire in a safety lamp?

6. Why do most acids decompose carbonates?

7. What effect would doubling the pressure have upon the solubility of carbon dioxide in water?

8. What compound would be formed by passing carbon dioxide into a solution of ammonium hydroxide? Write the equation.

9. Write equations for the preparation of K_2CO_3; of $BaCO_3$; of $MgCO_3$.

10. In what respects are carbonic and sulphurous acids similar?

11. Give three reasons why the reaction which takes place when a solution of calcium acid carbonate is heated, completes itself.

12. How could you distinguish between carbonates and sulphites?

13. How could you distinguish between oxygen, hydrogen, nitrogen, nitrous oxide, and carbon dioxide? [Pg 212]

14. Could a solution of sodium hydroxide be substituted for the solution of calcium hydroxide in testing for carbon dioxide?

15. What weight of sodium hydroxide is necessary to neutralize the carbonic acid formed by the action of hydrochloric acid on 100 g. of calcium carbonate?

16. What weight of calcium carbonate would be necessary to prepare sufficient carbon dioxide to saturate 10 l. of water at 15° and under ordinary pressure?

17. On the supposition that calcium carbide costs 12 cents a kilogram, what would be the cost of an amount sufficient to generate 100 l. of acetylene measured at 20° and 740 mm.?

18. How would the volume of a definite amount of carbon monoxide compare with the volume of carbon dioxide formed by its combustion, the measurements being made under the same conditions?

CHAPTER XVIII

FLAMES,—ILLUMINANTS

Conditions necessary for flames. It has been seen that when two substances unite chemically, with the production of light and heat, the act of union is called combustion. When one of the substances undergoing combustion remains solid at the temperature occasioned by the combustion, light may be given off, but there is no flame. Thus iron wire burning in oxygen throws off a shower of sparks and is brilliantly incandescent, but no flame is seen. When, however, both of the substances are gases or vapors at the temperature reached in the combustion, the act of union is accompanied by a flame.

Flames from burning liquids or solids. Many substances which are liquids or solids at ordinary temperatures burn with a flame because the heat of combustion vaporizes them slowly, and the flame is due to the union of this vapor with the gas supporting the combustion.

Supporter of combustion. That gas which surrounds the flame and constitutes the atmosphere in which the combustion occurs is said to support the combustion. The other gas which issues into this atmosphere is said to be the combustible gas. Thus, in the ordinary combustion of coal gas in the air the coal gas is said to be combustible, while the air is regarded as the supporter of combustion. These terms are entirely relative, however, for a jet of air issuing into an atmosphere of coal gas will burn when ignited, the coal gas supporting the combustion. [Pg 214] Ordinarily, when we say that a gas is combustible we mean that it is combustible in an atmosphere of air.

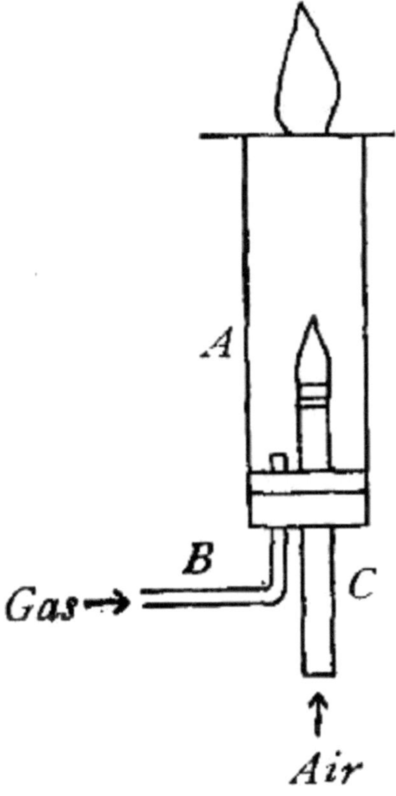

Fig. 62

Either gas may be the supporter of combustion. That the terms *combustible* and *supporter of combustion* are merely relative may be shown in the following way: A lamp chimney *A* is fitted with a cork and glass tubes, as shown in Fig. 62. The tube *C* should have a diameter of from 12 to 15 mm. A thin sheet of asbestos in which is cut a circular opening about 2 cm. in diameter is placed over the top of the chimney. The opening in the asbestos is closed with the palm of the hand, and gas is admitted to the chimney through the tube *B*. The air in the chimney is soon expelled through the tube *C*, and the gas itself is then lighted at the lower end of this tube. The hand is now removed from the opening in the asbestos, when the flame at the end of the tube at once rises and appears at the end within the chimney, as shown in the figure. The excess of coal gas now escapes from the opening in the asbestos and may be lighted. The flame at the top of the asbestos board is due to the combustion of coal gas in air, while the flame within the chimney is due to the combustion of air in coal gas, the air being drawn up through the tube by the escaping gas.

Appearance of flames. The flame caused by the union of hydrogen and oxygen is almost colorless and invisible. Chlorine and hydrogen combine with a pale violet flame, carbon monoxide burns in oxygen with a blue flame, while ammonia burns with a deep yellow flame. The color and appearance of flames are therefore often quite characteristic of the particular combustion which occasions them.

Structure of flames. When the gas undergoing combustion issues from a round opening into an atmosphere of the gas supporting combustion, as is the case with the burning Bunsen burner (Fig. 63), the flame is generally [Pg 215] conical in outline. It consists of several distinct cones, one within the other, the boundary between them being marked by differences of color or luminosity. In the simplest flame, of which hydrogen burning in oxygen is a good example, these cones are two in number,—an inner one, formed by unburned gas, and an outer one, usually more or less luminous, consisting of the combining gases. This outer one is in turn surrounded by a third envelope of the products of combustion; this envelope is sometimes invisible, as in the present case, but is sometimes faintly luminous. The lower part of the inner cone of the flame is quite cool and consists of unburned gas. Toward the top of the inner cone the gas has become heated to a high temperature by the burning envelope surrounding it. On reaching the supporter of combustion on the outside it is far above its kindling temperature, and combustion follows with the evolution of much heat. The region of combustion just outside the inner cone is therefore the hottest part of the flame.

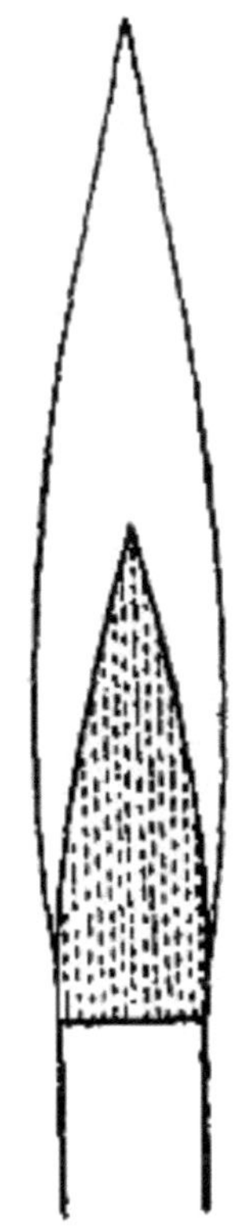

Fig. 63

Oxidizing and reducing flames. Since the tip of the outside cone consists of very hot products of combustion mixed with oxygen from the air, a substance capable of oxidation placed in this part of the flame becomes very hot and is easily oxidized. The oxygen with which it combines comes, of course, from the atmosphere, and not from the products of combustion. This outer tip of the flame is called the *oxidizing flame.*

At the tip of the inner cone the conditions are quite different. This region consists of a highly heated combustible gas, which has not yet reached a supply of oxygen. [Pg 216]

If a substance rich in oxygen, such as a metallic oxide, is placed in this region of the flame, the heated gases combine with its oxygen and the substance is reduced. This part of the flame is called the *reducing flame*. These flames are used in testing certain substances, especially minerals. For this purpose they are produced by blowing into a small luminous Bunsen flame from one side through a blow-pipe. This is a tube of the shape shown in Fig. 64. The flame is directed in any desired way and has the oxidizing and reducing regions very clearly marked (Fig. 65). It is non-luminous from the same causes which render the open Bunsen burner flame non-luminous, the gases from the lungs serving to furnish oxygen and to dilute the combustible gas.

Fig. 64

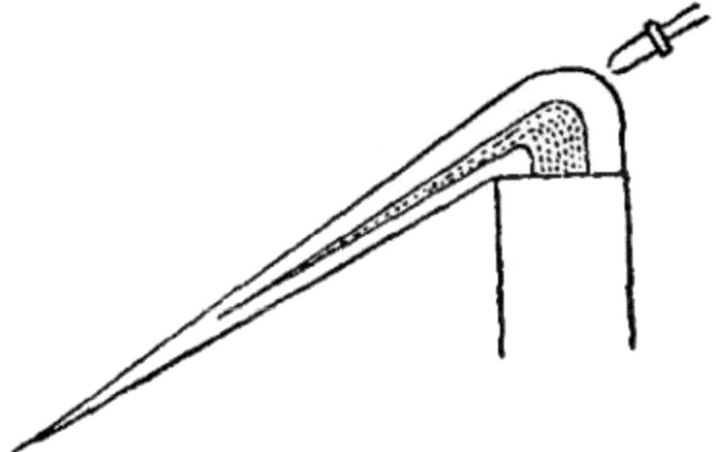

Fig. 65

Luminosity of flames. The luminosity of flames is due to a number of distinct causes, and may therefore be increased or diminished in several ways.

1. *Presence of solid matter.* The most obvious of these causes is the presence in the flame of incandescent solid matter. Thus chalk dust sifted into a non-luminous flame renders it luminous. When hydrocarbons form a part of the combustible gas, as they do in nearly all illuminating gases and oils, some carbon is usually set free in the process of combustion. This is made very hot by the flame and becomes incandescent, giving out light. In a well-regulated flame it is afterward burned up, but when the supply of oxygen is insufficient it escapes from the flame as lampblack or soot. That it is temporarily present in a well-burning luminous flame may be demonstrated by holding a cold object, such as a small evaporating dish, in the flame for a few seconds. This cold object cools the carbon below its kindling temperature, and it is deposited on the object as soot. [Pg 217]

2. *Pressure.* A second factor in the luminosity of flames is the pressure under which the gases are burning. Under increased pressure there is more matter in a given volume of a gas, and the chemical action is more energetic than when the gases are rarefied. Conse-

quently there is more heat and light. A candle burning on a high mountain gives less light than when it burns at the sea level.

If the gas is diluted with a non-combustible gas, the effect is the same as if it is rarefied, for under these conditions there is less combustible gas in a given volume.

3. *Temperature.* The luminosity also depends upon the temperature attained in the combustion. In general the hotter the flame the greater the luminosity; hence cooling the gases before combustion diminishes the luminosity of the flame they will make, because it diminishes the temperature attained in the combustion. Thus the luminosity of the Bunsen flame is largely diminished by the air drawn up with the gas. This is due in part to the fact that the burning gas is diluted and cooled by the air drawn in. The oxygen thus introduced into the flame also causes the combustion of the hot particles of carbon which would otherwise tend to make the flame luminous.

Illuminating and fuel gases. A number of mixtures of combustible gases, consisting largely of carbon compounds and hydrogen, find extensive use for the production of light and heat. The three chief varieties are coal gas, water gas, and natural gas. The use of acetylene gas has already been referred to.

Coal gas. Coal gas is made by heating bituminous coal in large retorts out of contact with the air. Soft or bituminous coal contains, in addition to large amounts of carbon, considerable quantities of compounds of hydrogen, oxygen, nitrogen, and sulphur. When distilled the nitrogen is liberated partly in the form of ammonia and cyanides and partly as free nitrogen gas; the sulphur is converted into hydrogen sulphide, carbon disulphide, and oxides of sulphur; the oxygen into water and oxides of carbon. The [Pg 218] remaining hydrogen is set free partly as hydrogen and partly in combination with carbon in the form of hydrocarbons. The most important of these is methane, with smaller quantities of many others, some of which are liquids or solids at ordinary temperatures. The great bulk of the carbon remains behind as coke and retort carbon.

The manufacture of coal gas. In the manufacture of coal gas it is necessary to separate from the volatile constituents formed by the heating of the coal all those substances which are either solid or liquid at ordinary temperature, since these would clog the gas pipes. Certain gaseous constituents, such as hydrogen sulphide and ammonia, must also be removed. The method used to accomplish this is shown in Fig. 66. The coal is heated in air-tight retorts illustrated by *A*. The volatile products escape through the pipe *X* and bubble into the tarry liquid in the large pipe *B*, known as the *hydraulic main*, which runs at right angles to the retorts. Here is deposited the greater portion of the solid and liquid products, forming a tarry mass known as *coal tar*. Much of the ammonia also remains dissolved in this liquid. The partially purified gas then passes into the pipes *C*, which serve to cool it and further remove the solid and liquid matter. The gas then passes into *D*, which is filled with coke over which a jet of water is sprayed. The water still further cools the gas and at the same time partially removes such gaseous products as hydrogen sulphide and ammonia, which are soluble in water. In *E* the gas passes over some material such as lime, which removes the last portions of the sulphur compounds as well as much of the carbon dioxide present. From *E* the gas passes into the large gas holder *F*, from which it is distributed through pipes to the places where it is burned.

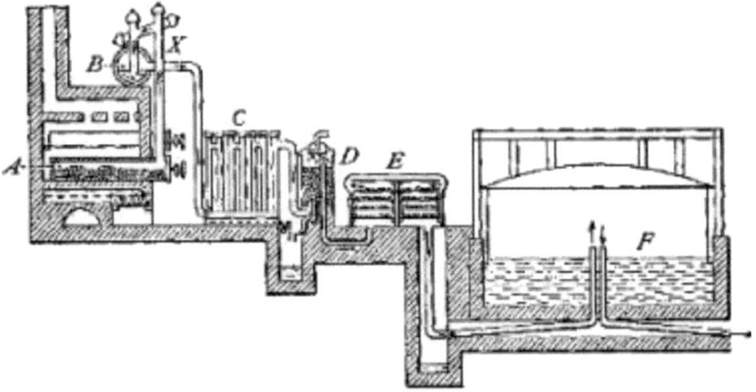

Fig. 66

[Pg 219]

One ton of good gas coal yields approximately 10,000 cu. ft. of gas, 1400 lb. of coke, 120 lb. of tar, and 20 gal. of ammoniacal liquor.

Not only is the ammonia obtained in the manufacture of the gas of great importance, but the coal tar also serves as the source of many very useful substances, as will be explained in Chapter XXXII.

Water gas. Water gas is essentially a mixture of carbon monoxide and hydrogen. It is made by passing steam over very hot anthracite coal, when the reaction shown in the following equation takes place:

$$C + H_2O = CO + 2H.$$

When required merely to produce heat the gas is at once ready for use. When made for illuminating purposes it must be enriched, that is, illuminants must be added, since both carbon monoxide and hydrogen burn with non-luminous flames. This is accomplished by passing it into heaters containing highly heated petroleum oils. The gas takes up hydrocarbon gases formed in the decomposition of the petroleum oils, which make it burn with a luminous flame.

Water gas is very effective as a fuel, since both carbon monoxide and hydrogen burn with very hot flames. It has little odor and is very poisonous. Its use is therefore attended with some risk, since leaks in pipes are very likely to escape notice.

Natural gas. This substance, so abundant in many localities, varies much in composition, but is composed principally of methane. When used for lighting purposes it is usually burned in a burner resembling an open Bunsen, the illumination being furnished by an incandescent mantle. This is the case in the familiar Welsbach burner. Contrary to statements frequently made, natural gas contains no free hydrogen. [Pg 220]

TABLE SHOWING COMPOSITION OF GASES

	PENNSYLVANIA NATURAL GAS	COAL GAS	WATER GAS	ENRICHED WATER GAS
Hydrogen		41.3	52.88	30.00
Methane	90.64	43.6	2.16	24.00

Illuminants		3.9		12.05
Carbon monoxide		6.4	36.80	29.00
Carbon dioxide	0.30	2.0	3.47	0.30
Nitrogen	9.06	1.2	4.69	2.50
Oxygen		0.3		1.50
Hydrocarbon vapors		1.5		1.50

These are analyses of actual samples, and may be taken as about the average for the various kinds of gases. Any one of these may vary considerably. The nitrogen and oxygen in most cases is due to a slight admixture of air which is difficult to exclude entirely in the manufacture and handling of gases.

Fuels. A variety of substances are used as fuels, the most important of them being wood, coal, and the various gases mentioned above. Wood consists mainly of compounds of carbon, hydrogen, and oxygen. The composition of coal and the fuel gases has been given. Since these fuels are composed principally of carbon and hydrogen or their compounds, the chief products of combustion are carbon dioxide and water. The practice of heating rooms with portable gas or oil stoves with no provision for removing the products of combustion is to be condemned, since the carbon dioxide is generated in sufficient quantities to render the air unfit for breathing. Rooms so heated also become very damp from the large amount of water vapor formed in the combustion, and which in [Pg 221] cold weather condenses on the window glass, causing the glass to "sweat." Both coal and wood contain a certain amount of mineral substances which constitute the ashes.

The electric furnace. In recent years electric furnaces have come into wide use in operations requiring a very high temperature. Temperatures as high as 3500° can be easily reached, whereas the hottest oxyhydrogen flame is not much above 2000°. These furnaces are constructed on one of two general principles.

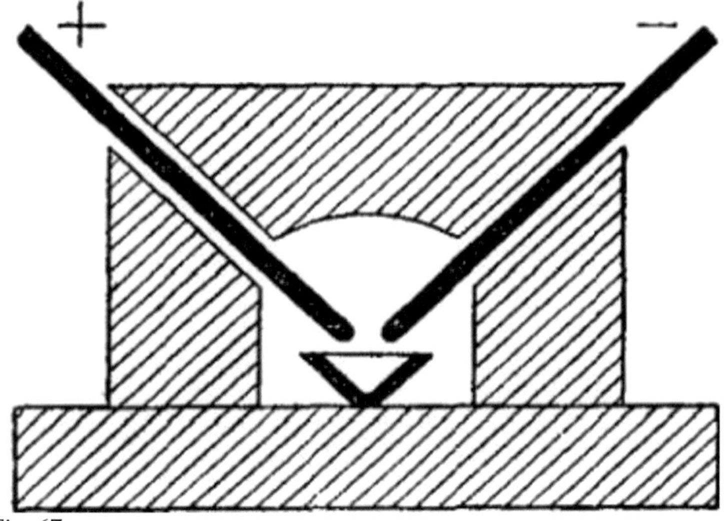

Fig. 67

1. *Arc furnaces.* In the one type the source of heat is an electric arc formed between carbon electrodes separated a little from each other, as shown in Fig. 67. The substance to be heated is placed in a vessel, usually a graphite crucible, just below the arc. The electrodes and crucible are surrounded by materials which fuse with great difficulty, such as magnesium oxide, the walls of the furnace being so shaped as to reflect the heat downwards upon the contents of the crucible.

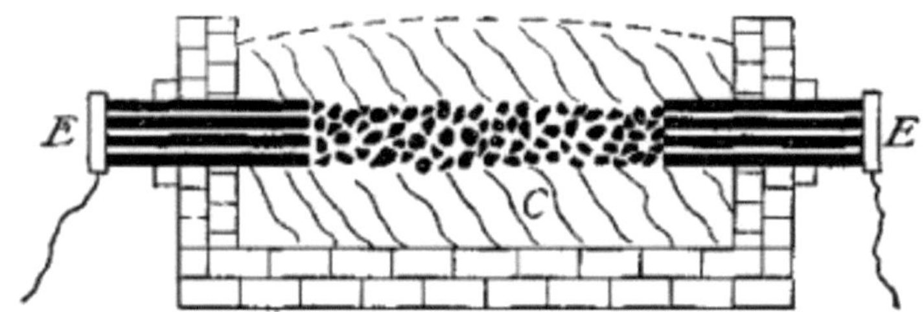

Fig. 68

2. *Resistance furnaces.* In the other type of furnace the heat is generated by the resistance offered to the current in its passage through the furnace. In its simplest form it may be represented by Fig. 68.

The furnace is merely a rectangular box built up of loose bricks. The electrodes E, each consisting of a bundle of carbon rods, are introduced through the sides of the furnace. The materials to be heated, C, are filled into the furnace up to the electrodes, and a layer of broken coke is arranged so as to extend from one electrode to the other. More of the charge is then placed on top of the coke. In passing through the broken coke the electrical current encounters great resistance. This generates great heat, and the charge surrounding the coke is brought to a very high temperature. The advantage of this type of furnace is that the temperature can be regulated to any desired intensity. [Pg 222]

EXERCISES

1. Why does charcoal usually burn with no flame? How do you account for the flame sometimes observed when it burns?

2. How do you account for the fact that a candle burns with a flame?

3. What two properties must the mantle used in the Welsbach lamp possess?

4. (*a*) In what respects does the use of the Welsbach mantle resemble that of lime in the calcium light? (*b*) If the mantle were made of carbon, would it serve the same purpose?

5. Would anthracite coal be suitable for the manufacture of coal gas?

6. How could you prove the formation of carbon dioxide and wa ter in the combustion of illuminating gases?

7. Suggest a probable way in which natural gas has been formed.

8. Coal frequently contains a sulphide of iron. (*a*) What two sulphur compounds are likely to be formed when gas is made from such coal? (*b*) Suggest some suitable method for the removal of these compounds.

9. Why does the use of the bellows on the blacksmith's forge cause a more intense heat?

10. What volume of oxygen is necessary to burn 100 l. of marsh gas and what volume of carbon dioxide would be formed, all of the gases being measured under standard conditions?

11. Suppose a cubic meter of Pennsylvania natural gas, measured under standard conditions, were to be burned. How much water by weight would result?

[Pg 223]

CHAPTER XIX

MOLECULAR WEIGHTS, ATOMIC WEIGHTS, FORMULAS

Introduction. In the chapter on The Atomic Theory, it was shown that if it were true that two elements uniting to form a compound always combined in the ratio of one atom of one element to one atom of the other element, it would be a very easy matter to decide upon figures which would represent the relative weights of the different atoms. It would only be necessary to select some one element as a standard and determine the weight of every element which combines with a definite weight (say 1 g.) of the standard element. The figures so obtained would evidently represent the relative weights of the atoms.

But the law of multiple proportion at once reminds us that two elements may unite in several proportions; and there is no simple way to determine the number of atoms present in the molecule of any compound. Consequently the problem of deciding upon the relative atomic weights is not an easy one. To the solution of this problem we must now turn.

Dalton's method of determining atomic weights. When Dalton first advanced the atomic theory he attempted to solve this problem by very simple methods. He thought that when only one compound of two elements is known it is reasonable to suppose that it contains one atom of each element. He therefore gave the formula HO to water, and HN to ammonia. When more than two compounds were known he assumed that the most familiar or the most stable one had the simple formula. He then determined the atomic weight as [Pg 224] explained above. The results he obtained were contradicto-

258

ry and very far from satisfactory, and it was soon seen that some other method, resting on much more scientific grounds, must be found to decide what compounds, if any, have a single atom of each element present.

Determination of atomic weights. Three distinct steps are involved in the determination of the atomic weight of an element: (1) determination of the equivalent, (2) determination of molecular weights of its compounds, and (3) deduction of the exact atomic weight from the equivalent and molecular weights.

1. Determination of the equivalent. By the equivalent of an element is meant the weight of the element which will combine with a fixed weight of some other element chosen as a standard. It has already been explained that oxygen has been selected as the standard element for atomic weights, with a weight of 16. This same standard will serve very well as a standard for equivalents. *The equivalent of an element is the weight of the element which will combine with 16 g. of oxygen.* Thus 16 g. of oxygen combines with 16.03 g. of sulphur, 65.4 g. of zinc, 215.86 g. of silver, 70.9 g. of chlorine. These figures, therefore, represent the equivalent weights of these elements.

Relation of atomic weights to equivalents. According to the atomic theory combination always takes place between whole numbers of atoms. Thus one atom unites with one other, or with two or three; or two atoms may unite with three, or three with five, and so on.

When oxygen combines with zinc the combination must be between definite numbers of the two kinds of atoms. Experiment shows that these two elements combine in the ratio of 16 g. of oxygen to 65.4 g. of zinc. If one atom of [Pg 225] oxygen combines with one atom of zinc, then this ratio must be the ratio between the weights of the two atoms. If one atom of oxygen combines with two atoms of zinc, then the ratio between the weights of the two atoms will be 16: 32.7. If two atoms of oxygen combine with one atom of zinc, the ratio by weight between the two atoms will be 8: 65.4. It is evident, therefore, that the real atomic weight of an element must be some multiple or submultiple of the equivalent; in other words, the equivalent multiplied by 1/2, 1, 2, or 3 will give the atomic weight.

Combining weights. A very interesting relation holds good between the equivalents of the various elements. We have just seen that the figures 16.03, 65.4, 215.86, and 70.9 are the equivalents respectively of sulphur, zinc, silver, and chlorine. These same figures represent the ratios by weight in which these elements combine among themselves. Thus 215.86 g. of silver combine with 70.9 g. of chlorine and with 2 × 16.03 g. of sulphur. 65.4 g. of zinc combine with 70.9 g. of chlorine and 2 × 16.03 g. of sulphur.

By taking the equivalent or some multiple of it a value can be obtained for each element which will represent its combining value, and for this reason is called its *combining weight.* It is important to notice that the fact that a combining weight can be obtained for each element is not a part of a theory, but is the direct result of experiment.

Elements with more than one equivalent. It will be remembered that oxygen combines with hydrogen in two ratios. In one case 16 g. of oxygen combine with 2.016 g. of hydrogen to form water; in the other 16 g. of oxygen combine with 1.008 g. of hydrogen to form hydrogen dioxide. The equivalents of hydrogen are therefore 2.016 and 1.008. Barium combines with oxygen in two proportions: in barium oxide the proportion is 16 g. of oxygen to 137.4 g. of barium; in barium dioxide the proportion is 16 g. of oxygen to 68.7 g. of barium. [Pg 226]

In each case one equivalent is a simple multiple of the other, so the fact that there may be two equivalents does not add to the uncertainty. All we knew before was that the true atomic weight is some multiple of the equivalent.

2. The determination of molecular weights. To decide the question as to which multiple of the equivalent correctly represents the atomic weight of an element, it has been found necessary to devise a method of determining the molecular weights of compounds containing the element in question. Since the molecular weight of a compound is merely the sum of the weights of all the atoms present in it, it would seem to be impossible to determine the molecular weight of a compound without first knowing the atomic weights of the constituent atoms, and how many atoms of each element are

260

present in the molecule. But certain facts have been discovered which suggest a way in which this can be done.

Avogadro's hypothesis. We have seen that the laws of Boyle, Charles, and Gay-Lussac apply to all gases irrespective of their chemical character. This would lead to the inference that the structure of gases must be quite simple, and that it is much the same in all gases.

In 1811 Avogadro, an Italian physicist, suggested that if we assume all gases under the same conditions of temperature and pressure to have the same number of molecules in a given volume, we shall have a probable explanation of the simplicity of the gas laws. It is difficult to prove the truth of this hypothesis by a simple experiment, but there are so many facts known which are in complete harmony with this suggestion that there is little doubt that it expresses the truth. Avogadro's hypothesis may be stated thus: *Equal volumes of all gases under the same conditions of temperature and pressure contain the same number of molecules.*

[Pg 227]

Avogadro's hypothesis and molecular weights. Assuming that Avogadro's hypothesis is correct, we have a very simple means for deciding upon the relative weights of molecules; for if equal volumes of two gases contain the same number of molecules, the weights of the two volumes must be in the same ratio as the weights of the individual molecules which they contain. If we adopt some one gas as a standard, we can express the weights of all other gases as compared with this one, and the same figures will express the relative weights of the molecules of which the gases are composed.

Oxygen as the standard. It is important that the same standard should be adopted for the determination of molecular weights as has been decided upon for atomic weights and equivalents, so that the three values may be in harmony with each other. Accordingly it is best to adopt oxygen as the standard element with which to compare the molecular weights of other gases, being careful to keep the oxygen atom equal to 16.

The oxygen molecule contains two atoms. One point must not be overlooked, however. We desire to have our unit, the oxygen *atom,* equal to 16. The method of comparing the weights of gases just suggested compares the molecules of the gases with the *molecule* of oxygen. Is the molecule and the atom of oxygen the same thing? This question is answered by the following considerations.

We have seen that when steam is formed by the union of oxygen and hydrogen, two volumes of hydrogen combine with one volume of oxygen to form two volumes of steam. Let us suppose that the one volume of oxygen contains 100 [Pg 228] molecules; then the two volumes of steam must, according to Avogadro's hypothesis, contain 200 molecules. But each of these 200 molecules must contain at least one atom of oxygen, or 200 in all, and these 200 atoms came from 100 molecules of oxygen. It follows that each molecule of oxygen must contain at least two atoms of oxygen.

Evidently this reasoning merely shows that there are *at least* two atoms in the oxygen molecule. There may be more than that, but as there is no evidence to this effect, we assume that the molecule contains two atoms only.

It is evident that if we wish to retain the value 16 for the atom of oxygen we must take twice this value, or 32, for the value of the oxygen molecule, when using it as a standard for molecular weights.

Determination of the molecular weights of gases from their weights compared with oxygen. Assuming the molecular weight of oxygen to be 32, Avogadro's hypothesis gives us a ready means for determining the molecular weight of any other gas, for all that is required is to know its weight compared with that of an equal volume of oxygen. For example, 1 l. of chlorine is found by experiment to weigh 2.216 times as much as 1 l. of oxygen. The molecular weight of chlorine must therefore be 2.216 ×32, or 70.91.

If, instead of comparing the relative weights of 1 l. of the two gases, we select such a volume of oxygen as will weigh 32 g., or the weight in grams corresponding to the molecular weight of the gas, the calculation is much simplified. It has been found that 32 g. of oxygen, under standard conditions, measure 22.4 l. This same volume of hydrogen weighs 2.019 g.; of chlorine 70.9 g.; of hydrochloric acid 36.458 g. The weights of these equal volumes must be proportional to their molecular weights, and since [Pg 229] the weight of the oxygen is the same as the value of its molecular weight, so too will the weights of the 22.4 l. of the other gases be equal to the value of their molecular weights.

As a summary we can then make the following statement: *The molecular weight of any gas may be determined by calculating the weight of 22.4 l. of the gas, measured under standard conditions.*

Determination of molecular weights from density of gases. In an actual experiment it is easier to determine the density of a gas than the weight of a definite volume of it. The density of a gas is usually defined as its weight compared with that of an equal volume of air. Having determined the density of a gas, its weight compared with oxygen may be determined by multiplying its density by the ratio between the weights of air and oxygen. This ratio is 0.9046. To compare it with our standard for atomic weights we must further multiply it by 32, since the standard is 1/32 the weight of oxygen molecules. The steps then are these:

1. Determine the density of the gas (its weight compared with air).

2. Multiply by 0.9046 to make the comparison with oxygen molecules.

3. Multiply by 32 to make the comparison with the unit for atomic weights.

We have, then, the formula:

$$\text{molecular weight} = \text{density} \times 0.9046 \times 32;$$

or, still more briefly,

$$\text{M.} = \text{D.} \times 28.9.$$

The value found by this method for the determination of molecular weights will of course agree with those found [Pg 230] by calculating the weight of 22.4 l. of the gas, since both methods depend on the same principles.

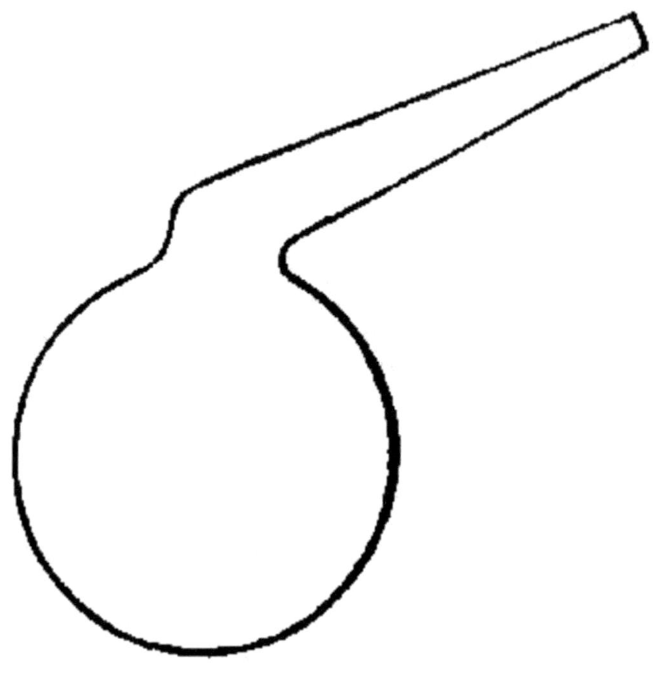

Fig. 69

Determination of densities of gases. The relative weights of equal volumes of two gases can be easily determined. The following is one of the methods used. A small flask, such as is shown in Fig. 69, is filled with one of the gases, and after the temperature and pressure have been noted the flask is sealed up and weighed. The tip of the sealed end is then broken off, the flask filled with the second gas, and its weight determined. If the weight of the empty flask is subtracted from these two weighings, the relative weights of the gases is readily found.

3. Deduction of atomic weights from molecular weights and equivalents. We have now seen how the equivalent of an element and the molecular weight of compounds containing the element can be obtained. Let us see how it is possible to decide which multiple of the equivalent really is the true atomic weight. As an example, let us suppose that the equivalent of nitrogen has been found to be 7.02 and that it is desired to obtain its atomic weight. The next step is to obtain the molecular weights of a large number of compounds containing nitrogen. The following will serve:

	DENSITY BY EXPERIMENT	APPROXIMATE MOLECULAR WEIGHT (D. × 28.9)	PERCENTAGE OF NITROGEN BY EXPERIMENT	PART OF MOLECULAR WEIGHT DUE TO NITROGEN
Nitrogen gas	0.9671	27.95	100.00	27.95
Nitrous oxide	1.527	44.13	63.70	27.11
Nitric oxide	1.0384	30.00	46.74	14.02
Nitrogen peroxide	1.580	45.66	30.49	13.90
Ammonia	0.591	17.05	82.28	14.03
Nitric acid	2.180	63.06	22.27	14.03
Hydrocyanic acid	0.930	26.87	51.90	13.94

[Pg 231]

Method of calculation. The densities of the various gases in the first column of this table are determined by experiment, and are fairly accurate but not entirely so. By multiplying these densities by 28.9 the molecular weights of the compounds as given in the second column are obtained. By chemical analysis it is possible to determine the percentage composition of these substances, and the percentages of nitrogen in them as determined by analysis are given in the third column. If each of these molecular weights is multiplied in turn by the percentage of nitrogen in the compound, the product will be the weight of the nitrogen in the molecular weight of the compound. This will be the sum of the weights of the nitrogen atoms in the molecule. These values are given in the fourth column in the table.

If a large number of compounds containing nitrogen are studied in this way, it is probable that there will be included in the list at least one substance whose molecule contains a single nitrogen atom. In this case the number in the fourth column will be the approximate atomic weight of nitrogen. On comparing the values for nitrogen in the table it will be seen that a number which is approximately 14 is the smallest, and that the others are multiples of this. These compounds of higher value, therefore, contain more than one nitrogen atom in the molecule.

Accurate determination of atomic weights. Molecular weights cannot be determined very accurately, and consequently the part in them due to nitrogen is a little uncertain, as will be seen in the table. All we can tell by this method is that the true weight is very near 14. The equivalent can however be determined very accurately, and we have seen that it is some multiple or submultiple [Pg 232] of the true atomic weight. Since molecular-weight determinations have shown that in the case of nitrogen the atomic weight is near 14, and we have found the equivalent to be 7.02, it is evident that the true atomic weight is twice the equivalent, or $7.02 \times 2 = 14.04$.

Summary. These, then, are the steps necessary to establish the atomic weight of an element.

1. Determine the equivalent accurately by analysis.

2. Determine the molecular weight of a large number of compounds of the element, and by analysis the part of the molecular weight due to the element. The smallest number so obtained will be approximately the atomic weight.

3. Multiply the equivalent by the small whole number (usually 1, 2, or 3), which will make a number very close to the approximate atomic weight. The figure so obtained will be the true atomic weight.

Molecular weights of the elements. It will be noticed that the molecular weight of nitrogen obtained by multiplying its density by 28.9 is 28.08. Yet the atomic weight of nitrogen as deduced from a study of its gaseous compounds is 14.04. The simplest explanation that can be given for this is that the gaseous nitrogen is made up of molecules, each of which contains two atoms. In this respect it resembles oxygen; for we have seen that an entirely different line of reasoning leads us to believe that the molecule of oxygen contains two atoms. When we wish to indicate molecules of these gases the symbols N_2 and O_2 should be used. When we desire to merely show the weights taking part in a reaction this is not necessary.

The vapor densities of many of the elements show that, like oxygen and nitrogen, their molecules consist of two atoms. In other cases, particularly among the metals, [Pg 233] the molecule and the atom are identical. Still other elements have four atoms in their molecules.

While oxygen contains two atoms in its molecules, a study of ozone has led to the conclusion that it has three. The formation of ozone from oxygen can therefore be represented by the equation

$$3O_2 = 2O_3.$$

Other methods of determining molecular weights. It will be noticed that Avogadro's law gives us a method by which we can determine the relative weights of the molecules of two gases because it enables us to tell when we are dealing with an equal number of the two kinds of molecules. If by any other means we can get this information, we can make use of the knowledge so gained to determine the molecular weights of the two substances.

Raoult's laws. Two laws have been discovered which give us just such information. They are known as Raoult's laws, and can be stated as follows:

1. *When weights of substances which are proportional to their molecular weights are dissolved in the same weight of solvent, the rise of the boiling point is the same in each case.*

2. *When weights of substances which are proportional to their molecular weights are dissolved in the same weight of solvent, the lowering of the freezing point is the same in each case.*

By taking advantage of these laws it is possible to determine when two solutions contain the same number of molecules of two dissolved substances, and consequently the relative molecular weights of the two substances.

Law of Dulong and Petit. In 1819 Dulong and Petit discovered a very interesting relation between the atomic [Pg 234] weight of an element and its specific heat, which holds true for elements in the solid state. If equal weights of two solids, say, lead and silver, are heated through the same range of temperature, as from 10° to 20°, it is found that very different amounts of heat are required. The amount of heat required to change the temperature of a solid or a liquid by a definite amount compared with the amount required to change the temperature of an equal weight of water by the same amount is called its specific heat. Dulong and Petit discovered the following law: *The specific heat of an element in the solid form multiplied by its atomic weight is approximately equal to the constant 6.25. That is,*

$$\text{at. wt.} \times \text{sp. ht.} = 6.25.$$

Consequently,

$$\text{at. wt.} = \frac{6.25}{\text{sp. ht.}}$$

This law is not very accurate, but it is often possible by means of it to decide upon what multiple of the equivalent is the real atomic weight. Thus the specific heat of iron is found by experiment to be 0.112, and its equivalent is 27.95. 6.25 ÷ 0.112 = 55.8. We see, therefore, that the atomic weight is twice the equivalent, or 55.9.

How formulas are determined. It will be well in connection with molecular weights to consider how the formula of a compound is decided upon, for the two subjects are very closely associated. Some examples will make clear the method followed.

The molecular weight of a substance containing hydrogen and chlorine was 36.4. By analysis 36.4 parts of the substance was found

268

to contain 1 part of hydrogen and 35.4 parts of chlorine. As these are the simple atomic [Pg 235] weights of the two elements, the formula of the compound must be HCl.

A substance consisting of oxygen and hydrogen was found to have a molecular weight of 34. Analysis showed that in 34 parts of the substance there were 2 parts of hydrogen and 32 parts of oxygen. Dividing these figures by the atomic weights of the two elements, we get $2 \div 1 = 2$ for H; $32 \div 16 = 2$ for O. The formula is therefore H_2O_2.

A substance containing 2.04% H, 32.6% S, and 65.3% O was found to have a molecular weight of 98. In these 98 parts of the substance there are $98 \times 2.04\% = 2$ parts of H, $98 \times 32.6\% = 32$ parts of S, and $98 \times 65.3\% = 64$ parts of O. If the molecule weighs 98, the hydrogen atoms present must together weigh 2, the sulphur atoms 32, and the oxygen atoms 64. Dividing these figures by the respective atomic weights of the three elements, we have, for H, $2 \div 1 = 2$ atoms; for S, $32 \div 32 = 1$ atom; for O, $64 \div 16 = 4$ atoms. Hence the formula is H_2SO_4.

We have, then, this general procedure: Find the percentage composition of the substance and also its molecular weight. Multiply the molecular weight successively by the percentage of each element present, to find the amount of the element in the molecular weight of the compound. The figures so obtained will be the respective parts of the molecular weight due to the several atoms. Divide by the atomic weights of the respective elements, and the quotient will be the number of atoms present.

Avogadro's hypothesis and chemical calculations. This law simplifies many chemical calculations.

1. *Application to volume relations in gaseous reactions.* Since equal volumes of gases contain an equal number of [Pg 236] molecules, it follows that when an equal number of gaseous molecules of two or more gases take part in a reaction, the reaction will involve equal volumes of the gases. In the equation

$$C_2H_2O_4 = H_2O + CO_2 + CO,$$

since 1 molecule of each of the gases CO_2 and CO is set free from each molecule of oxalic acid, the two substances must always be set free in equal volumes.

Acetylene burns in accordance with the equation

$$2C_2H_2 + 5O_2 = 4CO_2 + 2H_2O.$$

Hence 2 volumes of acetylene will react with 5 volumes of oxygen to form 4 volumes of carbon dioxide and 2 volumes of steam. That the volume relations may be correct a gaseous element must be given its molecular formula. Thus oxygen must be written O_2 and not $2O$.

2. *Application to weights of gases.* It will be recalled that the molecular weight of a gas is determined by ascertaining the weight of 22.4 l. of the gas. This weight in grams is called the *gram-molecular weight* of a gas. If the molecular weight of any gas is known, the weight of a liter of the gas under standard conditions may be determined by dividing its gram-molecular weight by 22.4. Thus the gram-molecular weight of a hydrochloric acid gas is 36.458. A liter of the gas will therefore weigh $36.458 \div 22.4 = 1.627$ g.

EXERCISES

1. From the following data calculate the atomic weight of sulphur. The equivalent, as obtained by an analysis of sulphur dioxide, is 16.03. The densities and compositions of a number of compounds containing sulphur are as follows: [Pg 237]

NAME	DENSITY	COMPOSITION BY PERCENTA-GE	
Hydrosulphuric acid	1.1791	S = 94.11	H = 5.89
Sulphur dioxide	2.222	S = 50.05	O = 49.95
Sulphur trioxide	2.74	S = 40.05	O = 59.95
Sulphur chloride	4.70	S = 47.48	Cl = 52.52

| Sulphuryl chloride | 4.64 | S = 23.75 | Cl = 52.53 | O = 23.70 |
| Carbon disulphide | 2.68 | S = 84.24 | C = 15.76 | |

2. Calculate the formulas for compounds of the following compositions:

			MOLECULAR WEIGHT
(1) S = 39.07%	O = 58.49%	H = 2.44%	81.0
(2) Ca = 29.40	S = 23.56	O = 47.04	136.2
(3) K = 38.67	N = 13.88	O = 47.45	101.2

3. The molecular weight of ammonia is 17.06; of sulphur dioxide is 64.06; of chlorine is 70.9. From the molecular weight calculate the weight of 1 l. of each of these gases. Compare your results with the table on the back cover of the book.

4. From the molecular weight of the same gases calculate the density of each, referred to air as a standard.

5. A mixture of 50 cc. of carbon monoxide and 50 cc. of oxygen was exploded in a eudiometer, (*a*) What gases remained in the tube after the explosion? (*b*) What was the volume of each?

6. In what proportion must acetylene and oxygen be mixed to produce the greatest explosion?

7. Solve Problem 18, Chapter XVII, without using molecular weights. Compare your results.

8. Solve Problem 10, Chapter XVIII, without using molecular weights. Compare your results.

9. The specific heat of aluminium is 0.214; of lead is 0.031. From these specific heats calculate the atomic weights of each of the elements.

[Pg 238]

CHAPTER XX

THE PHOSPHORUS FAMILY

	SYMBOL	ATOMIC WEIGHT	DENSITY	MELTING POINT
Phosphorus	P	31.0	1.8	43.3°
Arsenic	As	75.0	5.73	—
Antimony	Sb	120.2	6.7	432°
Bismuth	Bi	208.5	9.8	270°

The family. The elements constituting this family belong in the same group with nitrogen and therefore resemble it in a general way. They exhibit a regular gradation of physical properties, as is shown in the above table. The same general gradation is also found in their chemical properties, phosphorus being an acid-forming element, while bismuth is essentially a metal. The other two elements are intermediate in properties.

Compounds. In general the elements of the family form compounds having similar composition, as is shown in the following table:

PH_3	PCl_3	PCl_5	P_2O_3	P_2O_5
AsH_3	$AsCl_3$	$AsCl_5$	As_2O_3	As_2O_5
SbH_3	$SbCl_3$	$SbCl_5$	Sb_2O_3	Sb_2O_5
	$BiCl_3$	$BiCl_5$	Bi_2O_3	Bi_2O_5

In the case of phosphorus, arsenic, and antimony the oxides are acid anhydrides. Salts of at least four acids of each of these three elements are known, the free acid in [Pg 239] some instances being unstable. The relation of these acids to the corresponding anhydrides may be illustrated as follows, phosphorus being taken as an example:

$$P_2O_3 + 3H_2O = 2H_3PO_3 \text{ (phosphorous acid)}.$$

$$P_2O_5 + 3H_2O = 2H_3PO_4 \text{ (phosphoric acid)}.$$

$$P_2O_5 + 2H_2O = H_4P_2O_7 \text{ (pyrophosphoric acid)}.$$

$$P_2O_5 + H_2O = 2HPO_3 \text{ (metaphosphoric acid)}.$$

PHOSPHORUS

History. The element phosphorus was discovered by the alchemist Brand, of Hamburg, in 1669, while searching for the philosopher's stone. Owing to its peculiar properties and the secrecy which was maintained about its preparation, it remained a very rare and costly substance until the demand for it in the manufacture of matches brought about its production on a large scale.

Occurrence. Owing to its great chemical activity phosphorus never occurs free in nature. In the form of phosphates it is very abundant and widely distributed. *Phosphorite* and *sombrerite* are mineral forms of calcium phosphate, while *apatite* consists of calcium phosphate together with calcium fluoride or chloride. These minerals form very large deposits and are extensively mined for use as fertilizers. Calcium phosphate is a constituent of all fertile soil, having been supplied to the soil by the disintegration of rocks containing it. It is the chief mineral constituent of bones of animals, and bone ash is therefore nearly pure calcium phosphate.

Preparation. Phosphorus is now manufactured from bone ash or a pure mineral phosphate by heating the phosphate with sand and carbon in an electric furnace. The materials [Pg 240] are fed in at M (Fig. 70) by the feed screw F. The phosphorus vapor escapes at P and is condensed under water, while the calcium silicate is tapped off as a liquid at S. The phosphorus obtained in this way is quite impure, and is purified by distillation.

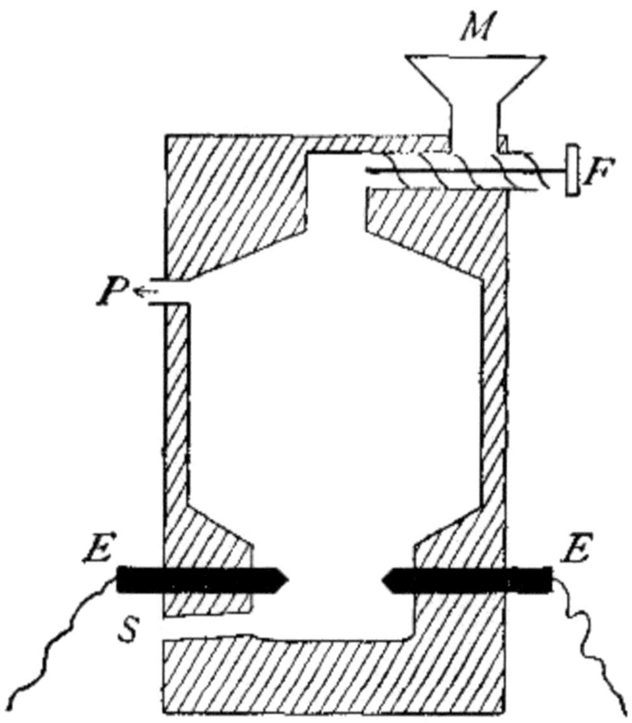

Fig. 70

Explanation of the reaction. To understand the reaction which occurs, it must be remembered that a volatile acid anhydride is expelled from its salts when heated with an anhydride which is not volatile. Thus, when sodium carbonate and silicon dioxide are heated together the following reaction takes place:

$$Na_2CO_3 + SiO_2 = Na_2SiO_3 + CO_2.$$

Silicon dioxide is a less volatile anhydride than phosphoric anhydride (P_2O_5), and when strongly heated with a phosphate the phosphoric anhydride is driven out, thus:

$$Ca_3(PO_4)_2 + 3SiO_2 = 3CaSiO_3 + P_2O_5.$$

If carbon is added before the heat is applied, the P_2O_5 is reduced to phosphorus at the same time, according to the equation

$$P_2O_5 + 5C = 2P + 5CO.$$

Physical properties. The purified phosphorus is a pale yellowish, translucent, waxy solid which melts at 43.3° and boils at 269°. It can therefore be cast into any convenient form under warm water, and is usually sold in the market in the form of sticks. It is quite soft and can be easily cut with a knife, but this must always be done while the element is covered with water, since it is extremely inflammable, and the friction of the knife blade is almost [Pg 241] sure to set it on fire if cut in the air. It is not soluble in water, but is freely soluble in some other liquids, notably in carbon disulphide. Its density is 1.8.

Chemical properties. Exposed to the air phosphorus slowly combines with oxygen, and in so doing emits a pale light, or phosphorescence, which can be seen only in a dark place. The heat of the room may easily raise the temperature to the kindling point of phosphorus, when it burns with a sputtering flame, giving off dense fumes of oxide of phosphorus. It burns with dazzling brilliancy in oxygen, and combines directly with many other elements, especially with sulphur and the halogens. On account of its great affinity for oxygen it is always preserved under water.

Phosphorus is very poisonous, from 0.2 to 0.3 gram being a fatal dose. Ground up with flour and water or similar substances, it is often used as a poison for rats and other vermin.

Precaution. The heat of the body is sufficient to raise phosphorus above its kindling temperature, and for this reason it should always be handled with forceps and never with the bare fingers. Burns occasioned by it are very painful and slow in healing.

Red phosphorus. On standing, yellow phosphorus gradually undergoes a remarkable change, being converted into a dark red powder which has a density of 2.1. It no longer takes fire easily, neither does it dissolve in carbon disulphide. It is not poisonous and, in fact, seems to be an entirely different substance. The velocity of this change increases with rise in temperature, and the red phosphorus is therefore prepared by heating the yellow just below the boiling point (250°-300°). When distilled and quickly condensed the red form changes back to the yellow. This is in accordance with the general rule that when a substance capable [Pg 242] of existing in several allotropic forms is condensed from a gas or crystallized from the liquid state, the more unstable variety forms first, and this then passes into the more stable forms.

Matches. The chief use of phosphorus is in the manufacture of matches. Common matches are made by first dipping the match sticks into some inflammable substance, such as melted paraffin, and afterward into a paste consisting of (1) phosphorus, (2) some oxidizing substance, such as manganese dioxide or potassium chlorate, and (3) a binding material, usually some kind of glue. On friction the phosphorus is ignited, the combustion being sustained by the oxidizing agent and communicated to the wood by the burning paraffin. In sulphur matches the paraffin is replaced by sulphur.

In safety matches *red* phosphorus, an oxidizing agent, and some gritty material such as emery is placed on the side of the box, while the match tip is provided as before with an oxidizing agent and an easily oxidized substance, usually antimony sulphide. The match cannot be ignited easily by friction, save on the prepared surface.

Compounds of phosphorus with hydrogen. Phosphorus forms several compounds with hydrogen, the best known of which is phosphine (PH_3) analogous to ammonia (NH_3).

Preparation of phosphine. Phosphine is usually made by heating phosphorus with a strong solution of potassium hydroxide, the reaction being a complicated one.

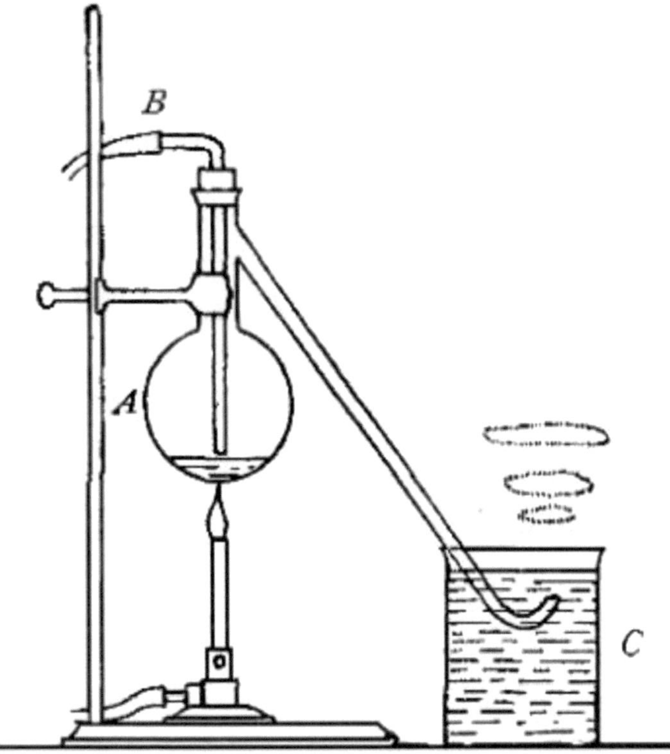

Fig. 71

The experiment can be conveniently made in the apparatus shown in Fig. 71. A strong solution of potassium hydroxide together with several small bits of phosphorus are placed in the flask A, and a current of coal gas is passed into the flask through the tube B until [Pg 243] all the air has been displaced. The gas is then turned off and the flask is heated. Phosphine is formed in small quantities and escapes through the delivery tube, the exit of which is just covered by the water in the vessel C. Each bubble of the gas as it escapes into the air takes fire, and the product of combustion (P_2O_5) forms beautiful small rings, which float unbroken for a considerable time in quiet air. The pure phosphine does not take fire spontaneously.

When prepared as directed above, impurities are present which impart this property.

Properties. Phosphine is a gas of unpleasant odor and is exceedingly poisonous. Like ammonia it forms salts with the halogen acids. Thus we have phosphonium chloride (PH_4Cl) analogous to ammonium chloride (NH_4Cl). The phosphonium salts are of but little importance.

Oxides of phosphorus. Phosphorus forms two well-known oxides,—the trioxide (P_2O_3) and the pentoxide (P_2O_5), sometimes called phosphoric anhydride. When phosphorus burns in an insufficient supply of air the product is partially the trioxide; in oxygen or an excess of air the pentoxide is formed. The pentoxide is much the better known of the two. It is a snow-white, voluminous powder whose most marked property is its great attraction for water. It has no chemical action upon most gases, so that they can be very thoroughly dried by allowing them to pass through properly arranged vessels containing phosphorus pentoxide.

Acids of phosphorus. The important acids of phosphorus are the following:

H_3PO_3	phosphorous acid.
H_3PO_4	phosphoric acid.
$H_4P_2O_7$	pyrophosphoric acid.
HPO_3	metaphosphoric acid.

These may be regarded as combinations of the oxides of phosphorus with water according to the equations given in the discussion of the characteristics of the family. [Pg 244]

1. *Phosphorous acid* (H_3PO_3). Neither the acid nor its salts are at all frequently met with in chemical operations. It can be easily obtained, however, in the form of transparent crystals when phosphorus trichloride is treated with water and the resulting solution is evaporated:

$$PCl_3 + 3H_2O = H_3PO_3 + 3HCl.$$

Its most interesting property is its tendency to take up oxygen and pass over into phosphoric acid.

2. *Orthophosphoric acid (phosphoric acid)* (H_3PO_4). This acid can be obtained by dissolving phosphorus pentoxide in boiling water, as represented in the equation

$$P_2O_5 + 3H_2O = 2H_3PO_4.$$

It is usually made by treating calcium phosphate with concentrated sulphuric acid. The calcium sulphate produced in the reaction is nearly insoluble, and can be filtered off, leaving the phosphoric acid in solution. Very pure acid is made by oxidizing phosphorus with nitric acid. It forms large colorless crystals which are exceedingly soluble in water. Being a tribasic acid, it forms acid as well as normal salts. Thus the following compounds of sodium are known:

NaH_2PO_4 monosodium hydrogen phosphate.

Na_2HPO_4 disodium hydrogen phosphate.

Na_3PO_4 normal sodium phosphate.

These salts are sometimes called respectively primary, secondary, and tertiary phosphates. They may be prepared by bringing together phosphoric acid and appropriate quantities of sodium hydroxide. Phosphoric acid also forms mixed salts, that is, salts containing two different metals. The most familiar compound of this kind is microcosmic [Pg 245] salt, which has the formula $Na(NH_4)HPO_4$.

Orthophosphates. The orthophosphates form an important class of salts. The normal salts are nearly all insoluble and many of them occur in nature. The secondary phosphates are as a rule insoluble, while most of the primary salts are soluble.

3. *Pyrophosphoric acid* ($H_4P_2O_7$). On heating orthophosphoric acid to about 225° pyrophosphoric acid is formed in accordance with the following equation:

$$2H_3PO_4 = H_4P_2O_7 + H_2O.$$

It is a white crystalline solid. Its salts can be prepared by heating a secondary phosphate:

$$2Na_2HPO_4 = Na_4P_2O_7 + H_2O.$$

4. *Metaphosphoric acid (glacial phosphoric acid)* (HPO_3). This acid is formed when orthophosphoric acid is heated above 400°:

$$H_3PO_4 = HPO_3 + H_2O.$$

It is also formed when phosphorus pentoxide is treated with cold water:

$$P_2O_5 + H_2O = 2HPO_3.$$

It is a white crystalline solid, and is so stable towards heat that it can be fused and even volatilized without decomposition. On cooling from the fused state it forms a glassy solid, and on this account is often called glacial phosphoric acid. It possesses the property of dissolving small quantities of metallic oxides, with the formation of compounds which, in the case of certain metals, have characteristic colors. It is therefore used in the detection of these metals.

While the secondary phosphates, on heating, give salts of pyrophosphoric acid, the primary phosphates yield salts of metaphosphoric acid. The equations representing these reactions are as follows:

$$2Na_2HPO_4 = Na_4P_3O_7 + H_2O,$$

$$NaH_2PO_4 = NaPO_3 + H_2O.$$

Fertilizers. When crops are produced year after year on the same field certain constituents of the soil essential to plant growth are removed, and the soil becomes impoverished and unproductive. To make the land once more [Pg 246] fertile these constituents must be replaced. The calcium phosphate of the mineral deposits or of bone ash serves well as a material for restoring phosphorus to soils exhausted of that essential element; but a more soluble substance, which the plants can more readily assimilate, is desirable. It is better, therefore, to convert the insoluble calcium phosphate into the soluble primary phosphate before it is applied as fertilizer. It will be seen by reference to the formulas for the orthophosphates (see page 244) that in a primary phosphate only one hydrogen atom of phosphoric acid is replaced by a metal. Since the calcium atom always replaces two hydrogen atoms, it might be thought that there could be no primary calcium phosphate; but if the calcium atom replaces one hydrogen atom from each of two molecules of phosphoric acid, the salt $Ca(H_2PO_4)_2$ will result, and this is a primary phosphate. It can be made by treatment of the normal phosphate with the necessary amount of sulphuric acid, calcium sulphate being formed at the same time, thus:

$$Ca_3(PO_4)_2 + 2H_2SO_4 = Ca(H_2PO_4)_2 + 2CaSO_4.$$

The resulting mixture is a powder, which is sold as a fertilizer under the name of "superphosphate of lime."

ARSENIC

Occurrence. Arsenic occurs in considerable quantities in nature as the native element, as the sulphides realgar (As_2S_2) and orpiment (As_2S_3), as oxide (As_2O_3), and as a constituent of many metallic sulphides, such as arsenopyrite (FeAsS).

Preparation. The element is prepared by purifying the native arsenic, or by heating the arsenopyrite in iron tubes, [Pg 247] out of contact with air, when the reaction expressed by the following equation occurs:

$$FeAsS = FeS + As.$$

The arsenic, being volatile, condenses in chambers connected with the heated tubes. It is also made from the oxide by reduction with carbon:

$$2As_2O_3 + 3C = 4As + 3CO_2.$$

Properties. Arsenic is a steel-gray, metallic-looking substance of density 5.73. Though resembling metals in appearance, it is quite brittle, being easily powdered in a mortar. When strongly heated it sublimes, that is, it passes into a vapor without melting, and condenses again to a crystalline solid when the vapor is cooled. Like phosphorus it can be obtained in several allotropic forms. It alloys readily with some of the metals, and finds its chief use as an alloy with lead, which is used for making shot, the alloy being harder than pure lead. When heated on charcoal with the blowpipe it is converted into an oxide which volatilizes, leaving the charcoal unstained by any oxide coating. It burns readily in chlorine gas, forming arsenic trichloride, —

$$As + 3Cl = AsCl_3.$$

Unlike most of its compounds, the element itself is not poisonous.

Arsine (AsH_3). When any compound containing arsenic is brought into the presence of nascent hydrogen, arsine (AsH_3), corresponding to phosphine and ammonia, is formed. The reaction when oxide of arsenic is so treated is

$$As_2O_3 + 12H = 2AsH_3 + 3H_2O.$$

[Pg 248]

Arsine is a gas with a peculiar garlic-like odor, and is intensely poisonous. A single bubble of pure gas has been known to prove fatal. It is an unstable compound, decomposing into its elements when heated to a moderate temperature. It is combustible, burning with a pale bluish-white flame to form arsenic trioxide and water when air is in excess:

$$2AsH_3 + 6O = As_2O_3 + 3H_2O.$$

When the supply of air is deficient water and metallic arsenic are formed:

$$2AsH_3 + 3O = 3H_2O + 2As.$$

These reactions make the detection of even minute quantities of arsenic a very easy problem.

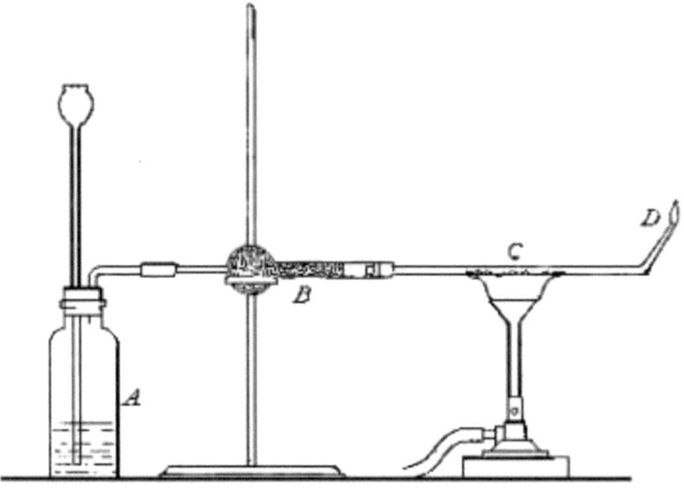

Fig. 72

Marsh's test for arsenic. The method devised by Marsh for detecting arsenic is most frequently used, the apparatus being shown in Fig. 72. Hydrogen is generated in the flask *A* by the action of dilute sulphuric acid on zinc, is dried by passing over calcium chloride in the tube *B*, and after passing through the hard-glass tube *C* is ignited at the jet *D*. If a substance containing arsenic is now introduced into the generator *A*, the arsenic is converted into arsine by the action of the nascent hydrogen, and [Pg 249] passes to the jet along with the hydrogen. If the tube *C* is strongly heated at some point near the middle, the arsine is decomposed while passing this point and the arsenic is deposited just beyond the heated point in the form of a shining, brownish-black mirror. If the tube is not heated, the arsine burns along with the hydrogen at the jet. Under these conditions a small porcelain dish crowded down into the flame is blackened by a spot of metallic arsenic, for the arsine is decomposed by the heat of the flame, and the arsenic, cooled below its kindling temperature by the cold porcelain, deposits upon it as a black spot. Antimony conducts itself in the same way as arsenic, but the antimony deposit is more sooty in appearance. The two can also be distinguished by the fact that sodium hypochlorite (NaClO) dissolves the arsenic deposit, but not that formed by antimony.

Oxides of arsenic. Arsenic forms two oxides, As_2O_3 and As_2O_5, corresponding to those of phosphorus. Of these arsenious oxide, or arsenic trioxide (As_2O_3), is much better known, and is the substance usually called white arsenic, or merely arsenic. It is found as a mineral, but is usually obtained as a by-product in burning pyrite in the sulphuric-acid industry. The pyrite has a small amount of arsenopyrite in it, and when this is burned arsenious oxide is formed as a vapor together with sulphur dioxide:

$$2FeAsS + 10O = Fe_2O_3 + As_2O_3 + 2SO_2.$$

The arsenious oxide is condensed in appropriate chambers. It is a rather heavy substance, obtained either as a crystalline powder or as large, vitreous lumps, resembling lumps of porcelain in appearance. It is very poisonous, from 0.2 to 0.3 g. being a fatal dose. It is frequently given as a poison, since it is nearly tasteless and does not act very rapidly. This slow action is due to the fact that it is not very soluble, and hence is absorbed slowly by the system. Arsenious oxide is also used as a chemical reagent in glass making and in the dye industry. [Pg 250]

284

Acids of arsenic. Like the corresponding oxides of phosphorus, the oxides of arsenic are acid anhydrides. In solution they combine with bases to form salts, corresponding to the salts of the acids of phosphorus. Thus we have salts of the following acids:

H_3AsO_3	arsenious acid.
H_3AsO_4	orthoarsenic acid.
$H_4As_2O_3$	pyroarsenic acid.
$HAsO_3$	metarsenic acid.

Several other acids of arsenic are also known. Not all of these can be obtained as free acids, since they tend to lose water and form the oxides. Thus, instead of obtaining arsenious acid (H_3AsO_3), the oxide As_2O_3 is obtained:

$$2H_3AsO_3 = As_2O_3 + 3H_2O.$$

Salts of all the acids are known, however, and some of them have commercial value. Most of them are insoluble, and some of the copper salts, which are green, are used as pigments. Paris green, which has a complicated formula, is a well-known insecticide.

Antidote for arsenical poisoning. The most efficient antidote for arsenic poisoning is ferric hydroxide. It is prepared as needed, according to the equation

$$Fe_2(SO_4)_3 + 3Mg(OH)_2 = 2Fe(OH)_3 + 3MgSO_4.$$

Sulphides of arsenic. When hydrogen sulphide is passed into an acidified solution containing an arsenic compound the arsenic is precipitated as a bright yellow sulphide, thus:

$$2H_3AsO_3 + 3H_2S = As_2S_3 + 6H_2O,$$

$$2H_3AsO_4 + 5H_2S = As_2S_5 + 8H_2O.$$

In this respect arsenic resembles the metallic elements, many of which produce sulphides under similar conditions. The sulphides of arsenic, both those produced artificially and those found in nature, are used as yellow pigments.

ANTIMONY

Occurrence. Antimony occurs in nature chiefly as the sulphide (Sb_2S_3), called stibnite, though it is also found as oxide and as a constituent of many complex minerals. [Pg 251]

Preparation. Antimony is prepared from the sulphide in a very simple manner. The sulphide is melted with scrap iron in a furnace, when the iron combines with the sulphur to form a slag, or liquid layer of melted iron sulphide, while the heavier liquid, antimony, settles to the bottom and is drawn off from time to time. The reaction involved is represented by the equation

$$Sb_2S_3 + 3Fe = 2Sb + 3FeS.$$

Physical properties. Antimony is a bluish-white, metallic-looking substance whose density is 6.7. It is highly crystalline, hard, and very brittle. It has a rather low melting point (432°) and expands very noticeably on solidifying.

Chemical properties. In chemical properties antimony resembles arsenic in many particulars. It forms the oxides Sb_2O_3 and Sb_2O_5, and in addition Sb_2O_4. It combines with the halogen elements with great energy, burning brilliantly in chlorine to form antimony trichloride ($SbCl_3$). When heated on charcoal with the blowpipe it is oxidized and forms a coating of antimony oxide on the charcoal which has a characteristic bluish-white color.

Stibine (SbH_3). The gas stibine (SbH_3) is formed under conditions which are very similar to those which produce arsine, and it closely resembles the latter compound, though it is still less stable. It is very poisonous.

Acids of antimony. The oxides Sb_2O_3 and Sb_2O_5 are weak acid anhydrides and are capable of forming two series of acids corresponding in formulas to the acids of phosphorus and arsenic. They are much weaker, however, and are of little practical importance.

Sulphides of antimony. Antimony resembles arsenic in that hydrogen sulphide precipitates it as a sulphide when conducted into an acidified solution containing an antimony compound:

$$2SbCl_3 + 3H_2S = Sb_2S_3 + 6HCl,$$

$$2SbCl_5 + 5H_2S = Sb_2S_5 + 10HCl.$$
[Pg 252]

The two sulphides of antimony are called the trisulphide and the pentasulphide respectively. When prepared in this way they are orange-colored substances, though the mineral stibnite is black.

Metallic properties of antimony. The physical properties of the element are those of a metal, and the fact that its sulphide is precipitated by hydrogen sulphide shows that it acts like a metal in a chemical way. Many other reactions show that antimony has more of the properties of a metal than of a non-metal. The compound $Sb(OH)_3$, corresponding to arsenious acid, while able to act as a weak acid is also able to act as a weak base with strong acids. For example, when treated with concentrated hydrochloric acid antimony chloride is formed:

$$Sb(OH)_3 + 3HCl = SbCl_3 + 3H_2O.$$

A number of elements act in this same way, their hydroxides under some conditions being weak acids and under others weak bases.

ALLOYS

Some metals when melted together thoroughly intermix, and on cooling form a homogeneous, metallic-appearing substance called an *alloy*. Not all metals will mix in this way, and in some cases definite chemical compounds are formed and separate out as the mixture solidifies, thus destroying the uniform quality of the alloy. In general the melting point of the alloy is below the average of the melting points of its constituents, and it is often lower than any one of them.

Antimony forms alloys with many of the metals, and its chief commercial use is for such purposes. It imparts to its alloys high density, rather low melting point, and the [Pg 253] property of expanding on solidification. Such an alloy is especially useful in type founding, where fine lines are to be reproduced on a cast. Type metal consists of antimony, lead, and tin. Babbitt metal, used for journal bearings in machinery, contains the same metals in a different proportion together with a small percentage of copper.

BISMUTH

Occurrence. Bismuth is usually found in the uncombined form in nature. It also occurs as oxide and sulphide. Most of the bismuth of commerce comes from Saxony, and from Mexico and Colorado, but it is not an abundant element.

Preparation. It is prepared by merely heating the ore containing the native bismuth and allowing the melted metal to run out into suitable vessels. Other ores are converted into oxides and reduced by heating with carbon.

Physical properties. Bismuth is a heavy, crystalline, brittle metal nearly the color of silver, but with a slightly rosy tint which distinguishes it from other metals. It melts at a low temperature (270°) and has a density of 9.8. It is not acted upon by the air at ordinary temperatures.

Chemical properties. When heated with the blowpipe on charcoal, bismuth gives a coating of the oxide Bi_2O_3. This has a yellowish-brown color which easily distinguishes it from the oxides formed by other metals. It combines very readily with the halogen elements, powdered bismuth burning readily in chlorine. It is not very easily acted upon by hydrochloric acid, but nitric and sulphuric acids act upon it in the same way that they do upon copper.

Uses. Bismuth finds its chief use as a constituent of alloys, particularly in those of low melting point. Some [Pg 254] of these melt in hot water. For example, Wood's metal, consisting of bismuth, lead, tin, and cadmium, melts at 60.5°.

Compounds of bismuth. Unlike the other elements of this group, bismuth has almost no acid properties. Its chief oxide, Bi_2O_3, is basic in its properties. It dissolves in strong acids and forms salts of bismuth:

$$Bi_2O_3 + 6HCl = 2BiCl_3 + 3H_2O,$$

$$Bi_2O_3 + 6HNO_3 = 2Bi(NO_3)_3 + 3H_2O.$$

The nitrate and chloride of bismuth can be obtained as well-formed colorless crystals. When treated with water the salts are decomposed in the manner explained in the following paragraph.

HYDROLYSIS

Many salts such as those of antimony and bismuth form solutions which are somewhat acid in reaction, and must therefore contain hydrogen ions. This is accounted for by the same principle suggested to explain the fact that solutions of potassium cyanide are alkaline in reaction (p. 210). Water forms an appreciable number of hydrogen and hydroxyl ions, and very weak bases such as bismuth hydroxide are dissociated to but a very slight extent. When Bi^{+++} ions from bismuth chloride, which dissociates very readily, are brought in contact with the OH^- ions from water, the two come to the equilibrium expressed in the equation

$$Bi^{+++} + 3OH^- \longleftrightarrow Bi(OH)_3.$$

For every hydroxyl ion removed from the solution in this way a hydrogen ion is left free, and the solution becomes acid in reaction. [Pg 255]

Reactions of this kind and that described under potassium cyanide are called *hydrolysis*.

DEFINITION: *Hydrolysis is the action of water upon a salt to form an acid and a base, one of which is very slightly dissociated.*

Conditions favoring hydrolysis. While hydrolysis is primarily due to the slight extent to which either the acid or the base formed is dissociated, several other factors have an influence upon the extent to which it will take place.

1. *Influence of mass.* Since hydrolysis is a reversible reaction, the relative masses of the reacting substances influence the point at which equilibrium will be reached. In the equilibrium

$$BiCl_3 + 3H_2O \longleftrightarrow Bi(OH)_3 + 3HCl$$

the addition of more water will result in the formation of more bismuth hydroxide and hydrochloric acid. The addition of more hydrochloric acid will convert some of the bismuth hydroxide into bismuth chloride.

2. *Formation of insoluble substances.* When one of the products of hydrolysis is nearly insoluble in water the solution will become saturated with it as soon as a very little has been formed. All in excess of this will precipitate, and the reaction will go on until the acid set free increases sufficiently to bring about an equilibrium. Thus a considerable amount of bismuth and antimony hydroxides are precipitated when water is added to the chlorides of these elements. The greater the dilution the more hydroxide precipitates. The addition of hydrochloric acid in considerable quantity will, however, redissolve the precipitate.

Partial hydrolysis. In many cases the hydrolysis of a salt is only partial, resulting in the formation of basic salts instead of the free base. Most of these basic salts are insoluble in water, which accounts for their ready formation. Thus bismuth chloride may hydrolyze by successive steps, as shown in the equations

$$BiCl_3 + H_2O = Bi(OH)Cl_2 + HCl,$$

$$BiCl_3 + 2H_2O = Bi(OH)_2Cl + 2HCl,$$

$$BiCl_3 + 3H_2O = Bi(OH)_3 + 3HCl.$$

The basic salt so formed may also lose water, as shown in the equation

$$Bi(OH)_2Cl = BiOCl + H_2O.$$

[Pg 256]

The salt represented in the last equation is sometimes called bismuth oxychloride, or bismuthyl chloride. The corresponding nitrate, $BiONO_3$, is largely used in medicine under the name of subnitrate of bismuth. In these two compounds the group of atoms, BiO, acts as a univalent metallic radical and is called *bismuthyl*. Similar basic salts are formed by the hydrolysis of antimony salts.

EXERCISES

1. Name all the elements so far studied which possess allotropic forms.

2. What compounds would you expect phosphorus to form with bromine and iodine? Write the equations showing the action of water on these compounds.

3. In the preparation of phosphine, why is coal gas passed into the flask? What other gases would serve the same purpose?

4. Give the formula for the salt which phosphine forms with hydriodic acid. Give the name of the compound.

5. Could phosphoric acid be substituted for sulphuric acid in the preparation of the common acids?

6. Write the equations for the preparation of the three sodium salts of orthophosphoric acid.

7. Why does a solution of disodium hydrogen phosphate react alkaline?

8. On the supposition that bone ash is pure calcium phosphate, what weight of it would be required in the preparation of 1 kg. of phosphorus?

9. If arsenopyrite is heated in a current of air, what products are formed?

10. (*a*) Write equations for the complete combustion of hydrosulphuric acid, methane, and arsine. (*b*) In what respects are the reactions similar?

11. Write the equations for all the reactions involved in Marsh's test for arsenic.

12. Write the names and formulas for the acids of antimony.

13. Write the equations showing the hydrolysis of antimony trichloride; of bismuth nitrate.

14. In what respects does nitrogen resemble the members of the phosphorus family?

[Pg 257]

CHAPTER XXI

SILICON, TITANIUM, BORON

	SYMBOL	ATOMIC WEIGHT	DENSITY	CHLORIDES	OXIDES
Silicon	Si	28.4	2.35	$SiCl_4$	SiO
Titanium	Ti	48.1	3.5	$TiCl_4$	TiO
Boron	B	11.0	2.45	BCl_3	B_2O_3

General. Each of the three elements, silicon, titanium, and boron, belongs to a separate periodic family, but they occur near together in the periodic grouping and are very similar in both physical and chemical properties. Since the other elements in their families are either so rare that they cannot be studied in detail, or are best understood in connection with other elements, it is convenient to consider these three together at this point.

The three elements are very difficult to obtain in the free state, owing to their strong attraction for other elements. They can be prepared by the action of aluminium or magnesium on their oxides and in impure state by reduction with carbon in an electric furnace. They are very hard and melt only at the highest temperatures. At ordinary temperatures they are not attacked by oxygen, but when strongly heated they burn with great brilliancy. Silicon and boron are not attacked by acids under ordinary conditions; titanium is easily dissolved by them. [Pg 258]

SILICON

Occurrence. Next to oxygen silicon is the most abundant element. It does not occur free in nature, but its compounds are very abundant and of the greatest importance. It occurs almost entirely in combination with oxygen as silicon dioxide (SiO_2), often called silica, or with oxygen and various metals in the form of salts of silicic acids, or silicates. These compounds form a large fraction of the earth's crust. Most plants absorb small amounts of silica from the soil, and it is also found in minute quantities in animal organisms.

Preparation. The element is most easily prepared by reducing pure powdered quartz with magnesium powder:

$$SiO_2 + 2Mg = 2MgO + Si.$$

Properties. As would be expected from its place in the periodic table, silicon resembles carbon in many respects. It can be obtained in several allotropic forms, corresponding to those of carbon. The crystallized form is very hard, and is inactive toward reagents. The amorphous variety has, in general, properties more similar to charcoal.

Compounds of silicon with hydrogen and the halogens. Silicon hydride (SiH_4) corresponds in formula to methane (CH_4), but its properties are more like those of phosphine (PH_3). It is a very inflammable gas of disagreeable odor, and, as ordinarily prepared, takes fire spontaneously on account of the presence of impurities.

Silicon combines with the elements of the chlorine family to form such compounds as $SiCl_4$ and SiF_4. Of these silicon fluoride is the most familiar and interesting. As stated in the discussion of fluorine, it is formed when [Pg 259] hydrofluoric acid acts upon silicon dioxide or a silicate. With silica the reaction is thus expressed:

$$SiO_2 + 4HF = SiF_4 + 2H_2O.$$

It is a very volatile, invisible, poisonous gas. In contact with water it is partially decomposed, as shown in the equation

$$SiF_4 + 4H_2O = 4HF + Si(OH)_4.$$

The hydrofluoric acid so formed combines with an additional amount of silicon fluoride, forming the complex fluosilicic acid (H_2SiF_6), thus:

$$2HF + SiF_4 = H_2SiF_6.$$

Silicides. As the name indicates, silicides are binary compounds consisting of silicon and some other element. They are very stable at high temperatures, and are usually made by heating the appropriate substances in an electric furnace. The most important one is *carborundum*, which is a silicide of carbon of the formula CSi. It is made by heating coke and sand, which is a form of silicon dioxide, in an electric furnace, the process being extensively carried on at Niagara Falls. The following equation represents the reaction

$$SiO_2 + 3C = CSi + 2CO.$$

The substance so prepared consists of beautiful purplish-black crystals, which are very hard. Carborundum is used as an abrasive, that is, as a material for grinding and polishing very hard substances. Ferrosilicon is a silicide of iron alloyed with an excess of iron, which finds extensive use in the manufacture of certain kinds of steel. [Pg 260]

Manufacture of carborundum. The mixture of materials is heated in a large resistance furnace for about thirty-six hours. After the reaction is completed there is left a core of graphite G. Surrounding this core is a layer of crystallized carborundum C, about 16 in. thick. Outside this is a shell of amorphous carborundum A. The remaining materials M are unchanged and are used for a new charge.

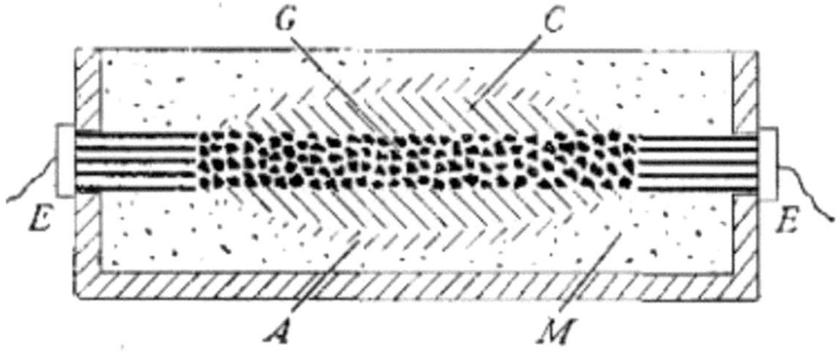

Fig. 73

Silicon dioxide (*silica*) (SiO_2). This substance is found in a great variety of forms in nature, both in the amorphous and in the crystalline condition. In the form of quartz it is found in beautifully formed six-sided prisms, sometimes of great size. When pure it is perfectly transparent and colorless. Some colored varieties are given special names, as amethyst (violet), rose quartz (pale pink), smoky or milky quartz (colored and opaque). Other varieties of silicon dioxide, some of which also contain water, are chalcedony, onyx, jasper, opal, agate, and flint. Sand and sandstone are largely silicon dioxide.

Properties. As obtained by chemical processes silicon dioxide is an amorphous white powder. In the crystallized state it is very hard and has a density of 2.6. It is insoluble in water and in most chemical reagents, and requires the hottest oxyhydrogen flame for fusion. Acids, excepting hydrofluoric acid, have little action on it, and it requires the most energetic reducing agents to deprive it of oxygen. It is the anhydride of an acid, and consequently it dissolves in fused alkalis to form silicates. Being nonvolatile, it will drive out most other anhydrides when heated [Pg 261] to a high temperature with their salts, especially when the silicates so formed are fusible. The following equations illustrate this property:

$$Na_2CO_3 + SiO_2 = Na_2SiO_3 + CO_2,$$

$$Na_2SO_4 + SiO_2 = Na_2SiO_3 + SO_3.$$

Silicic acids. Silicon forms two simple acids, orthosilicic acid (H_4SiO_4) and metasilicic acid (H_2SiO_3). Orthosilicic acid is formed as a jelly-like mass when orthosilicates are treated with strong acids such as hydrochloric. On attempting to dry this acid it loses water, passing into metasilicic or common silicic acid:

$$H_4SiO_4 = H_2SiO_3 + H_2O.$$

Metasilicic acid when heated breaks up into silica and water, thus:

$$H_2SiO_3 = H_2O + SiO_2.$$

Salts of silicic acids,—silicates. A number of salts of the orthosilicic and metasilicic acids occur in nature. Thus mica ($KAlSiO_4$) is a salt of orthosilicic acid.

Polysilicic acids. Silicon has the power to form a great many complex acids which may be regarded as derived from the union of several molecules of the orthosilicic acid, with the loss of water. Thus we have

$$3H_4SiO_4 = H_4Si_3O_8 + 4H_2O.$$

These acids cannot be prepared in the pure state, but their salts form many of the crystalline rocks in nature. Feldspar, for example, has the formula $KAlSi_3O_8$, and is a mixed salt of the acid $H_4Si_3O_8$, whose formation is represented in the equation above. Kaolin has the formula $Al_2Si_2O_7 \cdot 2H_2O$. Many other examples will be met in the study of the metals. [Pg 262]

Glass. When sodium and calcium silicates, together with silicon dioxide, are heated to a very high temperature, the mixture slowly fuses to a transparent liquid, which on cooling passes into the solid called glass. Instead of starting with sodium and calcium silicates it is more convenient and economical to heat sodium carbonate (or sulphate) and lime with an excess of clean sand, the silicates being formed during the heating:

$$Na_2CO_3 + SiO_2 = Na_2SiO_3 + CO_2,$$

$$CaO + SiO_2 = CaSiO_3.$$

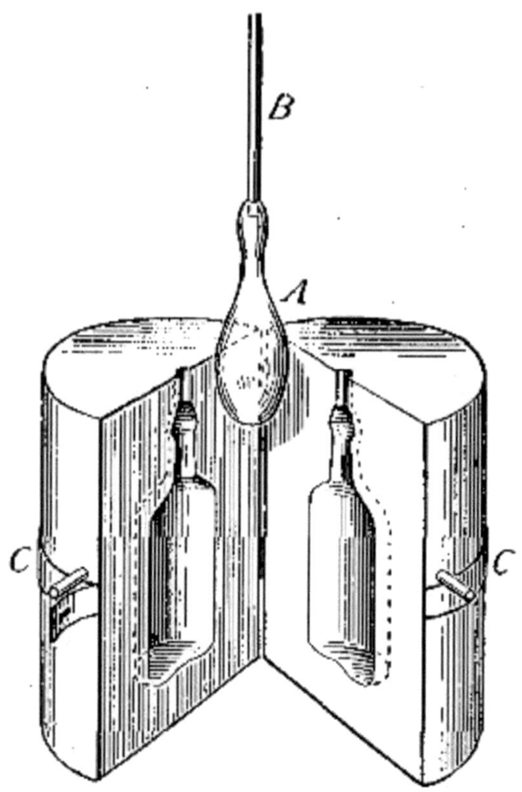

Fig. 74

The mixture is heated below the fusing point for some time, so that the escaping carbon dioxide may not spatter the hot liquid; the heat is then increased and the mixture kept in a state of fusion until all gases formed in the reaction have escaped.

Molding and blowing of glass. The way in which the melted mixture is handled in the glass factory depends upon the character of the article to be made. Many articles, such as bottles, are made by blowing the plastic glass into hollow molds of the desired shape. The mold is first opened, as shown in Fig. 74. A lump of plastic glass *A* on the hollow rod *B* is lowered into the mold, which is then closed by the handles *C*. By blowing into the tube the glass is blown into the shape of the mold. The mold is then opened and the bottle lifted out. The neck of the bottle must be cut off at the proper place and the sharp edges rounded off in a flame.

Other objects, such as lamp chimneys, are made by getting a lump of plastic glass on the end of a hollow iron rod and blowing it into the desired shape without the help of a mold, great skill being required in the manipulation of the glass. Window glass is made by blowing large hollow cylinders about 6 ft. long and 1-1/2 ft. in diameter. These are cut longitudinally, and are then placed in an oven and heated until they soften, when they are flattened out into plates (Fig. 75). Plate glass is cast into flat slabs, which are then ground and polished to perfectly plane surfaces. [Pg 263]

Varieties of glass. The ingredients mentioned above make a soft, easily fusible glass. If potassium carbonate is substituted for the sodium carbonate, the glass is much harder and less easily fused; increasing the amount of sand has somewhat the same effect. Potassium glass is largely used in making chemical glassware, since it resists the action of reagents better than the softer sodium glass. If lead oxide is substituted for the whole or a part of the lime, the glass is very soft, but has a high index of refraction and is valuable for making optical instruments and artificial jewels.

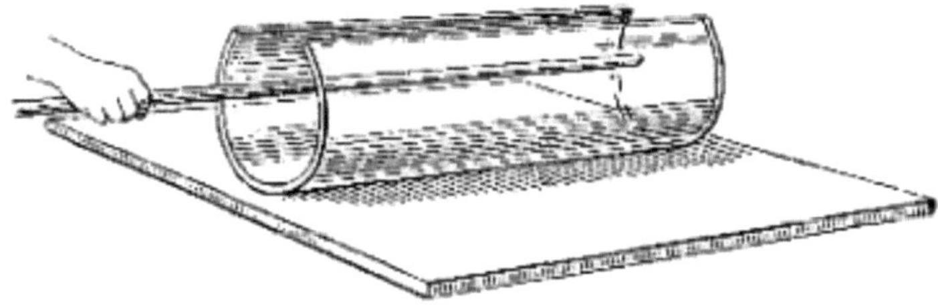

Fig. 75

Coloring of glass. Various substances fused along with the glass mixture give characteristic colors. The amber color of common bottles is due to iron compounds in the glass; in other cases iron colors the glass green. Cobalt compounds color it deep blue; those of manganese give it an amethyst tint and uranium compounds impart a peculiar yellowish green color. Since iron is nearly always present in the ingredients, glass is usually slightly yellow. This color can be removed by adding the proper amount of manganese dioxide, for the amethyst color of manganese and the yellow of iron together produce white light.

Nature of glass. Glass is not a definite chemical compound and its composition varies between wide limits. Fused glass is really a solution of various silicates, such as those of calcium and lead, in fused sodium or potassium silicate. A certain amount of silicon dioxide is also present. This solution is then allowed to solidify under such conditions of cooling that the dissolved substances do not separate from the solvent. The compounds which are used to color the glass are sometimes converted into silicates, which then dissolve in the glass, giving it a uniform color. In other cases, as in the milky glasses which resemble porcelain in appearance, the color or opaqueness is due to the finely divided color material evenly distributed throughout the glass, but not dissolved in it. Milky glass is made by mixing calcium fluoride, tin oxide, or some other insoluble substance in the melted glass. Copper or gold in metallic form scattered through glass gives it shades of red. [Pg 264]

TITANIUM

Titanium is a very widely distributed element in nature, being found in almost all soils, in many rocks, and even in plant and animal tissues. It is not very abundant in any one locality, and it possesses little commercial value save in connection with the iron industry. Its most common ore is rutile (TiO_2), which resembles silica in many respects.

In both physical and chemical properties titanium resembles silicon, though it is somewhat more metallic in character. This resemblance is most marked in the acids of titanium. It not only forms metatitanic and orthotitanic acids but a great variety of polytitanic acids as well.

BORON

Occurrence. Boron is never found free in nature. It occurs as boric acid (H_3BO_3), and in salts of polyboric acids, which usually have very complicated formulas.

Preparation and properties. Boron can be prepared from its oxide by reduction with magnesium, exactly as in the case of silicon. It resembles silicon very strikingly in its properties. It occurs in several allotropic forms, is very hard when crystallized, and is rather inactive toward reagents. It forms a hydride, BH_3, and combines directly with the elements of the chlorine family. Boron fluoride (BF_3) is very similar to silicon fluoride in its mode of formation and chemical properties.

Boric oxide (B_2O_3). Boron forms one well-known oxide, B_2O_3, called boric anhydride. It is formed as a glassy mass by heating boric acid to a high temperature. It absorbs water very readily, uniting with it to form boric acid again:

$$B_2O_3 + 3H_2O = 2H_3BO_3.$$

In this respect it differs from silicon dioxide, which will not combine directly with water. [Pg 265]

Boric acid (H_3BO_3). This is found in nature in considerable quantities and forms one of the chief sources of boron compounds. It is found dissolved in the water of hot springs in some localities, particularly in Italy. Being volatile with steam, the vapor which escapes from these springs has some boric acid in it. It is easily obtained from these sources by condensation and evaporation, the necessary heat being supplied by other hot springs.

Boric acid crystallizes in pearly flakes, which are greasy to the touch. In the laboratory it is easily prepared by treating a strong, hot solution of borax with sulphuric acid. Boric acid being sparingly soluble in water crystallizes out on cooling:

$$Na_2B_4O_7 + 5H_2O + H_2SO_4 = Na_2SO_4 + 4H_3BO_3.$$

The substance is a mild antiseptic, and on this account is often used in medicine and as a preservative for canned foods and milk.

Metaboric and polyboric acids. When boric acid is gently heated it is converted into metaboric acid (HBO_2):

$$H_3BO_3 = HBO_2 + H_2O.$$

On heating metaboric acid to a somewhat higher temperature tetraboric acid ($H_2B_4O_7$) is formed:

$$4HBO_2 = H_2B_4O_7 + H_2O.$$

Many other complex acids of boron are known.

Borax. Borax is the sodium salt of tetraboric acid, having the formula $Na_2B_4O_7 \cdot 10\ H_2O$. It is found in some arid countries, as southern California and Tibet, but is now made commercially from the mineral colemanite, which is the calcium salt of a complex boric acid. When this is treated with a solution of sodium carbonate, calcium [Pg 266] carbonate is precipitated and borax crystallizes from the solution.

When heated borax at first swells up greatly, owing to the expulsion of the water of crystallization, and then melts to a clear glass. This glass has the property of easily dissolving many metallic oxides, and on this account borax is used as a flux in soldering, for the purpose of removing from the metallic surfaces to be soldered the film of oxide with which they are likely to be covered. These oxides often give a characteristic color to the clear borax glass, and borax beads are therefore often used in testing for the presence of metals, instead of the metaphosphoric acid bead already described.

The reason that metallic oxides dissolve in borax is that borax contains an excess of acid anhydride, as can be more easily seen if its formula is written $2NaBO_2 + B_2O_3$. The metallic oxide combines with this excess of acid anhydride, forming a mixed salt of metaboric acid.

Borax is extensively used as a constituent of enamels and glazes for both metal ware and pottery. It is also used as a flux in soldering and brazing, and in domestic ways it serves as a mild alkali, as a preservative for meats, and in a great variety of less important applications.

EXERCISES

1. Account for the fact that a solution of borax in water is alkaline.

2. What weight of water of crystallization does 1 kg. of borax contain?

3. When a concentrated solution of borax acts on silver nitrate a borate of silver is formed. If the solution of borax is dilute, however, an hydroxide of silver forms. Account for this difference in behavior.

[Pg 267]

CHAPTER XXII

THE METALS

The metals. The elements which remain to be considered are known collectively as the metals. They are also called the base-forming elements, since their hydroxides are bases. A metal may therefore be defined as an element whose hydroxide is a base. When a base dissolves in water the hydroxyl groups form the anions, while the metallic element forms the cations. From this standpoint a metal can be defined as an element capable of forming simple cations in solution.

The distinction between a metal and a non-metal is not a very sharp one, since the hydroxides of a number of elements act as bases under some conditions and as acids under others. We have seen that antimony is an element of this kind.

Occurrence of metals in nature. A few of the metals are found in nature in the free state. Among these are gold, platinum, and frequently copper. They are usually found combined with other elements in the form of oxides or salts of various acids. Silicates, carbonates, sulphides, and sulphates are the most abundant salts. All inorganic substances occurring in nature, whether they contain a metal or not, are called *minerals*. Those minerals from which a useful substance can be extracted are called *ores* of the substance. These two terms are most frequently used in connection with the metals.
[Pg 268]

Extraction of metals,—metallurgy. The process of extracting a metal from its ores is called the metallurgy of the metal. The metallurgy of each metal presents peculiarities of its own, but there are several methods of general application which are very frequently employed.

1. *Reduction of an oxide with carbon.* Many of the metals occur in nature in the form of oxides. When these oxides are heated to a high temperature with carbon the oxygen combines with it and the metal is set free. Iron, for example, occurs largely in the form of the oxide Fe_2O_3. When this is heated with carbon the reaction expressed in the following equation takes place:

$$Fe_2O_3 + 3\,C = 2\,Fe + 3\,CO.$$

Many ores other than oxides may be changed into oxides which can then be reduced by carbon. The conversion of such ores into oxides is generally accomplished by heating, and this process is called *roasting*. Many carbonates and hydroxides decompose directly into the oxide on heating. Sulphides, on the other hand, must be heated in a current of air, the oxygen of the air entering into the reaction. The following equations will serve to illustrate these changes in the case of the ores of iron:

$$FeCO_3 = FeO + CO_2,$$

$$2Fe(OH)_3 = Fe_2O_3 + 3H_2O,$$

$$2FeS_2 + 11O = Fe_2O_3 + 4SO_2.$$

2. *Reduction of an oxide with aluminium.* Not all oxides, however, can be reduced by carbon. In such cases aluminium may be used. Thus chromium may be obtained in accordance with the following equation:

$$Cr_2O_3 + 2\,Al = 2\,Cr + Al_2O_3.$$

[Pg 269]

This method is a comparatively new one, having been brought into use by the German chemist Goldschmidt; hence it is sometimes called the Goldschmidt method.

3. *Electrolysis.* In recent years increasing use is being made of the electric current in the preparation of metals. In some cases the separation of the metal from its compounds is accomplished by passing the current through a solution of a suitable salt of the metal, the metal usually being deposited upon the cathode. In other cases the current is passed through a fused salt of the metal, the chloride being best adapted to this purpose.

Electro-chemical industries. Most of the electro-chemical industries of the country are carried on where water power is abundant, since this furnishes the cheapest means for the generation of electrical energy. Niagara Falls is the most important locality in this country for such industries, and many different electro-chemical products are manufactured there. Some industries depend upon electrolytic processes, while in others the electrical energy is used merely as a source of heat in electric furnaces.

Preparation of compounds of the metals. Since the compounds of the metals are so numerous and varied in character, there are many ways of preparing them. In many cases the properties of the substance to be prepared, or the material available for its preparation, suggest a rather unusual way. There are, however, a number of general principles which are constantly applied in the preparation of the compounds of the metals, and a clear understanding of them will save much time and effort in remembering the details in any given case. The most important of these general methods for the preparation of compounds are the following: [Pg 270]

1. *By direct union of two elements.* This is usually accomplished by heating the two elements together. Thus the sulphides, chlorides, and oxides of a metal can generally be obtained in this way. The following equations serve as examples of this method:

$$Fe + S = FeS,$$

$$Mg + O = MgO,$$

$$Cu + 2Cl = CuCl_2.$$

2. *By the decomposition of a compound.* This decomposition may be brought about either by heat alone or by the combined action of heat and a reducing agent. Thus when the nitrate of a metal is heated the oxide of the metal is usually obtained. Copper nitrate, for example, decomposes as follows:

$$Cu(NO_3)_2 = CuO + 2NO_2 + O.$$

Similarly the carbonates of the metals yield oxides, thus:

$$CaCO_3 = CaO + CO_2.$$

Most of the hydroxides form an oxide and water when heated:

$$2Al(OH)_3 = Al_2O_3 + 3H_2O.$$

When heated with carbon, sulphates are reduced to sulphides, thus:

$$BaSO_4 + 2C = BaS + 2CO_2.$$

3. *Methods based on equilibrium in solution.* In the preparation of compounds the first requisite is that the reactions chosen shall be of such a kind as will go on to completion. In the chapter on chemical equilibrium it was shown that reactions in solution may become complete in either of three ways: (1) a gas may be formed which escapes from solution; (2) an insoluble solid may be formed which

precipitates; (3) two different ions may combine to form [Pg 271] undissociated molecules. By the judicious selection of materials these principles may be applied to the preparation of a great variety of compounds, and illustrations of such methods will very frequently be found in the subsequent pages.

4. *By fusion methods.* It sometimes happens that substances which are insoluble in water and in acids, and which cannot therefore be brought into double decomposition in the usual way, are soluble in other liquids, and when dissolved in them can be decomposed and converted into other desired compounds. Thus barium sulphate is not soluble in water, and sulphuric acid, being less volatile than most other acids, cannot easily be driven out from this salt When brought into contact with melted sodium carbonate, however, it dissolves in it, and since barium carbonate is insoluble in melted sodium carbonate, double decomposition takes place:

$$Na_2CO_3 + BaSO_4 = BaCO_3 + Na_2SO_4.$$

On dissolving the cooled mixture in water the sodium sulphate formed in the reaction, together with any excess of sodium carbonate which may be present, dissolves. The barium carbonate can then be filtered off and converted into any desired salt by the processes already described.

5. *By the action of metals on salts of other metals.* When a strip of zinc is placed in a solution of a copper salt the copper is precipitated and an equivalent quantity of zinc passes into solution:

$$Zn + CuSO_4 = Cu + ZnSO_4.$$

In like manner copper will precipitate silver from its salts:

$$Cu + Ag_2SO_4 = 2Ag + CuSO_4.$$

[Pg 272]

It is possible to tabulate the metals in such a way that any one of them in the table will precipitate any one following it from its salts. The following is a list of some of the commoner metals arranged in this way:

Zinc
Iron
Tin
Lead
Copper
Bismuth
Mercury
Silver
Gold

According to this table copper will precipitate bismuth, mercury, silver, or gold from their salts, and will in turn be precipitated by zinc, iron, tin, or lead. Advantage is taken of this principle in the purification of some of the metals, and occasionally in the preparation of metals and their compounds.

Important insoluble compounds. Since precipitates play so important a part in the reactions which substances undergo, as well as in the preparation of many chemical compounds, it is important to know what substances are insoluble. Knowing this, we can in many cases predict reactions under certain conditions, and are assisted in devising ways to prepare desired compounds. While there is no general rule which will enable one to foretell the solubility of any given compound, nevertheless a few general statements can be made which will be of much assistance.

1. *Hydroxides.* All hydroxides are insoluble save those of ammonium, sodium, potassium, calcium, barium, and strontium.

2. *Nitrates.* All nitrates are soluble in water.

3. *Chlorides.* All chlorides are soluble save silver and mercurous chlorides. (Lead chloride is but slightly soluble.)

4. *Sulphates.* All sulphates are soluble save those of barium, strontium, and lead. (Sulphates of silver and calcium are only moderately soluble.) [Pg 273]

5. *Sulphides.* All sulphides are insoluble save those of ammonium, sodium, and potassium. The sulphides of calcium, barium, strontium, and magnesium are insoluble in water, but are changed by hydrolysis into acid sulphides which are soluble. On this account they cannot be prepared by precipitation.

6. *Carbonates, phosphates, and silicates.* All normal carbonates, phosphates, and silicates are insoluble save those of ammonium, sodium and potassium.

EXERCISES

1. Write equations representing four different ways for preparing $Cu(NO_3)_2$.

2. Write equations representing six different ways for preparing $ZnSO_4$.

3. Write equations for two reactions to illustrate each of the three ways in which reactions in solutions may become complete.

4. Give one or more methods for preparing each of the following compounds: $CaCl_2$, $PbCl_2$, $BaSO_4$, $CaCO_3$, $(NH_4)_2S$, Ag_2S, PbO, $Cu(OH)_2$ (for solubilities, see last paragraph of chapter). State in each case the general principle involved in the method of preparation chosen.

[Pg 274]

CHAPTER XXIII

THE ALKALI METALS

	SYMBOL	ATOMIC WEIGHT	DENSITY	MELTING POINT	FIRST PREPARED
Lithium	Li	7.03	0.59	186.°	Davy 1820
Sodium	Na	23.05	0.97	97.6°	" 1807

Potassium	K	39.15	0.87	62.5°	" 1807
Rubidium	Rb	85.5	1.52	38.5°	Bunsen 1861
Cæsium	Cs	132.9	1.88	26.5°	" 1860

The family. The metals listed in the above table constitute the even family in Group I in the periodic arrangement of the elements, and therefore form a natural family. The name alkali metals is commonly applied to the family for the reason that the hydroxides of the most familiar members of the family, namely sodium and potassium, have long been called alkalis.

1. *Occurrence.* While none of these metals occur free in nature, their compounds are very widely distributed, being especially abundant in sea and mineral waters, in salt beds, and in many rocks. Only sodium and potassium occur in abundance, the others being rarely found in any considerable quantity.

2. *Preparation.* The metals are most conveniently prepared by the electrolysis of their fused hydroxides or chlorides, though it is possible to prepare them by reducing their oxides or carbonates with carbon. [Pg 275]

3. *Properties.* They are soft, light metals, having low melting points and small densities, as is indicated in the table. Their melting points vary inversely with their atomic weights, while their densities (sodium excepted) vary directly with these. The pure metals have a silvery luster but tarnish at once when exposed to the air, owing to the formation of a film of oxide upon the surface of the metal. They are therefore preserved in some liquid, such as coal oil, which contains no oxygen. Because of their strong affinity for oxygen they decompose water with great ease, forming hydroxides and liberating hydrogen in accordance with the equation

$$M + H_2O = MOH + H,$$

where M stands for any one of these metals. These hydroxides are white solids; they are readily soluble in water and possess very

310

strong basic properties. These bases are nearly equal in strength, that is, they all dissociate in water to about the same extent.

4. *Compounds.* The alkali metals almost always act as univalent elements in the formation of compounds, the composition of which can be represented by such formulas as MH, MCl, MNO_3, M_2SO_4, M_3PO_4. These compounds, when dissolved in water, dissociate in such a way as to form simple, univalent metallic ions which are colorless. With the exception of lithium these metals form very few insoluble compounds, so that it is not often that precipitates containing them are obtained. Only sodium and potassium will be studied in detail, since the other metals of the family are of relatively small importance.

The compounds of sodium and potassium are so similar in properties that they can be used interchangeably for [Pg 276] most purposes. Other things being equal, the sodium compounds are prepared in preference to those of potassium, since they are cheaper. When a given sodium compound is deliquescent, or is so soluble that it is difficult to purify, the corresponding potassium compound is prepared in its stead, provided its properties are more desirable in these respects.

SODIUM

Occurrence in nature. Large deposits of sodium chloride have been found in various parts of the world, and the water of the ocean and of many lakes and springs contains notable quantities of it. The element also occurs as a constituent of many rocks and is therefore present in the soil formed by their disintegration. The mineral cryolite (Na_3AlF_6) is an important substance, and the nitrate, carbonate, and borate also occur in nature.

Preparation. In 1807 Sir Humphry Davy succeeded in preparing very small quantities of metallic sodium by the electrolysis of the fused hydroxide. On account of the cost of electrical energy it was for many years found more economical to prepare it by reducing the carbonate with carbon in accordance with the following equation:

$$Na_2CO_3 + 2C = 2Na + 3CO.$$

The cost of generating the electric current has been diminished to such an extent, however, that it is now more economical to prepare sodium by Davy's original method, namely, by the electrolysis of the fused hydroxide or chloride. When the chloride is used the process is difficult to manage, owing to the higher temperature required to keep the electrolyte fused, and because of the corroding action of the fused chloride upon the containing vessel.

SIR HUMPHRY DAVY (English) (1778-1829)

Isolated sodium, lithium, potassium, barium, strontium, and calcium by means of electrolysis; demonstrated the elementary nature of chlorine; invented the safety lamp; discovered the stupefying effects of nitrous oxide

[Pg 277]

Technical preparation. The sodium hydroxide is melted in a cylindrical iron vessel (Fig. 76) through the bottom of which rises the cathode K. The anodes A, several in number, are suspended around the cathode from above. A cylindrical vessel C floats in the fused alkali directly over the cathode, and under this cap the sodium and hydrogen liberated at the cathode collect. The hydrogen escapes by lifting the cover, and the sodium, protected from the air by the hydrogen, is skimmed or drained off from time to time. Oxygen is set free upon the anode and escapes into the air through the openings O without coming into contact with the sodium or hydrogen. This process is carried on extensively at Niagara Falls.

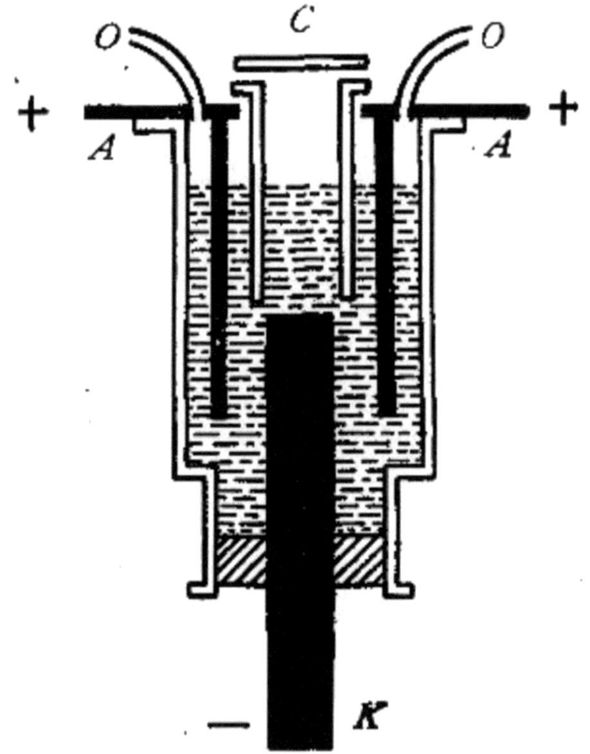

Fig. 76

Properties. Sodium is a silver-white metal about as heavy as water, and so soft that it can be molded easily by the fingers or pressed into wire. It is very active chemically, combining with most of the non-metallic elements, such as oxygen and chlorine, with great energy. It will often withdraw these elements from combination with other elements, and is thus able to decompose water and the oxides and chlorides of many metals.

Sodium peroxide (NaO). Since sodium is a univalent element we should expect it to form an oxide of the formula Na_2O. While such an oxide can be prepared, the peroxide (NaO) is much better known. It is a yellowish-white powder made by burning sodium in air. Its chief use is as an oxidizing agent. When heated with oxidizable substances it gives up a part of its oxygen, as shown in the equation

$$2NaO = Na_2O + O.$$

[Pg 278]

Water decomposes it in accordance with the equation

$$2NaO + 2H_2O = 2NaOH + H_2O_2.$$

Acids act readily upon it, forming a sodium salt and hydrogen peroxide:

$$2NaO + 2HCl = 2NaCl + H_2O_2.$$

In these last two reactions the hydrogen dioxide formed may decompose into water and oxygen if the temperature is allowed to rise:

$$H_2O_2 = H_2O + O.$$

Peroxides. It will be remembered that barium dioxide (BaO_{2}) yields hydrogen dioxide when treated with acids, and that manganese dioxide gives up oxygen when heated with sulphuric acid. Oxides which yield either hydrogen dioxide or oxygen when treated with water or an acid are called peroxides.

Sodium hydroxide (*caustic soda*) (NaOH). 1. *Preparation.* Sodium hydroxide is prepared commercially by several processes.

(*a*) In the older process, still in extensive use, sodium carbonate is treated with calcium hydroxide suspended in water. Calcium carbonate is precipitated according to the equation

$$Na_2CO_3 + Ca(OH)_2 = CaCO_3 + 2NaOH.$$

The dilute solution of sodium hydroxide, filtered from the calcium carbonate, is evaporated to a paste and is then poured into molds to solidify. It is sold in the form of slender sticks.

(*b*) The newer methods depend upon the electrolysis of sodium chloride. In the Castner process a solution of salt is electrolyzed, the reaction being expressed as follows:

$$NaCl + H_2O = NaOH + H + Cl.$$

[Pg 279]

The chlorine escapes as a gas, and by an ingenious mechanical device the sodium hydroxide is prevented from mixing with the salt in the solution.

In the Acker process the electrolyte is *fused* sodium chloride. The chlorine is evolved as a gas at the anode, while the sodium alloys with the melted lead which forms the cathode. When this alloy is treated with water the following reaction takes place:

$$Na + H_2O = NaOH + H.$$

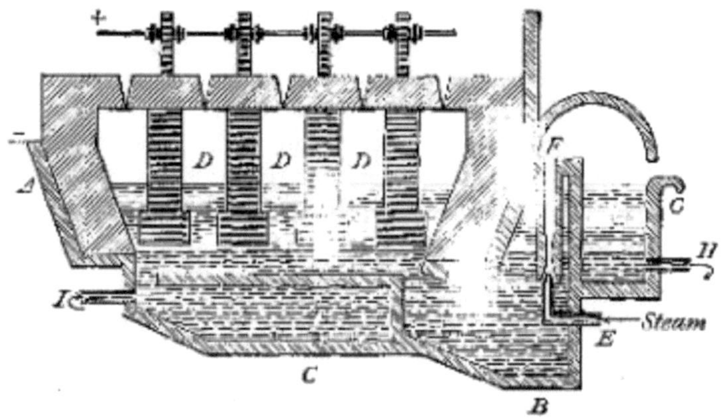

Fig. 77

Technical process. A sketch of an Acker furnace is represented in Fig. 77. The furnace is an irregularly shaped cast-iron box, divided into three compartments, *A*, *B*, and *C*. Compartment *A* is lined with magnesia brick. Compartments *B* and *C* are filled with melted lead, which also covers the bottom of *A* to a depth of about an inch. Above this layer in *A* is fused salt, into which dip carbon anodes *D*. The metallic box and melted lead is the cathode.

When the furnace is in operation chlorine is evolved at the anodes, and is drawn away through a pipe (not represented) to the bleaching-powder chambers. Sodium is set free at the surface of the melted lead in *A*, and at once alloys with it. Through the pipe *E* a powerful jet of steam is driven through the lead in *B* upwards [Pg 280] into the narrow tube *F*. This forces the lead alloy up through the tube and over into the chamber *G*.

In this process the steam is decomposed by the sodium in the alloy, forming melted sodium hydroxide and hydrogen. The melted lead and sodium hydroxide separate into two layers in *G*, and the sodium hydroxide, being on top, overflows into tanks from which it is drawn off and packed in metallic drums. The lead is returned to the other compartments of the furnace by a pipe leading from *H* to *I*. Compartment *C* serves merely as a reservoir for excess of melted lead.

2. *Properties.* Sodium hydroxide is a white, crystalline, brittle substance which rapidly absorbs water and carbon dioxide from the air. As the name (caustic soda) indicates, it is a very corrosive substance, having a disintegrating action on most animal and vegetable tissues. It is a strong base. It is used in a great many chemical industries, and under the name of lye is employed to a small extent as a cleansing agent for household purposes.

Sodium chloride (*common salt*) (NaCl). 1. *Preparation.* Sodium chloride, or common salt, is very widely distributed in nature. Thick strata, evidently deposited at one time by the evaporation of salt water, are found in many places. In the United States the most important localities for salt are New York, Michigan, Ohio, and Kansas. Sometimes the salt is mined, especially if it is in the pure form called rock salt. More frequently a strong brine is pumped from deep wells sunk into the salt deposit, and is then evaporated in large pans until the salt crystallizes out. The crystals are in the form of small cubes and contain no water of crystallization; some water is, however, held in cavities in the crystals and causes the salt to decrepitate when heated.

2. *Uses.* Since salt is so abundant in nature it forms the starting point in the preparation of all compounds [Pg 281] containing either sodium or chlorine. This includes many substances of the highest importance to civilization, such as soap, glass, hydrochloric acid, soda, and bleaching powder. Enormous quantities of salt are therefore produced each year. Small quantities are essential to the life of man and animals. Pure salt does not absorb moisture; the fact that ordinary salt becomes moist in air is not due to a property of the salt, but to impurities commonly occurring in it, especially calcium and magnesium chlorides.

Sodium sulphate (*Glauber's salt*) ($Na_2SO_4 \cdot 10H_2O$). This salt is prepared by the action of sulphuric acid upon sodium chloride, hydrochloric acid being formed at the same time:

$$2NaCl + H_2SO_4 = Na_2SO_4 + 2HCl.$$

Some sodium sulphate is prepared by the reaction represented in the equation

$$MgSO_4 + 2NaCl = Na_2SO_4 + MgCl_2.$$

The magnesium sulphate required for this reaction is obtained in large quantities in the manufacture of potassium chloride, and being of little value for any other purpose is used in this way. The reaction depends upon the fact that sodium sulphate is the least soluble of any of the four factors in the equation, and therefore crystallizes out when hot, saturated solutions of magnesium sulphate and sodium chloride are mixed together and the resulting mixture cooled.

Sodium sulphate forms large efflorescent crystals. The salt is extensively used in the manufacture of sodium carbonate and glass. Small quantities are used in medicine.

Sodium sulphite ($Na_2SO_3 \cdot 7H_2O$). Sodium sulphite is prepared by the action of sulphur dioxide upon solutions [Pg 282] of sodium hydroxide, the reaction being analogous to the action of carbon dioxide upon sodium hydroxide. Like the carbonate, the sulphite is readily decomposed by acids:

$$Na_2SO_3 + 2HCl = 2NaCl + H_2O + SO_2.$$

Because of this reaction sodium sulphite is used as a convenient source of sulphur dioxide. It is also used as a disinfectant and a preservative.

Sodium thiosulphate (*hyposulphite of soda or "hypo"*) ($Na_2S_2O_3 \cdot 5H_2O$). This salt, commonly called sodium hyposulphite, or merely hypo, is made by boiling a solution of sodium sulphite with sulphur:

$$Na_2SO_3 + S = Na_2S_2O_3.$$

It is used in photography and in the bleaching industry, to absorb the excess of chlorine which is left upon the bleached fabrics.

Thio compounds. The prefix "thio" means sulphur. It is used to designate substances which may be regarded as derived from oxygen compounds by replacing the whole or a part of their oxygen with sulphur. The thiosulphates may be regarded as sulphates in which one atom of oxygen has been replaced by an atom of sulphur. This may be seen by comparing the formula Na_2SO_4 (sodium sulphate) with the formula $Na_2S_2O_3$ (sodium thiosulphate).

Sodium carbonate (*sal soda*)($Na_2CO_3 \cdot 10H_2O$). There are two different methods now employed in the manufacture of this important substance.

1. *Le Blanc process.* This older process involves several distinct reactions, as shown in the following equations.

(*a*) Sodium chloride is first converted into sodium sulphate:

$$2NaCl + H_2SO_4 = Na_2SO_4 + 2HCl.$$

[Pg 283]

(*b*) The sodium sulphate is next reduced to sulphide by heating it with carbon:

$$Na_2SO_4 + 2C = Na_2S + 2CO_2.$$

(*c*) The sodium sulphide is then heated with calcium carbonate, when double decomposition takes place:

$$Na_2S + CaCO_3 = CaS + Na_2CO_3.$$

Technical preparation of sodium carbonate. In a manufacturing plant the last two reactions take place in one process. Sodium sulphate, coal, and powdered limestone are heated together to a rather high temperature. The coal reduces the sulphate to sulphide, which in turn reacts upon the calcium carbonate. Some limestone is decomposed by the heat, forming calcium oxide. When treated with water the calcium oxide is changed into hydroxide, and this prevents the water from decomposing the insoluble calcium sulphide.

The crude product of the process is a hard black cake called black ash. On digesting this mass with water the sodium carbonate passes into solution. The pure carbonate is obtained by evaporation of this solution, crystallizing from it in crystals of the formula $Na_2CO_3 \cdot 10H_2O$. Since over 60% of this salt is water, the crystals are sometimes heated until it is driven off. The product is called calcined soda, and is, of course, more valuable than the crystallized salt.

2. *Solvay process.* This more modern process depends upon the reactions represented in the equations

$$NaCl + NH_4HCO_3 = NaHCO_3 + NH_4Cl,$$

$$2NaHCO_3 = Na_2CO_3 + H_2O + CO_2.$$

The reason the first reaction takes place is that sodium hydrogen carbonate is sparingly soluble in water, while the other compounds are freely soluble. When strong solutions of sodium chloride and of ammonium hydrogen carbonate are brought together the sparingly soluble sodium hydrogen carbonate is precipitated. This is converted into the normal carbonate by heating, the reaction being represented in the second equation. [Pg 284]

Technical preparation. In the Solvay process a very concentrated solution of salt is first saturated with ammonia gas, and a current of carbon dioxide is then conducted into the solution. In this way ammonium hydrogen carbonate is formed:

$$NH_3 + H_2O + CO_2 = NH_4HCO_3.$$

This enters into double decomposition with the salt, as shown in the first equation under the Solvay process. After the sodium hydrogen carbonate has been precipitated the mother liquors containing ammonium chloride are treated with lime:

$$2NH_4Cl + CaO = CaCl_2 + 2\,NH_3 + H_2O.$$

The lime is obtained by burning limestone:

$$CaCO_3 = CaO + CO_2.$$

The ammonia and carbon dioxide evolved in the latter two reactions are used in the preparation of an additional quantity of ammonium hydrogen carbonate. It will thus be seen that there is no loss of ammonia. The only materials permanently used up are calcium carbonate and salt, while the only waste product is calcium chloride.

Historical. In former times sodium carbonate was made by burning seaweeds and extracting the carbonate from their ash. On this account the salt was called *soda ash*, and the name is still in common use. During the French Revolution this supply was cut off, and in behalf of the French government Le Blanc made a study of methods of preparing the carbonate directly from salt. As a result he devised the method which bears his name, and which was used exclusively for many years. It has been replaced to a large extent by the Solvay process, which has the advantage that the materials used are inexpensive, and that the ammonium hydrogen carbonate used can be regenerated from the products formed in the process. Much expense is also saved in fuel, and the sodium hydrogen carbonate, which is the first product of the process, has itself many commercial uses. The Le Blanc process is still used, however, since the hydrochloric acid generated is of value.

By-products. The substances obtained in a given process, aside from the main product, are called the by-products. The success of many processes depends upon the value of the by-products formed.

Thus hydrochloric acid, a by-product in the Le Blanc process, is valuable enough to make the process pay, even though sodium carbonate can be made cheaper in other ways.

[Pg 285]

Properties of sodium carbonate. Sodium carbonate forms large crystals of the formula $Na_2CO_3 \cdot 10\,H_2O$. It has a mild alkaline reaction and is used for laundry purposes under the name of washing soda. Mere mention of the fact that it is used in the manufacture of glass, soap, and many chemical reagents will indicate its importance in the industries. It is one of the few soluble carbonates.

Sodium hydrogen carbonate (*bicarbonate of soda*) ($NaHCO_3$). This salt, commonly called bicarbonate of soda, or baking soda, is made by the Solvay process, as explained above, or by passing carbon dioxide into strong solutions of sodium carbonate:

$$Na_2CO_3 + H_2O + CO_2 = 2NaHCO_3.$$

The bicarbonate, being sparingly soluble, crystallizes out. A mixture of the bicarbonate with some substance (the compound known as cream of tartar is generally used) which slowly reacts with it, liberating carbon dioxide, is used largely in baking. The carbon dioxide generated forces its way through the dough, thus making it porous and light.

Sodium nitrate (*Chili saltpeter*) ($NaNO_3$). This substance is found in nature in arid regions in a number of places, where it has been formed apparently by the decay of organic substances in the presence of air and sodium salts. The largest deposits are in Chili, and most of the nitrate of commerce comes from that country. Smaller deposits occur in California and Nevada. The commercial salt is prepared by dissolving the crude nitrate in water, [Pg 286] allowing the insoluble earthy materials to settle, and evaporating the clear solution so obtained to crystallization. The soluble impurities remain for the most part in the mother liquors.

Since this salt is the only nitrate found extensively in nature, it is the material from which other nitrates as well as nitric acid are prepared. It is used in enormous quantities in the manufacture of sulphuric acid and potassium nitrate, and as a fertilizer.

Sodium phosphate ($Na_2HPO_4 \cdot 12H_2O$). Since phosphoric acid has three replaceable hydrogen atoms, three sodium phosphates are possible,—two acid salts and one normal. All three can be made without difficulty, but disodium phosphate is the only one which is largely used, and is the salt which is commonly called sodium phosphate. It is made by the action of phosphoric acid on sodium carbonate:

$$Na_2CO_3 + H_3PO_4 = Na_2HPO_4 + CO_2 + H_2O.$$

It is interesting as being one of the few phosphates which are soluble in water, and is the salt commonly used when a soluble phosphate is needed.

Normal sodium phosphate (Na_3PO_4). Although this is a normal salt its solution has a strongly alkaline reaction. This is due to the fact that the salt hydrolyzes in solution into sodium hydroxide and disodium phosphate, as represented in the equation

$$Na_3PO_4 + H_2O = Na_2HPO_4 + NaOH.$$

Sodium hydroxide is strongly alkaline, while disodium phosphate is nearly neutral in reaction. The solution as a whole is therefore alkaline. The salt is prepared by adding a large excess of sodium hydroxide to a solution of disodium [Pg 287] phosphate and evaporating to crystallization. The excess of the sodium hydroxide reverses the reaction of hydrolysis and the normal salt crystallizes out.

Sodium tetraborate (*borax*) ($Na_2B_4O_7 \cdot 10H_2O$). The properties of this important compound have been discussed under the head of boron.

POTASSIUM

Occurrence in nature. Potassium is a constituent of many common rocks and minerals, and is therefore a rather abundant element, though not so abundant as sodium. Feldspar, which occurs both by itself and as a constituent of granite, contains considerable potassium. The element is a constituent of all clay and of mica and also occurs in very large deposits at Stassfurt, Germany, in the form of the chloride and sulphate, associated with compounds of sodium and magnesium. In small quantities it is found as nitrate and in many other forms.

The natural decomposition of rocks containing potassium gives rise to various compounds of the element in all fertile soils. Its soluble compounds are absorbed by growing plants and built up into complex vegetable substances; when these are burned the potassium remains in the ash in the form of the carbonate. Crude carbonate obtained from wood ashes was formerly the chief source of potassium compounds; they are now mostly prepared from the salts of the Stassfurt deposits.

Stassfurt salts. These salts form very extensive deposits in middle and north Germany, the most noted locality for working them being at Stassfurt. The deposits are very thick and rest upon an enormous layer of common salt. They are in the form of a series of strata, each consisting largely of a single mineral salt. A cross section of [Pg 288] these deposits is shown in Fig. 78. While these strata are salts from a chemical standpoint, they are as solid and hard as many kinds of stone, and are mined as stone or coal would be. Since the strata differ in general appearance, each can be mined separately, and the various minerals can be worked up by methods adapted to each particular case. The chief minerals of commercial importance in these deposits are the following:

Sylvine	KCl.
Anhydrite	$CaSO_4$.
Carnallite	$KCl \cdot MgCl_2 \cdot 6H_2O$.
Kainite	$K_2SO_4 \cdot MgSO_4 \cdot MgCl_2 \cdot 6H_2O$.
Polyhalite	$K_2SO_4 \cdot MgSO_4 \cdot 2CaSO_4 \cdot 2H_2O$.
Kieserite	$MgSO_4 \cdot H_2O$.
Schönite	$K_2SO_4 \cdot MgSO_4 \cdot 6H_2O$.

Preparation and properties. The metal is prepared by the same method used in the preparation of sodium. In most respects it is very similar to sodium, the chief difference being that it is even more energetic in its action upon other substances. The freshly cut, bright surface instantly becomes dim through oxidation by the air.

325

It decomposes water very vigorously, the heat of reaction being sufficient to ignite the hydrogen evolved. It is somewhat lighter than sodium and is preserved under gasoline.

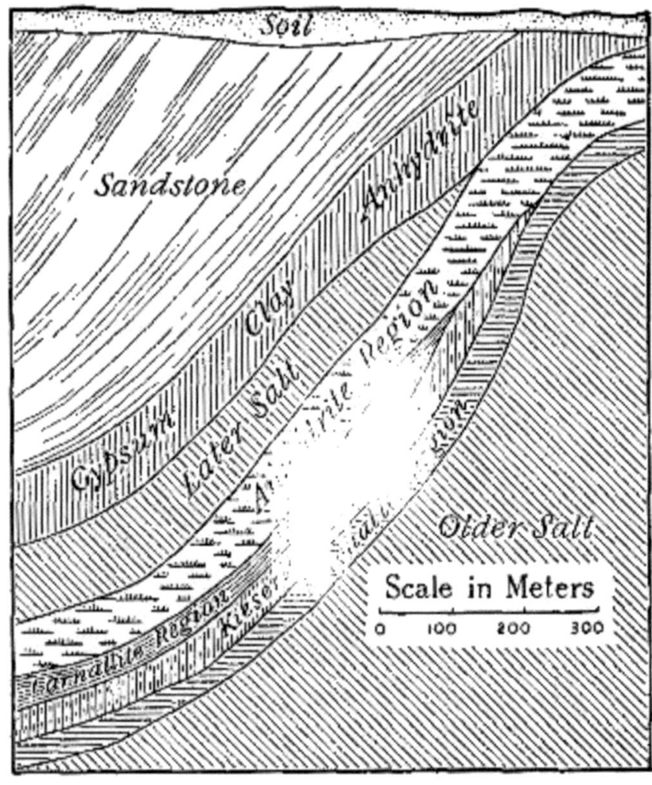

Fig. 78

Potassium hydroxide (*caustic potash*) (KOH). Potassium hydroxide is prepared by methods exactly similar to those [Pg 289] used in the preparation of sodium hydroxide, which compound it closely resembles in both physical and chemical properties. It is not used to any very great extent, being replaced by the cheaper sodium hydroxide.

Action of the halogen elements on potassium hydroxide. When any one of the three halogen elements—chlorine, bromine, and iodine—is added to a solution of potassium hydroxide a reaction takes place, the nature of which depends upon the conditions of the experiment. Thus, when chlorine is passed into a cold dilute solution of potassium hydroxide the reaction expressed by the following equation takes place:

$$(1)\ 2KOH + 2Cl = KCl + KClO + H_2O.$$

If the solution of hydroxide is concentrated and hot, on the other hand, the potassium hypochlorite formed according to equation (1) breaks down as fast as formed:

$$(2)\ 3KClO = KClO_3 + 2KCl.$$

Equation (1), after being multiplied by 3, may be combined with equation (2), giving the following:

$$(3)\ 6KOH + 6Cl = 5KCl + KClO_3 + 3H_2O.$$

This represents in a single equation the action of chlorine on hot, concentrated solutions of potassium hydroxide. By means of these reactions one can prepare potassium chloride, potassium hypochlorite, and potassium chlorate. By substituting bromine or iodine for chlorine the corresponding compounds of these elements are obtained. Some of these compounds can be obtained in cheaper ways.

If the halogen element is added to a solution of sodium hydroxide or calcium hydroxide, the reaction which takes place is exactly similar to that which takes place with [Pg 290] potassium hydroxide. It is possible, therefore, to prepare in this way the sodium and calcium compounds corresponding to the potassium compounds given above.

Potassium chloride (KCl). This salt occurs in nature in sea water, in the mineral sylvine, and, combined with magnesium chloride, as carnallite ($KCl \cdot MgCl_2\ 6H_2O$). It is prepared from carnallite by saturating boiling water with the mineral and allowing the solution to cool. The mineral decomposes while in solution, and the potassium chloride crystallizes out on cooling, while the very soluble magnesium chloride remains in solution. The salt is very similar to sodium chloride both in physical and chemical properties. It is used in the preparation of nearly all other potassium salts, and, together with potassium sulphate, is used as a fertilizer.

Potassium bromide (KBr). When bromine is added to a hot concentrated solution of potassium hydroxide there is formed a mixture of potassium bromide and potassium bromate in accordance with the reactions already discussed. There is no special use for the bromate, so the solution is evaporated to dryness, and the residue, consisting of a mixture of the bromate and bromide, is strongly heated. This changes the bromate to bromide, as follows:

$$KBrO_3 = KBr + 3O.$$

The bromide is then crystallized from water, forming large colorless crystals. It is used in medicine and in photography.

Potassium iodide (KI). Potassium iodide may be made by exactly the same method as has just been described for the bromide, substituting iodine for bromine. It is more frequently made as follows. Iron filings are [Pg 291] treated with iodine, forming the compound Fe_3I_8; on boiling this substance with potassium carbonate the reaction represented in the following equation occurs:

$$Fe_3I_8 + 4K_2CO_3 = Fe_3O_4 + 8KI + 4CO_2.$$

Potassium iodide finds its chief use in medicine.

Potassium chlorate (KClO$_3$). This salt, as has just been explained, can be made by the action of chlorine on strong potassium hydroxide solutions. The chief use of potassium chlorate is as an oxidizing agent in the manufacture of matches, fireworks, and explosives; it is also used in the preparation of oxygen and in medicine.

Commercial preparation. By referring to the reaction between chlorine and hot concentrated solutions of potassium hydroxide, it will be seen that only one molecule of potassium chlorate is formed from six molecules of potassium hydroxide. Partly because of this poor yield and partly because the potassium hydroxide is rather expensive, this process is not an economical one for the preparation of potassium chlorate. The commercial method is the following. Chlorine is passed into hot solutions of calcium hydroxide, a compound which is very cheap. The resulting calcium chloride and chlorate are both very soluble. To the solution of these salts potassium chloride is added, and as the solution cools the sparingly soluble potassium chlorate crystallizes out:

$$Ca(ClO_3)_2 + 2KCl = 2KClO_3 + CaCl_2.$$

Electro-chemical processes are also used.

Potassium nitrate (*saltpeter*) (KNO_3). This salt was formerly made by allowing animal refuse to decompose in the open air in the presence of wood ashes or earthy materials containing potassium. Under these conditions the nitrogen in the organic matter is in part converted into potassium nitrate, which was obtained by extracting the mass with water and evaporating to crystallization. This crude and [Pg 292] slow process is now almost entirely replaced by a manufacturing process in which the potassium salt is made from Chili saltpeter:

$$NaNO_3 + KCl = NaCl + KNO_3.$$

This process has been made possible by the discovery of the Chili niter beds and the potassium chloride of the Stassfurt deposits.

The reaction depends for its success upon the apparently insignificant fact that sodium chloride is almost equally soluble in cold and hot water. All four factors in the equation are rather soluble in cold water, but in hot water sodium chloride is far less soluble than the other three. When hot saturated solutions of sodium nitrate and potassium chloride are brought together, sodium chloride precipitates and can be filtered off, leaving potassium nitrate in solution, together with some sodium chloride. On cooling, potassium nitrate crystallizes out, leaving small amounts of the other salts in solution.

Potassium nitrate is a colorless salt which forms very large crystals. It is stable in the air, and when heated is a good oxidizing agent, giving up oxygen quite readily. Its chief use is in the manufacture of gunpowder.

Gunpowder. The object sought for in the preparation of gunpowder is to secure a solid substance which will remain unchanged under ordinary conditions, but which will explode readily when ignited, evolving a large volume of gas. When a mixture of carbon and potassium nitrate is ignited a great deal of gas is formed, as will be seen from the equation

$$2KNO_3 + 3C = CO_2 + CO + N_2 + K_2CO_3.$$

By adding sulphur to the mixture the volume of gas formed in the explosion is considerably increased:

$$2KNO_3 + 3C + S = 3CO_2 + N_2 + K_2S.$$

Gunpowder is simply a mechanical mixture of these three substances in the proportion required for the above reaction. While the equation represents the principal reaction, other reactions also take place. [Pg 293] The gases formed in the explosion, when measured under standard conditions, occupy about two hundred and eighty times the volume of the original powder. Potassium sulphide (K_2S) is a solid substance, and it is largely due to it that gunpowder gives off smoke and soot when it explodes. Smokeless powder consists of organic substances which, on explosion, give only colorless gases, and hence produce no smoke. Sodium nitrate is cheaper than potassium nitrate, but it is not adapted to the manufacture of the best grades of powder, since it is somewhat deliquescent and does not give up its oxygen so readily as does potassium nitrate. It is used, however, in the cheaper grades of powder, such as are employed for blasting.

Potassium cyanide (KCN). When animal matter containing nitrogen is heated with iron and potassium carbonate, complicated changes occur which result in the formation of a substance commonly called yellow prussiate of potash, which has the formula

$K_4FeC_6N_6$. When this substance is heated with potassium, potassium cyanide is formed:

$$K_4FeC_6N_6 + 2\,K = 6KCN + Fe.$$

Since sodium is much cheaper than potassium it is often used in place of it:

$$K_4FeC_6N_6 + 2Na = 4KCN + 2NaCN + Fe.$$

The mixture of cyanides so resulting serves most of the purposes of the pure salt. It is used very extensively in several metallurgical processes, particularly in the extraction of gold. Potassium cyanide is a white solid characterized by its poisonous properties, and must be used with extreme caution.

Potassium carbonate (*potash*) (K_2CO_3). This compound occurs in wood ashes in small quantities. It cannot be prepared by the Solvay process, since the acid carbonate is quite soluble in water, but is made by the Le Blanc process. Its chief use is in the manufacture of other potassium salts. [Pg 294]

Other salts of potassium. Among the other salts of potassium frequently met with are the sulphate (K_2SO_4), the acid carbonate ($KHCO_3$), the acid sulphate ($KHSO_4$), and the acid sulphite ($KHSO_3$). These are all white solids.

LITHIUM, RUBIDIUM, CÆSIUM

Of the three remaining elements of the family—lithium, rubidium, and cæsium—lithium is by far the most common, the other two being very rare. Lithium chloride and carbonate are not infrequently found in natural mineral waters, and as these substances are supposed to increase the medicinal value of the water, they are very often added to artificial mineral waters in small quantities.

COMPOUNDS OF AMMONIUM

General. As explained in a previous chapter, when ammonia is passed into water the two compounds combine to form the base NH_4OH, known as ammonium hydroxide. When this base is neutralized with acids there are formed the corresponding salts, known as the ammonium salts. Since the ammonium group is univalent, ammonium salts resemble those of the alkali metals in formulas; they also resemble the latter salts very much in their chemical properties, and may be conveniently described in connection with them. Among the ammonium salts the chloride, sulphate, carbonate, and sulphide are the most familiar.

Ammonium chloride (*sal ammoniac*) (NH_4Cl). This substance is obtained by neutralizing ammonium hydroxide with hydrochloric acid. It is a colorless substance crystallizing in fine needles, and, like most ammonium salts, is very soluble in water. When placed in a tube and heated strongly it decomposes into hydrochloric acid and ammonia. When these gases reach a cooler portion of the tube they [Pg 295] at once recombine, and the resulting ammonium chloride is deposited on the sides of the tube. In this way the salt can be separated from nonvolatile impurities. Ammonium chloride is sometimes used in preparation of ammonia; it is also used in making dry batteries and in the laboratory as a chemical reagent.

Ammonium sulphate ($(NH_4)_2SO_4$). This salt resembles the chloride very closely, and, being cheaper, is used in place of it when possible. It is used in large quantity as a fertilizer, the nitrogen which it contains being a very valuable food for plants.

Ammonium carbonate ($(NH_4)_2CO_3$). This salt, as well as the acid carbonate (NH_4HCO_3), is used as a chemical reagent. They are colorless solids, freely soluble in water. The normal carbonate is made by heating ammonium chloride with powdered limestone (calcium carbonate), the ammonium carbonate being obtained as a sublimate in compact hard masses:

$$2NH_4Cl + CaCO_3 = (NH_4)_2CO_3 + CaCl_2.$$

The salt always smells of ammonia, since it slowly decomposes, as shown in the equation

$$(NH_4)_2CO_3 = NH_4HCO_3 + NH_3.$$

The acid carbonate, or bicarbonate, is prepared by saturating a solution of ammonium hydroxide with carbon dioxide:

$$NH_4OH + CO_2 = NH_4HCO_3.$$

It is a well-crystallized stable substance.

Ammonium sulphide $((NH_4)_2S)$. Ammonium sulphide is prepared by the action of hydrosulphuric acid upon ammonium hydroxide:

$$2NH_4OH + H_2S = (NH_4)_2S + 2H_2O.$$

[Pg 296]

If the action is allowed to continue until no more hydrosulphuric acid is absorbed, the product is the acid sulphide, sometimes called the hydrosulphide:

$$NH_4OH + H_2S = NH_4HS + H_2O.$$

If equal amounts of ammonium hydroxide and ammonium acid sulphide are brought together, the normal sulphide is formed:

$$NH_4OH + NH_4HS = (NH_4)_2S + H_2O$$

It has been obtained in the solid state, but only with great difficulty. As used in the laboratory it is always in the form of a solution. It is much used in the process of chemical analysis because it is a soluble sulphide and easily prepared. On exposure to the air ammonium sulphide slowly decomposes, being converted into ammonia, water, and sulphur:

$$(NH_4)_2S + O = 2NH_3 + H_2O + S.$$

As fast as the sulphur is liberated it combines with the unchanged sulphide to form several different ammonium sulphides in which there are from two to five sulphur atoms in the molecule, thus: $(NH_4)_2S_2$, $(NH_4)_2S_3$, $(NH_4)_2S_5$. These sulphides in turn decompose by further action of oxygen, so that the final products of the reaction are those given in the equation. A solution of these compounds is yellow and is sometimes called *yellow ammonium sulphide*.

FLAME REACTION — SPECTROSCOPE

When compounds of either sodium or potassium are brought into the non-luminous flame of a Bunsen burner the flame becomes colored. Sodium compounds color it intensely yellow, while those of potassium color it pale violet. When only one of these elements is present it is [Pg 297] easy to identify it by this simple test, but when both are present the intense color of the sodium flame entirely conceals the pale tint characteristic of potassium compounds.

It is possible to detect the potassium flame in such cases, however, in the following way. When light is allowed to shine through a very small hole or slit in some kind of a screen, such as a piece of metal, upon a triangular prism of glass, the light is bent or refracted out of its course instead of passing straight through the glass. It thus comes out of the prism at some angle to the line at which it entered. Yellow light is bent more than red, and violet more than yellow. When light made up of the yellow of sodium and the violet of potassium shines through a slit upon such a prism, the yellow and the violet lights come out at somewhat different angles, and so two colored lines of light—a yellow line and a violet line—are seen on looking into the prism in the proper direction. The instrument used for separating the rays of light in this way is called a *spectroscope* (Fig. 79). The material to be tested is placed on a platinum wire and held in the colorless Bunsen flame. The resulting light passes through the slit in the end of tube *B*, and then through *B* to the prism. The resulting lines of light are seen by looking into the tube *A*, which contains a magnifying lens. Most elements give more than one image of the slit, each having a different color, and the series of colored lines due to an element is called its spectrum.

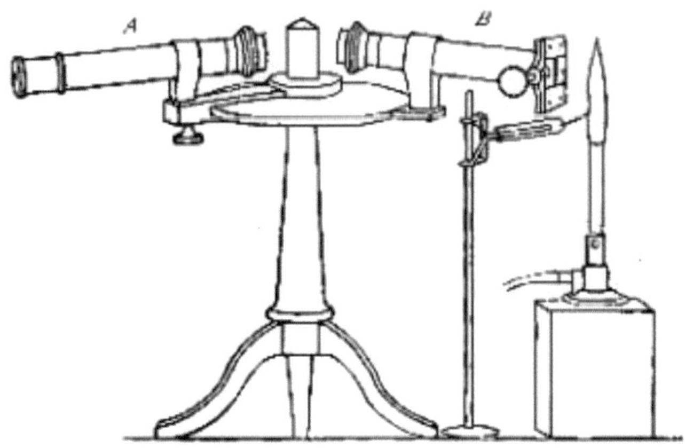

Fig. 79

[Pg 298]

The spectra of the known elements have been carefully studied, and any element which imparts a characteristic color to a flame, or has a spectrum of its own, can be identified even when other elements are present. Through the spectroscopic examination of certain minerals a number of elements have been discovered by the observation of lines which did not belong to any known element. A study of the substance then brought to light the new element. Rubidium and cæsium were discovered in this way, rubidium having bright red lines and cæsium a very intense blue line. Lithium colors the flame deep red, and has a bright red line in its spectrum.

EXERCISES

1. What is an alkali? Can a metal itself be an alkali?

2. Write equations showing how the following changes may be brought about, giving the general principle involved in each change: NaCl --> Na_2SO_3, Na_2SO_3 --> NaCl, NaCl --> NaBr, Na_2SO_4 -> $NaNO_3$, $NaNO_3$ --> $NaHCO_3$.

3. What carbonates are soluble?

4. State the conditions under which the reaction represented by the following equation can be made to go in either direction:

$$Na_2CO_3 + H_2O + CO_2 \longleftrightarrow 2\,NaHCO_3.$$

5. Account for the fact that solutions of sodium carbonate and potassium carbonate are alkaline.

6. What non-metallic element is obtained from the deposits of Chili saltpeter?

7. Supposing concentrated hydrochloric acid (den. = 1.2) to be worth six cents a pound, what is the value of the acid generated in the preparation of 1 ton of sodium carbonate by the Le Blanc process?

8. What weight of sodium carbonate crystals will 1 kg. of the anhydrous salt yield?

9. Write equations for the preparation of potassium hydroxide by three different methods.

10. What would take place if a bit of potassium hydroxide were left exposed to the air?

11. Write the equations for the reactions between sodium hydroxide and bromine; between potassium hydroxide and iodine.

12. Write equations for the preparation of potassium sulphate; of potassium acid carbonate.

ROBERT WILHELM BUNSEN (German) (1811-1899)

Invented many lecture-room and laboratory appliances (Bunsen
burner); invented the spectroscope and with it discovered rubidium
and cæsium; greatly perfected methods of electrolysis, inventing a
new battery; made many investigations among metallic and organic
substances

[Pg 299]

13. What weight of carnallite would be necessary in the preparation of 1 ton of potassium carbonate?

14. Write the equations showing how ammonium chloride, ammonium sulphate, ammonium carbonate, and ammonium nitrate may be prepared from ammonium hydroxide.

15. Write an equation to represent the reaction involved in the preparation of ammonia from ammonium chloride.

16. What substances already studied are prepared from the following compounds? ammonium chloride; ammonium nitrate; ammonium nitrite; sodium nitrate; sodium chloride.

17. How could you prove that the water in crystals of common salt is not water of crystallization?

18. How could you distinguish between potassium chloride and potassium iodide? between sodium chloride and ammonium chloride? between sodium nitrate and potassium nitrate?

[Pg 300]

CHAPTER XXIV

THE ALKALINE-EARTH FAMILY

	SYMBOL	ATOMIC WEIGHT	DENSITY	MILLIGRAMS SOLUBLE IN 1 L OF WATER AT 18°		CARBON. DECOMP
				SULPHATE	HYDROXIDE	
lcium	Ca	40.1	1.54	2070.00	1670.	At dull red heat
rontium	Sr	87.6	2.50	170.00	7460.	At white h
rium	Ba	137.4	3.75	2.29	36300.	Scarcely a

The family. The alkaline-earth family consists of the very abundant element calcium and the much rarer elements strontium and barium. They are called the alkaline-earth metals because their properties are between those of the alkali metals and the earth metals. The earth metals will be discussed in a later chapter. The family is also frequently called the calcium family.

1. *Occurrence.* These elements do not occur free in nature. Their most abundant compounds are the carbonates and sulphates; calcium also occurs in large quantities as the phosphate and silicate.

2. *Preparation.* The metals were first prepared by Davy in 1808 by electrolysis. This method has again come into use in recent years. Strontium and barium have as yet been obtained only in small quantities and in the impure state, and many of their physical properties, [Pg 301] such as their densities and melting points, are therefore imperfectly known.

3. *Properties.* The three metals resemble each other very closely. They are silvery-white in color and are about as hard as lead. Their densities increase with their atomic weights, as is shown in the table on opposite page. Like the alkali metals they have a strong affinity for oxygen, tarnishing in the air through oxidation. They decompose water at ordinary temperatures, forming hydroxides and liberating hydrogen. When ignited in the air they burn with brilliancy, forming oxides of the general formula MO. These oxides readily combine with water, according to the equation

$$MO + H_2O = M(OH)_2.$$

Each of the elements has a characteristic spectrum, and the presence of the metals can easily be detected by the spectroscope.

4. *Compounds.* The elements are divalent in almost all of their compounds, and these compounds in solution give simple, divalent, colorless ions. The corresponding salts of the three elements are very similar to each other and show a regular variation in properties in passing from calcium to strontium and from strontium to barium. This is seen in the solubility of the sulphate and hydroxide, and in the ease of decomposition of the carbonates, as given in the table. Unlike the alkali metals, their normal carbonates and phosphates are insoluble in water.

CALCIUM

Occurrence. The compounds of calcium are very abundant in nature, so that the total amount of calcium in the earth's crust is very large. A great many different compounds [Pg 302] containing the clement are known, the most important of which are the following:

Calcite (marble)	$CaCO_3$.
Phosphorite	$Ca_3(PO_4)_2$.
Fluorspar	CaF_2.
Wollastonite	$CaSiO_3$.
Gypsum	$CaSO_4 \cdot 2H_2O$.
Anhydrite	$CaSO_4$.

Preparation. Calcium is now prepared by the electrolysis of the melted chloride, the metal depositing in solid condition on the cathode. It is a gray metal, considerably heavier and harder than sodium. It acts upon water, forming calcium hydroxide and hydrogen, but the action does not evolve sufficient heat to melt the metal. It promises to become a useful substance, though no commercial applications for it have as yet been found.

Calcium oxide (*lime, quicklime*) (CaO). Lime is prepared by strongly heating calcium carbonate (limestone) in large furnaces called kilns:

$$CaCO_3 = CaO + CO_2.$$

When pure, lime is a white amorphous substance. Heated intensely, as in the oxyhydrogen flame, it gives a brilliant light called the lime light. Although it is a very difficultly fusible substance, yet in the electric furnace it can be made to melt and even boil. Water acts upon lime with the evolution of a great deal of heat,—hence the name quicklime, or live lime,—the process being called slaking. The equation is

$$CaO + H_2O = Ca(OH)_2.$$

Lime readily absorbs moisture from the air, and is used to dry moist gases, especially ammonia, which cannot be [Pg 303] dried by the usual desiccating agents. It also absorbs carbon dioxide, forming the carbonate

$$CaO + CO_2 = CaCO_3.$$

Lime exposed to air is therefore gradually converted into hydroxide and carbonate, and will no longer slake with water. It is then said to be air-slaked.

Limekilns. The older kiln, still in common use, consists of a large cylindrical stack in which the limestone is loosely packed. A fire is built at the base of the stack, and when the burning is complete it is allowed to die out and the lime is removed from the kiln. The newer kilns are constructed as shown in Fig. 80. A number of fire boxes are built around the lower part of the kiln, one of which is shown at B. The fire is built on the grate F and the hot products of combustion are drawn up through the stack, decomposing the limestone. The kiln is charged at C, and sometimes fuel is added with the limestone to cause combustion throughout the contents of the kiln. The burned lime is raked out through openings in the bottom of the stack, one of which is shown at D. The advantage of this kind of a kiln over the older form is that the process is continuous, limestone being charged in at the top as fast as the lime is removed at the bottom.

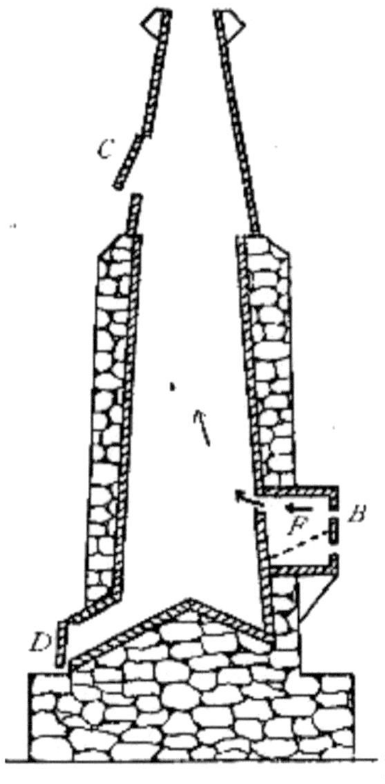

Fig. 80

Calcium hydroxide (*slaked lime*) (Ca(OH)$_2$). Pure calcium hydroxide is a light white powder. It is sparingly soluble in water, forming a solution called *limewater*, which is often used in medicine as a mild alkali. Chemically, calcium hydroxide is a moderately strong base, though not so strong as sodium hydroxide. Owing to its cheapness it is much used in the [Pg 304] industries whenever an alkali is desired. A number of its uses have already been mentioned. It is used in the preparation of ammonia, bleaching powder, and potassium hydroxide. It is also used to remove carbon dioxide and sulphur compounds from coal gas, to remove the hair from hides in the tanneries (this recalls the caustic or corrosive properties of sodium hydroxide), and for making mortar.

Mortar is a mixture of calcium hydroxide and sand. When it is exposed to the air or spread upon porous materials moisture is removed from it partly by absorption in the porous materials and

342

partly by evaporation, and the mortar becomes firm, or *sets*. At the same time carbon dioxide is slowly absorbed from the air, forming hard calcium carbonate:

$$Ca(OH)_2 + CO_2 = CaCO_3 + H_2O.$$

By this combined action the mortar becomes very hard and adheres firmly to the surface upon which it is spread. The sand serves to give body to the mortar and makes it porous, so that the change into carbonate can take place throughout the mass. It also prevents too much shrinkage.

Cement. When limestone to which clay and sand have been added in certain proportions is burned until it is partly fused (some natural marl is already of about the right composition), and the clinker so produced is ground to powder, the product is called cement. When this material is moistened it sets to a hard stone-like mass which retains its hardness even when exposed to the continued action of water. It can be used for under-water work, such as bridge piers, where mortar would quickly soften. Several varieties of cement are made, the best known of which is Portland cement. [Pg 305]

Growing importance of cement. Cement is rapidly coming into use for a great variety of purposes. It is often used in place of mortar in the construction of brick buildings. Mixed with crushed stone and sand it forms concrete which is used in foundation work. It is also used in making artificial stone, terra-cotta trimmings for buildings, artificial stone walks and floors, and the like. It is being used more and more for making many articles which were formerly made of wood or stone, and the entire walls of buildings are sometimes made of cement blocks or of concrete.

Calcium carbonate ($CaCO_3$). This substance is found in a great many natural forms to which various names have been given. They may be classified under three heads:

1. *Amorphous carbonate.* This includes those forms which are not markedly crystalline. Limestone is the most familiar of these and is a grayish rock usually found in hard stratified masses. Whole

mountain ranges are sometimes made up of this material. It is always impure, usually containing magnesium carbonate, clay, silica, iron and aluminium compounds, and frequently fossil remains. Marl is a mixture of limestone and clay. Pearls, chalk, coral, and shells are largely calcium carbonate.

2. *Hexagonal carbonate.* Calcium carbonate crystallizes in the form of rhomb-shaped crystals which belong to the hexagonal system. When very pure and transparent the substance is called Iceland spar. Calcite is a similar form, but somewhat opaque or clouded. Mexican onyx is a massive variety, streaked or banded with colors due to impurities. Marble when pure is made up of minute calcite crystals. Stalactites and stalagmites are icicle-like forms sometimes found in caves.

3. *Rhombic carbonate.* Calcium carbonate sometimes crystallizes in needle-shaped crystals belonging to the rhombic system. This is the unstable form and tends to [Pg 306] go over into the other variety. Aragonite is the most familiar example of this form.

Preparation and uses of calcium carbonate. In the laboratory pure calcium carbonate can be prepared by treating a soluble calcium salt with a soluble carbonate:

$$Na_2CO_3 + CaCl_2 = CaCO_3 + 2NaCl.$$

When prepared in this way it is a soft white powder often called precipitated chalk, and is much used as a polishing powder. It is insoluble in water, but dissolves in water saturated with carbon dioxide, owing to the formation of the acid calcium carbonate which is slightly soluble:

$$CaCO_3 + H_2CO_3 = Ca(HCO_3)_2.$$

The natural varieties of calcium carbonate find many uses, such as in the preparation of lime and carbon dioxide; in metallurgical operations, especially in the blast furnaces; in the manufacture of soda, glass, and crayon (which, in addition to chalk, usually contains clay and calcium sulphate); for building stone and ballast for roads.

Calcium chloride ($CaCl_2$). This salt occurs in considerable quantity in sea water. It is obtained as a by-product in many technical processes, as in the Solvay soda process. When crystallized from its saturated solutions it forms colorless needles of the composition $CaCl_2 \cdot 6H_2O$. By evaporating a solution to dryness and heating to a moderate temperature calcium chloride is obtained anhydrous as a white porous mass. In this condition it absorbs water with great energy and is a valuable drying agent.

Bleaching powder ($CaOCl_2$). When chlorine acts upon a solution of calcium hydroxide the reaction is similar to that which occurs between chlorine and potassium hydroxide:

$$2Ca(OH)_2 + 4Cl = CaCl_2 + Ca(ClO)_2 + 2H_2O.$$

[Pg 307]

If, however, chlorine is conducted over calcium hydroxide in the form of a dry powder, it is absorbed and a substance is formed which appears to have the composition represented in the formula $CaOCl_2$. This substance is called bleaching powder, or hypochlorite of lime. It is probably the calcium salt of both hydrochloric and hypochlorous acids, so that its structure is represented by the formula

    /ClO
 Ca
    \Cl.

In solution this substance acts exactly like a mixture of calcium chloride ($CaCl_2$) and calcium hypochlorite ($Ca(ClO)_2$), since it dissociates to form the ions Ca^{++}, Cl^-, and ClO^-.

Bleaching powder undergoes a number of reactions which make it an important substance.

1. When treated with an acid it evolves chlorine:

$$Ca\begin{cases}ClO\\Cl\end{cases} + H_2SO_4 = CaSO_4 + HCl + HClO,$$

$$HCl + HClO = H_2O + 2Cl.$$

This reaction can be employed in the preparation of chlorine, or the nascent chlorine may be used as a bleaching agent.

2. It is slowly decomposed by the carbon dioxide of the air, yielding calcium carbonate and chlorine:

$$CaOCl_2 + CO_2 = CaCO_3 + 2Cl.$$

Owing to this slow action the substance is a good disinfectant.

3. When its solution is boiled the substance breaks down into calcium chloride and chlorate:

$$6CaOCl_2 = 5CaCl_2 + Ca(ClO_3)_2.$$

This reaction is used in the preparation of potassium chlorate. [Pg 308]

Calcium fluoride (*fluorspar*) (CaF_2). Fluorspar has already been mentioned as the chief natural compound of fluorine. It is found in large quantities in a number of localities, and is often crystallized in perfect cubes of a light green or amethyst color. It can be melted easily in a furnace, and is sometimes used in the fused condition in metallurgical operations to protect a metal from the action of the air during its reduction. It is used as the chief source of fluorine compounds, especially hydrofluoric acid.

Calcium sulphate (*gypsum*) ($CaSO_4 \cdot 2H_2O$). This abundant substance occurs in very perfectly formed crystals or in massive deposits. It is often found in solution in natural waters and in the sea water. Salts deposited from sea water are therefore likely to contain this substance (see Stassfurt salts).

It is very sparingly soluble in water, and is thrown down as a fine white precipitate when any considerable amounts of a calcium salt and a soluble sulphate (or sulphuric acid) are brought together in solution. Its chief use is in the manufacture of plaster of Paris and of hollow tiles for fireproof walls. Such material is called *gypsite*. It is also used as a fertilizer.

Calcium sulphate, like the carbonate, occurs in many forms in nature. Gypsum is a name given to all common varieties. Granular or massive specimens are called alabaster, while all those which are well crystallized are called selenite. Satin spar is still another variety often seen in mineral collections.

Plaster of Paris. When gypsum is heated to about 115° it loses a portion of its water of crystallization in accordance with the equation

$$2(CaSO_4 \cdot 2H_2O) = 2CaSO_4 \cdot H_2O + 2H_2O.$$

[Pg 309]

The product is a fine white powder called *plaster of Paris*. On being moistened it again takes up this water, and in so doing first forms a plastic mass, which soon becomes very firm and hard and regains its crystalline structure. These properties make it very valuable as a material for forming casts and stucco work, for cementing glass to metals, and for other similar purposes. If overheated so that all water is driven off, the process of taking up water is so slow that the material is worthless. Such material is said to be dead burned. Plaster of Paris is very extensively used as the finishing coat for plastered walls.

Hard water. Waters containing compounds of calcium and magnesium in solution are called hard waters because they feel harsh to the touch. The hardness of water may be of two kinds, — (1) temporary hardness and (2) permanent hardness.

1. *Temporary hardness.* We have seen that when water charged with carbon dioxide comes in contact with limestone a certain amount of the latter dissolves, owing to the formation of the soluble acid carbonate of calcium. The hardness of such waters is said to be temporary, since it may be removed by boiling. The heat changes the acid carbonate into the insoluble normal carbonate which then precipitates, rendering the water soft:

$$Ca(HCO_3)_2 = CaCO_3 + H_2O + CO_2.$$

Such waters may also be softened by the addition of sufficient lime or calcium hydroxide to convert the acid carbonate of calcium into the normal carbonate. The equation representing the reaction is

$$Ca(HCO_3)_2 + Ca(OH)_2 = 2CaCO_3 + 2H_2O.$$

[Pg 310]

2. *Permanent hardness.* The hardness of water may also be due to the presence of calcium and magnesium sulphates or chlorides. Boiling the water does not affect these salts; hence such waters are said to have permanent hardness. They may be softened, however, by the addition of sodium carbonate, which precipitates the calcium and magnesium as insoluble carbonates:

$$CaSO_4 + Na_2CO_3 = CaCO_3 + Na_2SO_4.$$

This process is sometimes called "breaking" the water.

Commercial methods for softening water. The average water of a city supply contains not only the acid carbonates of calcium and magnesium but also the sulphates and chlorides of these metals, together with other salts in smaller quantities. Such waters are softened on a commercial scale by the addition of the proper quantities of calcium hydroxide and sodium carbonate. The calcium hydroxide is added first to precipitate all the acid carbonates. After a short time the sodium carbonate is added to precipitate the other soluble salts of calcium and magnesium, together with any excess of calcium hydroxide which may have been added. The quantity of calcium hydroxide and sodium carbonate required is calculated from a chemical analysis of the water. It will be noticed that the water softened in this way will contain sodium sulphate and chloride, but the presence of these salts is not objectionable.

Calcium carbide (CaC_2). This substance is made by heating well-dried coke and lime in an electrical furnace. The equation is

$$CaO + 3C = CaC_2 + CO.$$

The pure carbide is a colorless, transparent, crystalline substance. In contact with water it is decomposed with the evolution of pure acetylene gas, having a pleasant ethereal odor. The commercial article is a dull gray porous substance which contains many impurities. The acetylene prepared from this substance has a very characteristic odor [Pg 311] due to impurities, the chief of these being phosphine. It is used in considerable quantities as a source of acetylene gas for illuminating purposes.

Technical preparation. Fig. 81 represents a recent type of a carbide furnace. The base of the furnace is provided with a large block of carbon A, which serves as one of the electrodes. The other electrodes B, several in number, are arranged horizontally at some distance above this. A mixture of coal and lime is fed into the furnace through the trap top C, and in the lower part of the furnace this mixture becomes intensely heated, forming liquid carbide. This is drawn off through the taphole D.

The carbon monoxide formed in the reaction escapes through the pipes E and is led back into the furnace. The pipes F supply air, so that the monoxide burns as it reënters the furnace and assists in heating the charge. The carbon dioxide so formed, together with the nitrogen entering as air, escape at G. An alternating current is used.

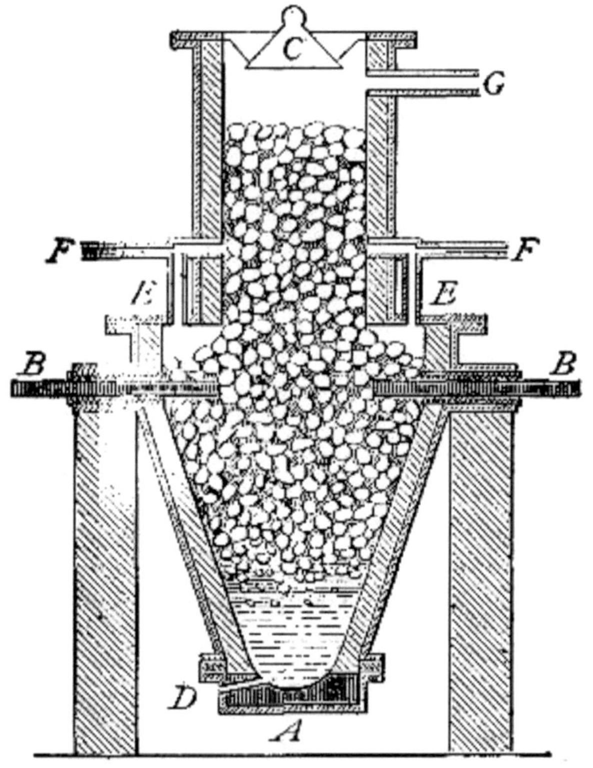

Fig. 81

Calcium phosphate ($Ca_3(PO_4)_2$). This important substance occurs abundantly in nature as a constituent of apatite ($3\ Ca_3(PO_4)_2 \cdot CaF_2$), in phosphate rock, and as the chief mineral constituent of bones. Bone ash is therefore nearly pure calcium phosphate. It is a white powder, insoluble in water, although it readily dissolves in acids, being decomposed by them and converted into soluble acid phosphates, as explained in connection with the acids of phosphorus. [Pg 312]

STRONTIUM

Occurrence. Strontium occurs sparingly in nature, usually as strontianite ($SrCO_3$) and as celestite ($SrSO_4$). Both minerals form beautiful colorless crystals, though celestite is sometimes colored a faint blue. Only a few of the compounds of strontium have any commercial applications.

Strontium hydroxide ($Sr(OH)_2 \cdot 8H_2O$). The method of preparation of strontium hydroxide is analogous to that of calcium hydroxide. The substance has the property of forming an insoluble compound with sugar, which can easily be separated again into its constituents. It is therefore sometimes used in the sugar refineries to extract sugar from impure mother liquors from which the sugar will not crystallize.

Strontium nitrate ($Sr(NO_3)_2 \cdot 4H_2O$). This salt is prepared by treating the native carbonate with nitric acid. When ignited with combustible materials it imparts a brilliant crimson color to the flame, and because of this property it is used in the manufacture of red lights.

BARIUM

Barium is somewhat more abundant than strontium, occurring in nature largely as barytes, or heavy spar ($BaSO_4$), and witherite ($BaCO_3$). Like strontium, it closely resembles calcium both in the properties of the metal and in the compounds which it forms.

Oxides of barium. Barium oxide (BaO) can be obtained by strongly heating the nitrate:

$$Ba(NO_3)_2 = BaO + 2NO_2 + O.$$

Heated to a low red heat in the air, the oxide combines with oxygen, [Pg 313] forming the peroxide (BaO_2). If the temperature is raised still higher, or the pressure is reduced, oxygen is given off and the oxide is once more formed. The reaction

$$BaO_2 \longleftrightarrow BaO + O$$

is reversible and has been used as a means of separating oxygen from the air. Treated with acids, barium peroxide yields hydrogen peroxide:

$$BaO_2 + 2HCl = BaCl_2 + H_2O_2.$$

Barium chloride ($BaCl_2 \cdot 2H_2O$). Barium chloride is a white well-crystallized substance which is easily prepared from the native carbonate. It is largely used in the laboratory as a reagent to detect the presence of sulphuric acid or soluble sulphates.

Barium sulphate *(barytes)* ($BaSO_4$). Barium sulphate occurs in nature in the form of heavy white crystals. It is precipitated as a crystalline powder when a barium salt is added to a solution of a sulphate or sulphuric acid:

$$BaCl_2 + H_2SO_4 = BaSO_4 + 2HCl.$$

This precipitate is used, as are also the finely ground native sulphate and carbonate, as a pigment in paints. On account of its low cost it is sometimes used as an adulterant of white lead, which is also a heavy white substance.

Barium compounds color the flame green, and the nitrate ($Ba(NO_3)_2$) is used in the manufacture of green lights. Soluble barium compounds are poisonous.

RADIUM

Historical. In 1896 the French scientist Becquerel observed that the mineral pitchblende possesses certain remarkable properties. It affects photographic plates even in complete darkness, and discharges [Pg 314] a gold-leaf electroscope when brought close to it. In 1898 Madam Curie made a careful study of pitchblende to see if these properties belong to it or to some unknown substance contained in it. She succeeded in extracting from it a very small quantity of a substance containing a new element which she named radium.

In 1910 Madam Curie succeeded in obtaining radium itself by the electrolysis of radium chloride. It is a silver-white metal melting at about 700°. It blackens in the air, forming a nitride, and decomposes water. Its atomic weight is about 226.5.

Properties. Compounds of radium affect a photographic plate or electroscope even through layers of paper or sheets of metal. They also bring about chemical changes in substances placed near them. Investigation of these strange properties has suggested that the radium atoms are unstable and undergo a decomposition. As a result of this decomposition very minute bodies, to which the name corpuscles has been given, are projected from the radium atom with exceedingly great velocity. It is to these corpuscles that the strange properties of radium are due. It seems probable that the gas helium is in some way formed during the decomposition of radium.

Two or three other elements, particularly uranium and thorium, have been found to possess many of the properties of radium in smaller degree.

Radium and the atomic theory. If these views in regard to radium should prove to be well founded, it will be necessary to modify in some respects the conception of the atom as developed in a former chapter. The atom would have to be regarded as a compound unit made up of several parts. In a few cases, as in radium and uranium, it would appear that this unit is unstable and undergoes transformation into more stable combinations. This modification would not, in any essential way, be at variance with the atomic theory as propounded by Dalton.

EXERCISES

1. What properties have the alkaline-earth metals in common with the alkali metals? In what respects do they differ?

2. Write the equation for the reaction between calcium carbide and water.

3. For what is calcium chlorate used? [Pg 315]

4. Could limestone be completely decomposed if heated in a closed vessel?

5. Caves often occur in limestone. Account for their formation.

6. What is the significance of the term fluorspar? (Consult dictionary.)

7. Could calcium chloride be used in place of barium chloride in testing for sulphates?

8. What weight of water is necessary to slake the lime obtained from 1 ton of pure calcium carbonate?

9. What weight of gypsum is necessary in the preparation of 1 ton of plaster of Paris?

10. Write equations to represent the reactions involved in the preparation of strontium hydroxide and strontium nitrate from strontianite.

11. Write equations to represent the reactions involved in the preparation of barium chloride from heavy spar.

12. Could barium hydroxide be used in place of calcium hydroxide in testing for carbon dioxide?

[Pg 316]

CHAPTER XXV

THE MAGNESIUM FAMILY

	SYMBOL	ATOMIC WEIGHT	DENSITY	MELTING POINT	BOILING POINT	OXIDE
Magnesium	Mg	24.36	1.75	750°	920°	MgO
Zinc	Zn	65.4	7.00	420°	950°	ZnO
Cadmium	Cd	112.4	8.67	320°	778°	CdO

The family. In the magnesium family are included the four elements: magnesium, zinc, cadmium, and mercury. Between the first three of these metals there is a close family resemblance, such as has been traced between the members of the two preceding families. Mercury in some respects is more similar to copper and will be studied in connection with that metal.

1. *Properties.* When heated to a high temperature in the air each of these metals combines with oxygen to form an oxide of the general formula MO, in which M represents the metal. Magnesium decomposes boiling water slowly, while zinc and cadmium have but little action on it.

2. *Compounds.* The members of this group are divalent in nearly all their compounds, so that the formulas of their salts resemble those of the alkaline-earth metals. Like the alkaline-earth metals, their carbonates and phosphates are insoluble in water. Their sulphates, however, are readily soluble. Unlike both the alkali and alkaline-earth [Pg 317] metals, their hydroxides are nearly insoluble in water. Most of their compounds dissociate in such a way as to give a simple, colorless, metallic ion.

MAGNESIUM

Occurrence. Magnesium is a very abundant element in nature, ranking a little below calcium in this respect. Like calcium, it is a constituent of many rocks and also occurs in the form of soluble salts.

Preparation. The metal magnesium, like most metals whose oxides are difficult to reduce with carbon, was formerly prepared by heating the anhydrous chloride with sodium:

$$MgCl_2 + 2Na = 2NaCl + Mg.$$

It is now made by electrolysis, but instead of using as the electrolyte the melted anhydrous chloride, which is difficult to obtain, the natural mineral carnallite is used. This is melted in an iron pot which also serves as the cathode in the electrolysis. A rod of carbon dipping into the melted salt serves as the anode. The apparatus is very similar to the one employed in the preparation of sodium.

Properties. Magnesium is a rather tough silvery-white metal of small density. Air does not act rapidly upon it, but a thin film of oxide forms upon its surface, dimming its bright luster. The common acids dissolve it with the formation of the corresponding salts. It can be ignited readily and in burning liberates much heat and gives a brilliant white light. This light is very rich in the rays which affect photographic plates, and the metal in the form of fine powder is extensively used in the production of flash lights and for white lights in pyrotechnic displays. [Pg 318]

Magnesium oxide (*magnesia*) (MgO). Magnesium oxide, sometimes called magnesia or magnesia usta, resembles lime in many respects. It is much more easily formed than lime and can be made in the same way,—by igniting the carbonate. It is a white powder, very soft and light, and is unchanged by heat even at very high temperatures. For this reason it is used in the manufacture of crucibles, for lining furnaces, and for other purposes where a refractory substance is needed. It combines with water to form magnesium hydroxide, but much more slowly and with the production of much less heat than in the case of calcium oxide.

Magnesium hydroxide ($Mg(OH)_2$). The hydroxide formed in this way is very slightly soluble in water, but enough dissolves to give the water an alkaline reaction. Magnesium hydroxide is therefore a fairly strong base. It is an amorphous white substance. Neither magnesia nor magnesium salts have a very marked effect upon the system; and for this reason magnesia is a very suitable antidote for poisoning by strong acids, since any excess introduced into the system will have no injurious effect.

Magnesium cement. A paste of magnesium hydroxide and water slowly absorbs carbon dioxide from the air and becomes very hard. The hardness of the product is increased by the presence of a considerable amount of magnesium chloride in the paste. The hydroxide, with or without the chloride, is used in the preparation of cements for some purposes.

Magnesium carbonate ($MgCO_3$). Magnesium carbonate is a very abundant mineral. It occurs in a number of localities as magnesite, which is usually amorphous, but sometimes forms pure crystals resembling calcite. More commonly it is found associated with calcium carbonate. [Pg 319] The mineral dolomite has the composition $CaCO_3 \cdot MgCO_3$. Limestone containing smaller amounts of magnesium carbonate is known as dolomitic limestone. Dolomite is one of the most common rocks, forming whole mountain masses. It is harder and less readily attacked by acids than limestone. It is valuable as a building stone and as ballast for roadbeds and foundations. Like calcium carbonate, magnesium carbonate is insoluble in water, though easily dissolved by acids.

Basic carbonate of magnesium. We should expect to find magnesium carbonate precipitated when a soluble magnesium salt and a soluble carbonate are brought together:

$$Na_2CO_3 + MgCl_2 = MgCO_3 + 2NaCl.$$

Instead of this, some carbon dioxide escapes and the product is found to be a basic carbonate. The most common basic carbonate of magnesium has the formula $4MgCO_3 \cdot Mg(OH)_2$, and is sometimes called magnesia alba. This compound is formed by the partial hydrolysis of the normal carbonate at first precipitated:

$$5MgCO_3 + 2H_2O = 4MgCO_3 \cdot Mg(OH)_2 + H_2CO_3.$$

Magnesium chloride ($MgCl_2 \cdot 6H_2O$). Magnesium chloride is found in many natural waters and in many salt deposits (see Stassfurt salts). It is obtained as a by-product in the manufacture of potassium chloride from carnallite. As there is no very important use for it, large quantities annually go to waste. When heated to drive off the water of crystallization the chloride is decomposed as shown in the equation

$$MgCl_2 \cdot 6H_2O = MgO + 2HCl + 5H_2O.$$

[Pg 320]

Owing to the abundance of magnesium chloride, this reaction is being used to some extent in the preparation of both magnesium oxide and hydrochloric acid.

Boiler scale. When water which contains certain salts in solution is evaporated in steam boilers, a hard insoluble material called *scale* deposits in the boiler. The formation of this scale may be due to several distinct causes.

1. *To the deposit of calcium sulphate.* This salt, while sparingly soluble in cold water, is almost completely insoluble in superheated water. Consequently it is precipitated when water containing it is heated in a boiler.

2. *To decomposition of acid carbonates.* As we have seen, calcium and magnesium acid carbonates are decomposed on heating, forming insoluble normal carbonates:

$$Ca(HCO_3)_2 = CaCO_3 + H_2O + CO_2.$$

3. *To hydrolysis of magnesium salts.* Magnesium chloride, and to some extent magnesium sulphate, undergo hydrolysis when superheated in solution, and the magnesium hydroxide, being sparingly soluble, precipitates:

$$MgCl_2 + 2H_2O \longleftrightarrow Mg(OH)_2 + 2HCl.$$

This scale adheres tightly to the boiler in compact layers and, being a non-conductor of heat, causes much waste of fuel. It is very difficult to remove, owing to its hardness and resistance to reagents. Thick scale sometimes cracks, and the water coming in contact with the overheated iron occasions an explosion. Moreover, the acids set free in the hydrolysis of the magnesium salts attack the iron tubes and rapidly corrode them. These causes combine to make the formation of scale a matter which occasions much trouble in cases where hard water is used in steam boilers. Water containing such salts should be softened, therefore, before being used in boilers.

Magnesium sulphate (*Epsom salt*) ($MgSO_4 \cdot 7H_2O$). Like the chloride, magnesium sulphate is found rather commonly in springs and in salt deposits. A very large deposit of the almost pure salt has been found in Wyoming. Its name [Pg 321] was given to it because of its abundant occurrence in the waters of the Epsom springs in England.

Magnesium sulphate has many uses in the industries. It is used to a small extent in the preparation of sodium and potassium sulphates, as a coating for cotton cloth, in the dye industry, in tanning, and in the manufacture of paints and laundry soaps. To some extent it is used in medicine.

Magnesium silicates. Many silicates containing magnesium are known and some of them are important substances. Serpentine, asbestos, talc, and meerschaum are examples of such substances.

ZINC

Occurrence. Zinc never occurs free in nature. Its compounds have been found in many different countries, but it is not a constituent of common rocks and minerals, and its occurrence is rather local and confined to definite deposits or pockets. It occurs chiefly in the following ores:

Sphalerite (zinc blende)	ZnS.
Zincite	ZnO.
Smithsonite	$ZnCO_3$.
Willemite	Zn_2SiO_4.
Franklinite	$ZnO \cdot Fe_2O_3$.

One fourth of the world's output of zinc comes from the United States, Missouri being the largest producer.

Metallurgy. The ores employed in the preparation of zinc are chiefly the sulphide, oxide, and carbonate. They are first roasted in the air, by which process they are changed into oxide:

$$ZnCO_3 = ZnO + CO_2,$$
$$ZnS + 3O = ZnO + SO_2.$$

[Pg 322]

The oxide is then mixed with coal dust, and the mixture is heated in earthenware muffles or retorts, natural gas being used as fuel in many cases. The oxide is reduced by this means to the metallic state, and the zinc, being volatile at the high temperature reached, distills and is collected in suitable receivers. At first the zinc collects in the form of fine powder, called zinc dust or flowers of zinc, recalling the formation under similar conditions of flowers of sulphur. Later, when the whole apparatus has become warm, the zinc condenses to a liquid in the receiver, from which it is drawn off into molds. Commercial zinc often contains a number of impurities, especially carbon, arsenic, and iron.

Physical properties. Pure zinc is a rather heavy bluish-white metal with a high luster. It melts at about 420°, and if heated much above this temperature in the air takes fire and burns with a very bright bluish flame. It boils at about 950° and can therefore be purified by distillation.

Many of the physical properties of zinc are much influenced by the temperature and previous treatment of the metal. When cast into ingots from the liquid state it becomes at ordinary temperatures quite hard, brittle, and highly crystalline. At 150° it is malleable and can be rolled into thin sheets; at higher temperatures it again becomes very brittle. When once rolled into sheets it retains its softness and malleability at ordinary temperatures. When melted and poured into water it forms thin brittle flakes, and in this condition is called granulated or mossy zinc.

Chemical properties. Zinc is tarnished superficially by moist air, but beyond this is not affected by it. It does not decompose even boiling water. When the metal is quite pure, sulphuric and hydrochloric acids have scarcely any action upon it; when, however, it contains small [Pg 323] amounts of other metals such as magnesium or arsenic, or when it is merely in contact with metallic platinum, brisk action takes place and hydrogen is evolved. For this reason, when pure zinc is used in the preparation of hydrogen a few drops of platinum chloride are often added to the solution to assist the chemical action. Nitric acid dissolves the metal readily, with the formation of zinc nitrate and various reduction products of nitric acid. The strong alkalis act upon zinc and liberate hydrogen:

$$Zn + 2KOH = Zn(OK)_2 + 2H.$$

The product of this reaction, potassium zincate, is a salt of zinc hydroxide, which is thus seen to have acid properties, though it usually acts as a base.

Uses of zinc. The metal has many familiar uses. Rolled into sheets, it is used as a lining for vessels which are to contain water. As a thin film upon the surface of iron (galvanized iron) it protects the iron from rust. Iron is usually galvanized by dipping it into a bath of melted zinc, but electrical methods are also employed. Zinc plates are used in many forms of electrical batteries. In the laboratory zinc is used in the preparation of hydrogen, and in the form of zinc dust as a reducing agent.

One of the largest uses of zinc is in the manufacture of alloys. Brass, an alloy of zinc and copper, is the most important of these; German silver, consisting of copper, zinc, and nickel, has many uses; various bronzes, coin metals, and bearing metals also contain zinc. Its ability to alloy with silver finds application in the separation of silver from lead (see silver).

Compounds of zinc. In general, the compounds of zinc are similar in formula and appearance to those of magnesium, [Pg 324] but in other properties they often differ markedly. A number of them have value in commercial ways.

Zinc oxide (*zinc white*) (ZnO). Zinc oxide occurs in impure form in nature, being colored red by manganese and iron compounds. It can be prepared just like magnesium oxide, but is more often made by burning the metal.

Zinc oxide is a pure white powder which becomes yellow on heating and regains its white color when cold. It is much used as a white pigment in paints, under the name of zinc white, and has the advantage over white lead in that it is not changed in color by sulphur compounds, while lead turns black. It is also used in the manufacture of rubber goods.

Commercial preparation of zinc oxide. Commercially it is often made from franklinite in the following way. The franklinite is mixed with coal and heated to a high temperature in a furnace, by which process the zinc is set free and converted into vapor. As the vapor leaves the furnace through a conduit it meets a current of air and takes fire in it, forming zinc oxide. The oxide passes on and is filtered from the air through canvas bags, which allow the air to pass but retain the oxide. It is thus made by burning the metal, though the metal is not actually isolated in the process.

Soluble salts. The soluble salts of zinc can be made by dissolving the metal or the oxide in the appropriate acid. They are all somewhat poisonous. The sulphate and chloride are the most familiar.

Zinc sulphate (*white vitriol*) ($ZnSO_4 \cdot 7H_2O$). This salt is readily crystallized from strong solutions in transparent colorless crystals. It is prepared commercially by careful roasting of the sulphide:

$$ZnS + 4O = ZnSO_4.$$

[Pg 325]

Zinc chloride ($ZnCl_2 \cdot H_2O$). When a solution of zinc chloride is slowly evaporated a salt of the composition $ZnCl_2 \cdot H_2O$ crystallizes out. If the water is completely expelled by heat and the residue distilled, the anhydrous chloride is obtained and may be cast into sticks or broken into lumps. In this distillation, just as in heating magnesium chloride, some of the chloride is decomposed:

$$ZnCl_2 \cdot H_2O = ZnO + 2HCl.$$

The anhydrous chloride has a great affinity for water, and is used as a dehydrating agent. It is also a germicide, and wood which is to be exposed to conditions which favor decay, as, for example, railroad ties, is often soaked in solutions of this salt.

Insoluble compounds. The insoluble compounds of zinc can be prepared by precipitation. The most important are the sulphide, carbonate, and hydroxide.

Zinc sulphide (ZnS). This substance occurs as the mineral sphalerite, and is one of the most valued ores of zinc. Very large deposits occur in southwestern Missouri. The natural mineral is found in large crystals or masses, resembling resin in color and luster. When prepared by precipitation the sulphide is white.

CADMIUM

The element. This element occurs in small quantities in some zinc ores. In the course of the metallurgy of zinc the cadmium compounds undergo chemical changes quite similar to those of the zinc compounds, and the cadmium distills along with the zinc. Being more volatile, it comes over with the first of the zinc and is prepared from the first portions of the distillate by special methods of purification. [Pg 326] The element very closely resembles zinc in most respects. Some of its alloys are characterized by having low melting points.

Compounds of cadmium. Among the compounds of cadmium may be mentioned the chloride ($CdCl_2 \cdot 2H_2O$), the sulphate ($3CdSO_4 \cdot 8H_2O$), and the nitrate ($Cd(NO_3)_2 \cdot 4H_2O$). These are white solids soluble in water. The sulphide (CdS) is a bright yellow substance which is insoluble in water and in dilute acids. It is valuable as a pigment in fine paints.

EXERCISES

1. What properties have the metals of the magnesium family in common with the alkali metals; with the alkaline-earth metals?

2. Compare the action of the metals of the magnesium group on water with that of the other metals studied.

3. What metals already studied are prepared by electrolysis?

4. Write the equations representing the reactions between magnesium and hydrochloric acid; between magnesium and dilute sulphuric acid.

5. What property of magnesium was taken advantage of in the isolation of argon?

6. With phosphoric acid magnesium forms salts similar to those of calcium. Write the names and formulas of the corresponding magnesium salts.

7. How could you distinguish between magnesium chloride and magnesium sulphate? between Glauber's salts and Epsom salts?

8. What weight of carnallite is necessary in the preparation of 500 g. of magnesium?

9. Account for the fact that paints made of zinc oxide are not colored by hydrosulphuric acid.

10. What hydroxide studied, other than zinc hydroxide, has both acid and basic properties?

11. Write equations showing how the following compounds of zinc may be obtained from metallic zinc: the oxide, chloride, nitrate, carbonate, sulphate, sulphide, hydroxide.

[Pg 327]

CHAPTER XXVI

THE ALUMINIUM FAMILY

The family. The element aluminium is the most abundant member of the group of elements known as the aluminium family; indeed, the other members of the family — gallium, indium, and thallium — are of such rare occurrence that they need not be separately described. The elements of the family are ordinarily trivalent, so that the formulas for their compounds differ from those of the elements so far studied. Their hydroxides are practically insoluble in water and are very weak bases; indeed, the bases are so weak that their salts are often hydrolyzed into free base and free acid in solution. The salts formed from these bases usually contain water of crystallization, which cannot be driven off without decomposing them more or less.

The trivalent metals, which in addition to aluminium include also iron and chromium, are sometimes called the *earth metals*. The name refers to the earthy appearance of the oxides of these metals, and to the fact that many earths, soils, and rocks are composed in part of these substances.

ALUMINIUM

Occurrence. Aluminium never occurs in the free state in nature, owing to its great affinity for oxygen. In combined form, as oxides, silicates, and a few other salts, it is both abundant and widely distributed, being an essential [Pg 328] constituent of all soils and of most rocks excepting limestone and sandstone. Cryolite (Na_3AlF_6), found in Greenland, and bauxite, which is an aluminium hydroxide usually mixed with some iron hydroxide, are important minerals. It is estimated that aluminium composes about 8% of the earth's crust. In the industries the metal is called aluminum, but its chemical name is aluminium.

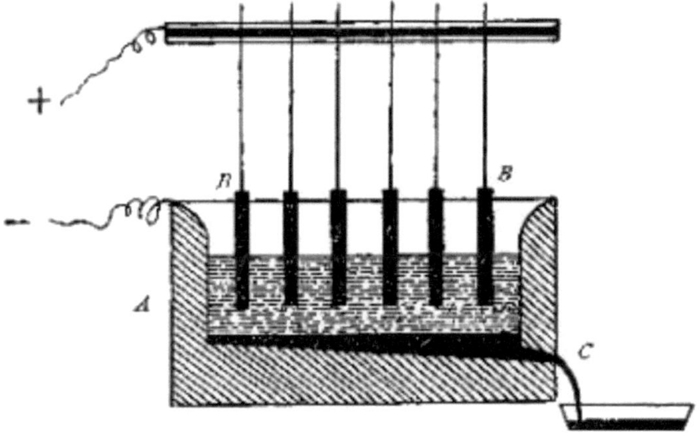

Fig. 82

Preparation. Aluminium was first prepared by Wöhler, in 1827, by heating anhydrous aluminium chloride with potassium:

$$AlCl_3 + 3K = 3KCl + Al.$$

This method was tried after it was found impossible to reduce the oxide of aluminium with carbon. The metal possessed such interesting properties and promised to be so useful that many efforts were made to devise a cheap way of preparing it. The method which has proved most successful consists in the electrolysis of the oxide dissolved in melted cryolite.

[Pg 329]

Metallurgy. An iron box *A* (Fig. 82) about eight feet long and six feet wide is connected with a powerful generator in such a way as to serve as the cathode upon which the aluminium is deposited. Three or four rows of carbon rods *B* dip into the box and serve as the anodes. The box is partially filled with cryolite and the current is turned on, generating enough heat to melt the cryolite. Aluminium oxide is then added, and under the influence of the electric current it decomposes into aluminium and oxygen. The temperature is maintained above the melting point of aluminium, and the liquid metal, being heavier than cryolite, sinks to the bottom of the vessel, from which it is tapped off from time to time through the tap hole *C*. The oxygen in part escapes as gas, and in part combines with the carbon of the anode, the combustion being very brilliant. The process is carried on at Niagara Falls.

The largest expense in the process, apart from the cost of electrical energy, is the preparation of aluminium oxide free from other oxides, for most of the oxide found in nature is too impure to serve without refining. Bauxite is the principal ore used as a source of the aluminium because it is converted into pure oxide without great difficulty. Since common clay is a silicate of aluminium and is everywhere abundant, it might be expected that this would be utilized in the preparation of aluminium. It is, however, very difficult to extract the aluminium from a silicate, and no practical method has been found which will accomplish this.

Physical properties. Aluminium is a tin-white metal which melts at 640° and is very light, having a density of 2.68. It is stiff and strong, and with frequent annealing can be rolled into thin foil. It is a good conductor of heat and electricity, though not so good as copper for a given cross section of wire.

Chemical properties. Aluminium is not perceptibly acted on by boiling water, and moist air merely dims its luster. Further action is prevented in each case by the formation of an extremely thin film of oxide upon the surface of the metal. It combines directly with chlorine, and when heated in oxygen burns with great energy and the liberation of much heat. It is therefore a good reducing agent. Hydrochloric acid acts upon it, forming aluminium chloride: [Pg 330] nitric acid and dilute sulphuric acid have almost no action on it, but hot, concentrated sulphuric acid acts upon it in the same way as upon copper:

$$2Al + 6H_2SO_4 = Al_2(SO_4)_3 + 6H_2O + 3SO_2.$$

Alkalis readily attack the metal, liberating hydrogen, as in the case of zinc:

$$Al + 3KOH = Al(OK)_3 + 3H.$$

Salt solutions, such as sea water, corrode the metal rapidly. It alloys readily with other metals.

Uses of aluminium. These properties suggest many uses for the metal. Its lightness, strength, and permanence make it well adapted for many construction purposes. These same properties have led to its extensive use in the manufacture of cooking utensils. The fact that it is easily corroded by salt solutions is, however, a disadvantage. Owing to its small resistance to electrical currents, it is replacing copper to some extent in electrical construction, especially for trolley and power wires. Some of its alloys have very valuable properties, and a considerable part of the aluminium manufactured is used for this purpose. Aluminium bronze, consisting of about 90% copper and 10% aluminium, has a pure golden color, is strong and malleable, is easily cast, and is permanent in the air. Considerable amounts of aluminium steel are also made.

Goldschmidt reduction process. Aluminium is frequently employed as a powerful reducing agent, many metallic oxides which resist reduction by carbon being readily reduced by it. The aluminium in the form of a fine powder is mixed with the metallic oxide, together with some substance such as fluorspar to act as a flux. The mixture is ignited, and the aluminium unites with the [Pg 331] oxygen of the metallic oxide, liberating the metal. This collects in a fused condition under the flux.

An enormous quantity of heat is liberated in this reaction, and a temperature as high as 3500° can be reached. The heat of the reaction is turned to practical account in welding car rails, steel castings, and in similar operations where an intense local heat is required. A mixture of aluminium with various metallic oxides, ready prepared for such purposes, is sold under the name of *thermite*.

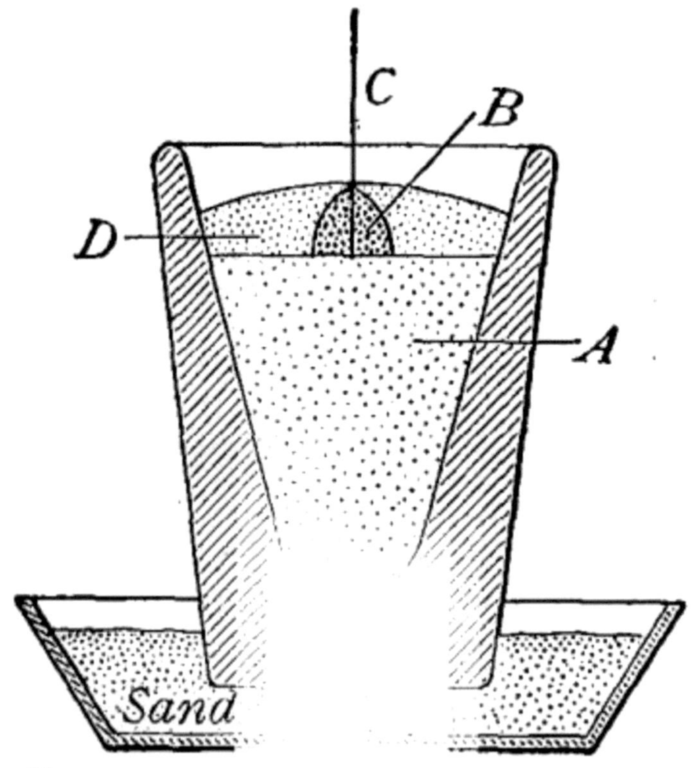

Fig. 83

Preparation of chromium by the Goldschmidt method. A mixture of chromium oxide and aluminium powder is placed in a Hessian crucible (*A*, Fig. 83), and on top of it is placed a small heap *B* of a mixture of sodium peroxide and aluminium, into which is stuck a piece of magnesium ribbon C. Powdered fluorspar D is placed around the sodium peroxide, after which the crucible is set on a pan of sand and the magnesium ribbon ignited. When the flame reaches the sodium peroxide mixture combustion of the aluminium begins with almost explosive violence, so that great care must be taken in the experiment. The heat of this combustion starts the reaction in the chromium oxide mixture, and the oxide is reduced to metallic chromium. When the crucible has cooled a button of chromium will be found in the bottom.

Aluminium oxide (Al_2O_3). This substance occurs in several forms in nature. The relatively pure crystals are called corundum, while emery is a variety colored dark gray or black, usually with iron compounds. In transparent crystals, tinted different colors by traces of impurities, it forms such precious stones as the sapphire, oriental ruby, topaz, and amethyst. All these varieties are very [Pg 332] hard, falling little short of the diamond in this respect. Chemically pure aluminium oxide can be made by igniting the hydroxide, when it forms an amorphous white powder:

$$2Al(OH)_3 = Al_2O_3 + 3H_2O.$$

The natural varieties, corundum and emery, are used for cutting and grinding purposes; the purest forms, together with the artificially prepared oxide, are largely used in the preparation of aluminium.

Aluminium hydroxide ($Al(OH)_3$). The hydroxide occurs in nature as the mineral hydrargyllite, and in a partially dehydrated form called bauxite. It can be prepared by adding ammonium hydroxide to any soluble aluminium salt, forming a semi-transparent precipitate which is insoluble in water but very hard to filter. It dissolves in most acids to form soluble salts, and in the strong bases to form aluminates, as indicated in the equations

$$Al(OH)_3 + 3HCl = AlCl_3 + 3H_2O,$$
$$Al(OH)_3 + 3NaOH = Al(ONa)_3 + 3H_2O.$$

It may act, therefore, either as a weak base or as a weak acid, its action depending upon the character of the substances with which it is in contact. When heated gently the hydroxide loses part of its hydrogen and oxygen according to the equation

$$Al(OH)_3 = AlO \cdot OH + H_2O.$$

This substance, the formula of which is frequently written $HAlO_2$, is a more pronounced acid than is the hydroxide, and its salts are frequently formed when aluminium compounds are fused with alkalis. The magnesium salt $Mg(AlO_2)_2$ is called spinel, and many other of its salts, called aluminates, are found in nature. [Pg 333]

When heated strongly the hydroxide is changed into oxide, which will not again take up water on being moistened.

Mordants and dyeing. Aluminium hydroxide has the peculiar property of combining with many soluble coloring materials and forming insoluble products with them. On this account it is often used as a filter to remove objectionable colors from water. This property also leads to its wide use in the dye industry. Many dyes will not adhere to natural fibers such as cotton and wool, that is, will not "dye fast." If, however, the cloth to be dyed is soaked in a solution of aluminium compounds and then treated with ammonia, the aluminium salts which have soaked into the fiber will be converted into the hydroxide, which, being insoluble, remains in the body of it. If the fiber is now dipped into a solution of the dye, the aluminium hydroxide combines with the color material and fastens, or "fixes," it upon the fiber. A substance which serves this purpose is called a *mordant*, and aluminium salts, particularly the acetate, are used in this way.

Aluminium chloride ($AlCl_3 \cdot 6\,H_2O$). This substance is prepared by dissolving the hydroxide in hydrochloric acid and evaporating to crystallization. When heated it is converted into the oxide, resembling magnesium in this respect:

$$2(AlCl_3 \cdot 6\,H_2O) = Al_2O_3 + 6HCl + 9H_2O.$$

The anhydrous chloride, which has some important uses, is made by heating aluminium turnings in a current of chlorine.

Alums. Aluminium sulphate can be prepared by the action of sulphuric acid upon aluminium hydroxide. It has the property of combining with the sulphates of the alkali metals to form compounds called *alums*. Thus, with potassium sulphate the reaction is expressed by the equation

$$K_2SO_4 + Al_2(SO_4)_3 + 24H_2O = 2(KAl(SO_4)_2 \cdot 12H_2O).$$

[Pg 334]

Under similar conditions ammonium sulphate yields ammonium alum:

$$(NH_4)_2SO_4 + Al_2(SO_4)_3 + 24H_2O = 2(NH_4Al(SO_4)_2 \cdot 12H_2O).$$

Other trivalent sulphates besides aluminium sulphate can form similar compounds with the alkali sulphates, and these compounds are also called alums, though they contain no aluminium. They all crystallize in octahedra and contain twelve molecules of water of crystallization. The alums most frequently prepared are the following:

Potassium alum	$KAl(SO_4)_2 \cdot 12H_2O.$
Ammonium alum	$NH_4Al(SO_4)_2 \cdot 12H_2O.$
Ammonium iron alum	$NH_4Fe(SO_4)_2 \cdot 12H_2O.$
Potassium chrome alum	$KCr(SO_4)_2 \cdot 12H_2O.$

An alum may therefore be regarded as a compound derived from two molecules of sulphuric acid, in which one hydrogen atom has been displaced by the univalent alkali atom, and the other three hydrogen atoms by an atom of one of the trivalent metals, such as aluminium, iron, or chromium.

Very large, well-formed crystals of an alum can be prepared by suspending a small crystal by a thread in a saturated solution of the alum, as shown in Fig. 84. The small crystal slowly grows and assumes a very perfect form.

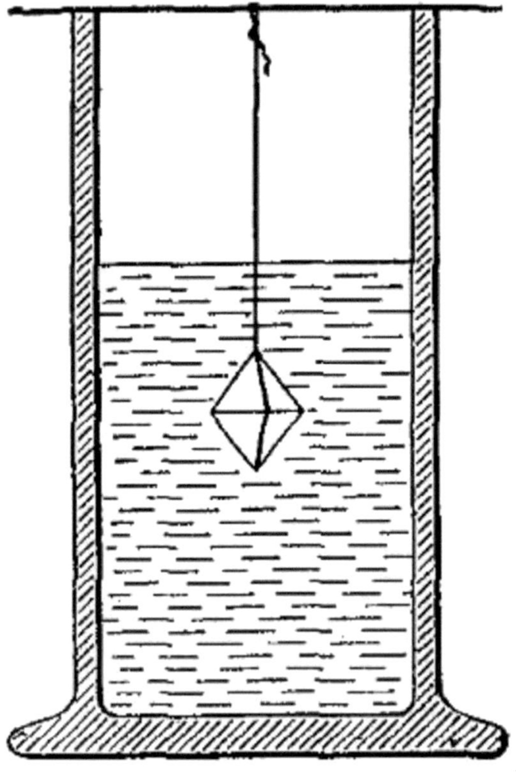

Fig. 84

Other salts of aluminium. While aluminium hydroxide forms fairly stable salts with strong acids, it is such a weak base that its salts with weak acids are readily hydrolyzed. Thus, when an aluminium salt and a soluble carbonate are [Pg 335] brought together in solution we should expect to have aluminium carbonate precipitated according to the equation

$$3Na_2CO_3 + 2AlCl_3 = Al_2(CO_3)_3 + 6NaCl.$$

But if it is formed at all, it instantly begins to hydrolyze, the products of the hydrolysis being aluminium hydroxide and carbonic acid,

$$Al_2(CO_3)_3 + 6H_2O = 2Al(OH)_3 + 3H_2CO_3.$$

Similarly a soluble sulphide, instead of precipitating aluminium sulphide (Al_2S_3), precipitates aluminium hydroxide; for hydrogen sulphide is such a weak acid that the aluminium sulphide at first formed hydrolyzes at once, forming aluminium hydroxide and hydrogen sulphide:

$$3Na_2S + 2AlCl_3 + 6H_2O = 2Al(OH)_3 + 6NaCl + 3H_2S.$$

Alum baking powders. It is because of the hydrolysis of aluminium carbonate that alum is used as a constituent of some baking powders. The alum baking powders consist of a mixture of alum and sodium hydrogen carbonate. When water is added the two compounds react together, forming aluminium carbonate, which hydrolyzes into aluminium hydroxide and carbonic acid. The carbon dioxide from the latter escapes through the dough and in so doing raises it into a porous condition, which is the end sought in the use of a baking powder.

Aluminium silicates. One of the most common constituents of rocks is feldspar ($KAlSi_3O_8$), a mixed salt of potassium and aluminium with the polysilicic acid ($H_4Si_3O_8$). Under the influence of moisture, carbon dioxide, and changes of temperature this substance is constantly being broken down into soluble potassium compounds and hydrated aluminium silicate. This compound has the formula $Al_2Si_2O_7 \cdot 2H_2O$. In relatively pure condition it is called kaolin; in the impure state, mixed with sand and other [Pg 336] substances, it forms common clay. Mica is another very abundant mineral, having varying composition, but being essentially of the formula $KAlSiO_4$. Serpentine, talc, asbestos, and meerschaum are important complex silicates of aluminium and magnesium, and granite is a mechanical mixture of quartz, feldspar, and mica.

Ceramic industries. Many articles of greatest practical importance, ranging from the roughest brick and tile to the finest porcelain and chinaware, are made from some form of kaolin, or clay. No very precise classification of such ware can be made, as the products vary greatly in properties, depending upon the materials used and the treatment during manufacture.

Porcelain is made from the purest kaolin, to which must be added some less pure, plastic kaolin, since the pure substance is not sufficiently plastic. There is also added some more fusible substance, such as feldspar, gypsum, or lime, together with some pure quartz. The constituents must be ground very fine, and when thoroughly mixed and moistened must make a plastic mass which can be molded into any desired form. The article molded from such materials is then burned. In this process the article is slowly heated to a point at which it begins to soften and almost fuse, and then it is allowed to cool slowly. At this stage, a very thin vessel will be translucent and have an almost glassy fracture; if, however, it is somewhat thicker, or has not been heated quite so high, it will still be porous, and partly on this account and partly to improve its appearance it is usually glazed.

Glazing is accomplished by spreading upon the object a thin layer of a more fusible mixture of the same materials as compose the body of the object itself, and again heating until the glaze melts to a transparent glassy coating upon the surface of the vessel. In some cases fusible mixtures of quite different composition from that used in fashioning the vessel may be used as a glaze. Oxides of lead, zinc, and barium are often used in this way.

When less carefully selected materials are used, or quite thick vessels are made, various grades of stoneware are produced. The inferior grades are glazed by throwing a quantity of common salt into the kiln towards the end of the first firing. In the form of vapor [Pg 337] the salt attacks the surface of the baked ware and forms an easily fusible sodium silicate upon it, which constitutes a glaze.

Vitrified bricks, made from clay or ground shale, are burned until the materials begin to fuse superficially, forming their own glaze.

Other forms of brick and tile are not glazed at all, but are left porous. The red color of ordinary brick and earthenware is due to an oxide of iron formed in the burning process.

The decorations upon china are sometimes painted upon the baked ware and then glazed over, and sometimes painted upon the glaze and burned in by a third firing. Care must be taken to use such pigments as are not affected by a high heat and do not react chemically with the constituents of the baked ware or the glaze.

EXERCISES

1. What metals and compounds studied are prepared by electrolysis?

2. Write the equation for the reaction between aluminium and hydrochloric acid; between aluminium and sulphuric acid (in two steps).

3. What hydroxides other than aluminium hydroxide have both acid and basic properties?

4. Write equations showing the methods used for preparing aluminium hydroxide and sulphate.

5. Write the general formula of an alum, representing an atom of an alkali metal by X and an atom of a trivalent metal by Y.

6. What is meant by the term polysilicic acid, as used in the discussion of aluminium silicates?

7. Compare the properties of the hydroxides of the different groups of metals so far studied.

8. In what respects does aluminium oxide differ from calcium oxide in properties?

9. Supposing bauxite to be 90% aluminium hydroxide, what weight of it is necessary for the preparation of 100 kg. of aluminium?

[Pg 338]

CHAPTER XXVII

THE IRON FAMILY

	SYMBOL	ATOMIC WEIGHT	DENSITY	APPROXIMATE MELTING POINT	OXIDES
Iron	Fe	55.9	7.93	1800°	FeO, Fe_2O_3
Cobalt	Co	59.0	8.55	1800°	CoO, Co_2O_3
Nickel	Ni	58.7	8.9	1600°	NiO, Ni_2O_3

The family. The elements iron, cobalt, and nickel form a group in the eighth column of the periodic table. The atomic weights of the three are very close together, and there is not the same gradual gradation in the properties of the three elements that is noticed in the families in which the atomic weights differ considerably in magnitude. The elements are very similar in properties, the similarity being so great in the case of nickel and cobalt that it is difficult to separate them by chemical analysis.

The elements occur in nature chiefly as oxides and sulphides, though they have been found in very small quantities in the native state, usually in meteorites. Their sulphides, carbonates, and phosphates are insoluble in water, the other common salts being soluble. Their salts are usually highly colored, those of iron being yellow or light green as a rule, those of nickel darker green, while cobalt salts are usually rose colored. The metals are obtained by reducing the oxides with carbon. [Pg 339]

IRON

Occurrence. The element iron has long been known, since its ores are very abundant and it is not difficult to prepare the metal from them in fairly pure condition. It occurs in nature in many forms of combination,—in large deposits as oxides, sulphides, and carbonates, and in smaller quantities in a great variety of minerals. Indeed, very few rocks or soils are free from small amounts of iron, and it is assimilated by plants and animals playing an important part in life processes.

Metallurgy. It will be convenient to treat of the metallurgy of iron under two heads, — Materials Used and Process.

Materials used. Four distinct materials are used in the metallurgy of iron:

1. *Iron ore.* The ores most frequently used in the metallurgy of iron are the following:

Hematite	Fe_2O_3.
Magnetite	Fe_3O_4.
Siderite	$FeCO_3$.
Limonite	$2Fe_2O_2 \cdot 3H_2O$.

These ores always contain impurities, such as silica, sulphides, and earthy materials. All ores, with the exception of the oxides, are first roasted to expel any water and carbon dioxide present and to convert any sulphide into oxide.

2. *Carbon.* Carbon in some form is necessary both as a fuel and as a reducing agent. In former times wood charcoal was used to supply the carbon, but now anthracite coal or coke is almost universally used.

3. *Hot air.* To maintain the high temperature required for the reduction of iron a very active combustion of fuel [Pg 340] is necessary. This is secured by forcing a strong blast of hot air into the lower part of the furnace during the reduction process.

4. *Flux.* (a) *Purpose of the flux.* All the materials which enter the furnace must leave it again either in the form of gases or as liquids. The iron is drawn off as the liquid metal after its reduction. To secure the removal of the earthy matter charged into the furnace along with the ore, materials are added to the charge which will, at the high temperature of the furnace, combine with the impurities in the

ore, forming a liquid. The material added for this purpose is called the *flux*; the liquid produced from the flux and the ore is called *slag*.

(*b*) *Function of the slag.* While the main purpose of adding flux to the charge is to remove from the furnace in the form of liquid slag the impurities originally present in the ore, the slag thus produced serves several other functions. It keeps the contents of the furnace in a state of fusion, thus preventing clogging, and makes it possible for the small globules of iron to run together with greater ease into one large liquid mass.

(*c*) *Character of the slag.* The slag is really a kind of readily fusible glass, being essentially a calcium-aluminium silicate. The ore usually contains silica and some aluminium compounds, so that limestone (which also contains some silica and aluminium) is added to furnish the calcium required for the slag. If the ore and the limestone do not contain a sufficient amount of silica and aluminium for the formation of the slag, these ingredients are added in the form of sand and feldspar. In the formation of slag from these materials the ore is freed from the silica and aluminium which it contained.

[Pg 341]

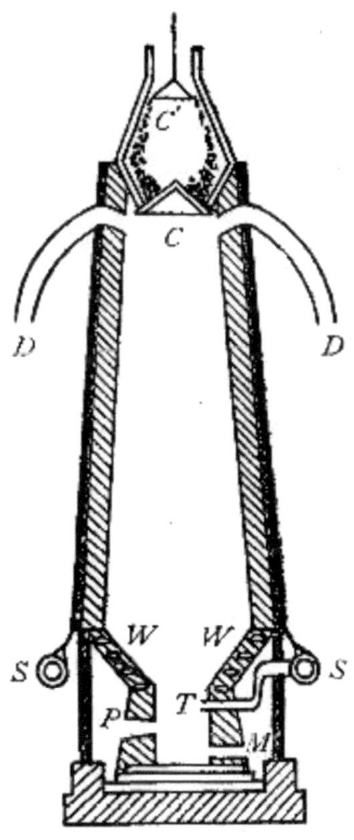

Fig. 85

Process. The reduction of iron is carried out in large towers called blast furnaces. The blast furnace (Fig. 85) is usually about 80 ft. high and 20 ft. in internal diameter at its widest part, narrowing somewhat both toward the top and toward the bottom. The walls are built of steel and lined with fire-brick. The base is provided with a number of pipes T, called tuyers, through which hot air can be forced into the furnace. The tuyers are supplied from a large pipe S, which circles the furnace as a girdle. The base has also an opening M, through which the liquid metal can be drawn off from time to time, and a second opening P, somewhat above the first, through which the excess of slag overflows. The top is closed by a movable trap C and C', called the cone, and through this the materials to be used are introduced. The gases produced by the combustion of the fuel and the reduction of the ore, together with the nitrogen of the air forced in through the tuyers, escape through pipes D, called downcomer pipes, which leave the furnace near the top. These gases are very hot and contain combustible substances, principally carbon monoxide; they are therefore utilized as fuel for the engines and also to heat the blast admitted through the tuyers. The lower part of the furnace is often furnished with a water jacket. This consists of a series of pipes W built into the walls, through which water can be circulated to reduce their temperature. [Pg 342]

Charges consisting of coke (or anthracite coal), ore, and flux in proper proportions are introduced into the furnace at intervals through the trap top. The coke burns fiercely in the hot-air blast, giving an intense heat and forming carbon monoxide. The ore, working down in the furnace as the coke burns, becomes very hot, and by the combined reducing action of the carbon and carbon monoxide is finally reduced to metal and collects as a liquid in the bottom of the furnace, the slag floating on the molten iron. After a considerable amount of the iron has collected the slag is drawn off through the opening P. The molten iron is then drawn off into large ladles and taken to the converters for the manufacture of steel, or it is run out into sand molds, forming the bars or ingots called "pigs." The process is a continuous one, and when once started it is kept in operation for months or even years without interruption.

It seems probable that the first product of combustion of the carbon, at the point where the tuyers enter the furnace, is carbon dioxide. This is at once reduced to carbon monoxide by the intensely heated carbon present, so that no carbon dioxide can be found at that point. For practical purposes, therefore, we may consider that carbon monoxide is the first product of combustion.

Varieties of iron. The iron of commerce is never pure, but contains varying amounts of other elements, such as carbon, silicon, phosphorus, sulphur, and manganese. These elements may either be alloyed with the iron or may be combined with it in the form of definite chemical compounds. In some instances, as in the case of graphite, the mixture may be merely mechanical.

The properties of iron are very much modified by the presence of these elements and by the form of the combination between them and the iron; the way in which the [Pg 343] metal is treated during its preparation has also a marked influence on its properties. Owing to these facts many kinds of iron are recognized in commerce, the chief varieties being cast iron, wrought iron, and steel.

Cast iron. The product of the blast furnace, prepared as just described, is called cast iron. It varies considerably in composition, usually containing from 90 to 95% iron, the remainder being largely carbon and silicon with smaller amounts of phosphorus and sulphur. When the melted metal from the blast furnace is allowed to cool rapidly most of the carbon remains in chemical combination with the iron, and the product is called white cast iron. If the cooling goes on slowly, the carbon partially separates as flakes of graphite which remain scattered through the metal. This product is softer and darker in color and is called gray cast iron.

Properties of cast iron. Cast iron is hard, brittle, and rather easily melted (melting point about 1100°). It cannot be welded or forged into shape, but is easily cast in sand molds. It is strong and rigid but not elastic. It is used for making castings and in the manufacture of other kinds of iron. Cast iron, which contains the metal manganese up to the extent of 20%, together with about 3% carbon, is called spiegel iron; when more than this amount of manganese is present the product is called ferromanganese. The ferromanganese may contain as much as 80% manganese. These varieties of cast iron are much used in the manufacture of steel.

Wrought iron. Wrought iron is made by burning out from cast iron most of the carbon, silicon, phosphorus, and sulphur which it contains. The process is called *puddling*, and is carried out in a furnace constructed as represented [Pg 344] in Fig. 86. The floor of the furnace *F* is somewhat concave and is made of iron covered with a layer of iron oxide. A long flame produced by burning fuel upon the grate *G* is directed downward upon the materials placed upon the floor, and the draught is maintained by the stack *S*. *A* is the ash box and *T* a trap to catch the solid particles carried into the stack by the draught. Upon the floor of the furnace is placed the charge of cast iron, together with a small amount of material to make a slag. The iron is soon melted by the flame directed upon it, and the sulphur, phosphorus, and silicon are oxidized by the iron oxide, forming oxides which are anhydrides of acids. These combine with the flux, which is basic in character, or with the iron oxide, to form a slag. The carbon is also oxidized and escapes as carbon dioxide. As the iron is freed from other elements it becomes pasty, owing to the higher melting point of the purer iron, and in this condition forms small lumps which are raked together into a larger one. The large lump is then removed from the furnace and rolled or hammered into bars, the slag; being squeezed out in this process. The product has a stranded or fibrous structure. *The product of a puddling furnace is called wrought iron.*

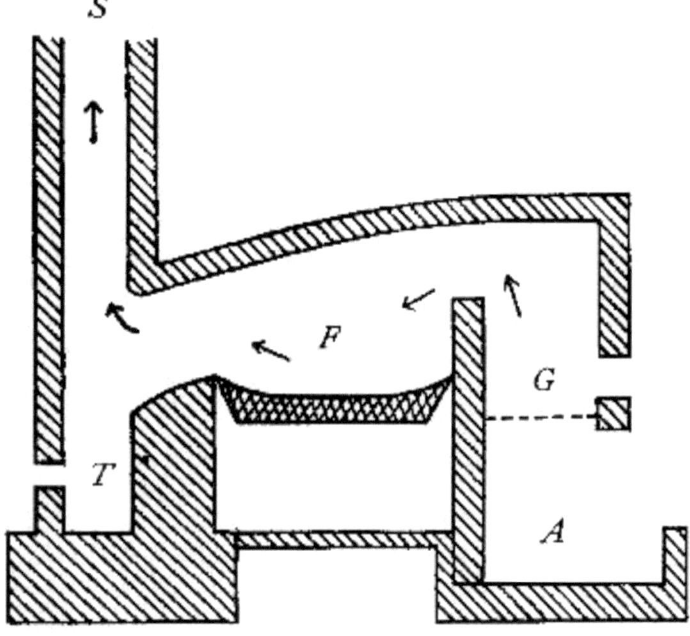

Fig. 86

Properties of wrought iron. Wrought iron is nearly pure iron, usually containing about 0.3% of other substances, chiefly carbon. It is tough, malleable, and fibrous [Pg 345] in structure. It is easily bent and is not elastic, so it will not sustain pressure as well as cast iron. It can be drawn out into wire of great tensile strength, and can also be rolled into thin sheets (sheet iron). It melts at a high temperature (about 1600°) and is therefore forged into shape rather than cast. If melted, it would lose its fibrous structure and be changed into a low carbon steel.

Steel. Steel, like wrought iron, is made by burning out from cast iron a part of the carbon, silicon, phosphorus, and sulphur which it contains; but the process is carried out in a very different way, and usually, though not always, more carbon is found in steel than in wrought iron. A number of processes are in use, but nearly all the steel of commerce is made by one of the two following methods.

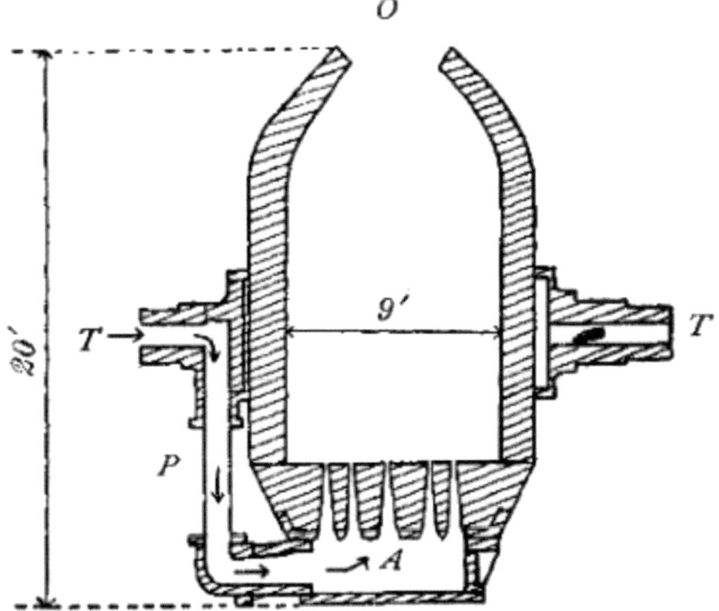

Fig. 87

1. *Bessemer process.* This process, invented about 1860, is by far the most important. It is carried out in great egg-shaped crucibles called converters (Fig. 87), each one of which will hold as much as 15 tons of steel. The converter is built of steel and lined with silica. It is mounted on trunnions *T*, so that it can be tipped over on its side for filling and emptying. One of the trunnions is hollow and a pipe *P* connects it with an air chamber *A*, which forms a false bottom to the

converter. The true bottom is perforated, so that air [Pg 346] can be forced in by an air blast admitted through the trunnion and the air chamber.

White-hot, liquid cast iron from a blast furnace is run into the converter through its open necklike top *O*, the converter being tipped over to receive it; the air blast is then turned on and the converter rotated to a nearly vertical position. The elements in the iron are rapidly oxidized, the silicon first and then the carbon. The heat liberated in the oxidation, largely due to the combustion of silicon, keeps the iron in a molten condition. When the carbon is practically all burned out cast iron or spiegel iron, containing a known percentage of carbon, is added and allowed to mix thoroughly with the fluid. The steel is then run into molds, and the ingots so formed are hammered or rolled into rails or other forms. By this process any desired percentage of carbon can be added to the steel. Low carbon steel, which does not differ much from wrought iron in composition, is now made in this way and is replacing the more expensive wrought iron for many purposes.

The basic lining process. When the cast iron contains phosphorus and sulphur in appreciable quantities, the lining of the converter is made of dolomite. The silicon and carbon burn, followed by the phosphorus and sulphur, and the anhydrides of acids so formed combine with the basic oxides of the lining, forming a slag. This is known as the basic lining process.

2. *Open-hearth process.* In this process a furnace very similar to a puddling furnace is used, but it is lined with silica or dolomite instead of iron oxide. A charge consisting in part of old scrap iron of any kind and in part of cast iron is melted in the furnace by a gas flame. The silicon and carbon are slowly burned away, and when a test shows that the desired percentage of carbon is present the steel [Pg 347] is run out of the furnace. *Steel may therefore be defined as the product of the Bessemer or open-hearth processes.*

Properties of steel. Bessemer and open-hearth steel usually contain only a few tenths of a per cent of carbon, less than 0.1% silicon, and a very much smaller quantity of phosphorus and sulphur. Any considerable amount of the latter elements makes the steel brittle, the sulphur affecting it when hot, and the phosphorus when cold.

This kind of steel is used for structural purposes, for rails, and for nearly all large steel articles. It is hard, malleable, ductile, and melts at a lower temperature than wrought iron. It can be forged into shape, rolled into sheets, or cast in molds.

Relation of the three varieties of iron. It will be seen that wrought iron is usually very nearly pure iron, while steel contains an appreciable amount of alloy material, chiefly carbon, and cast iron still more of the same substances. It is impossible, however, to assign a given sample of iron to one of these three classes on the basis of its chemical composition alone. A low carbon steel, for example, may contain less carbon than a given sample of wrought iron. The real distinction between the three is the process by which they are made. The product of the blast furnace is cast iron; that of the puddling furnace is wrought iron; that of the Bessemer and open-hearth methods is steel.

Tool steel. Steel designed for use in the manufacture of edged tools and similar articles should be relatively free from silicon and phosphorus, but should contain from 0.5 to 1.5% carbon. The percentage of carbon should be regulated by the exact use to which the steel is to be put. Steel of this character is usually made in small lots from either Bessemer or open-hearth steel in the following way. [Pg 348]

A charge of melted steel is placed in a large crucible and the calculated quantity of pure carbon is added. The carbon dissolves in the steel, and when the solution is complete the metal is poured out of the crucible. This is sometimes called crucible steel.

Tempering of steel. Steel containing from 0.5 to 1.5% carbon is characterized by the property of "taking temper." When the hot steel is suddenly cooled by plunging it into water or oil it becomes very hard and brittle. On carefully reheating this hard form it gradually becomes less brittle and softer, so that by regulating the temperature to which steel is reheated in tempering almost any condition of temper demanded for a given purpose, such as for making springs or cutting tools, can be obtained.

Steel alloys. It has been found that small quantities of a number of different elements when alloyed with steel very much improve its quality for certain purposes, each element having a somewhat different effect. Among the elements most used in this connection are manganese, silicon, chromium, nickel, tungsten, and molybdenum.

The usual method for adding these elements to the steel is to first prepare a very rich alloy of iron with the element to be added, and then add enough of this alloy to a large quantity of the steel to bring it to the desired composition. A rich alloy of iron with manganese or silicon can be prepared directly in a blast furnace, and is called ferromanganese or ferrosilicon. Similar alloys of iron with the other elements mentioned are made in an electric furnace by reducing the mixed oxides with carbon.

Pure iron. Perfectly pure iron is rarely prepared and is not adapted to commercial uses. It can be made by reducing pure oxide of iron in a current of hydrogen at a [Pg 349] high temperature. Prepared in this way it forms a black powder; when melted it forms a tin-white metal which is less fusible and more malleable than wrought iron. It is easily acted upon by moist air.

Compounds of iron. Iron differs from the metals so far studied in that it is able to form two series of compounds in which the iron has two different valences. In the one series the iron is divalent and forms compounds which in formulas and many chemical properties are similar to the corresponding zinc compounds. It can also act as a trivalent metal, and in this condition forms salts similar to those of aluminium. Those compounds in which the iron is divalent are known as *ferrous* compounds, while those in which it is trivalent are known as *ferric*.

Oxides of iron. Iron forms several oxides. Ferrous oxide (FeO) is not found in nature, but can be prepared artificially in the form of a black powder which easily takes up oxygen, forming ferric oxide:

$$2FeO + O = Fe_2O_3.$$

Ferric oxide is the most abundant ore of iron and occurs in great deposits, especially in the Lake Superior region. It is found in many mineral varieties which vary in density and color, the most abundant being hematite, which ranges in color from red to nearly black. When prepared by chemical processes it forms a red powder which is used as a paint pigment (Venetian red) and as a polishing powder (rouge).

Magnetite has the formula Fe_3O_4 and is a combination of FeO and Fe_2O_3. It is a very valuable ore, but is less abundant than hematite. It is sometimes called magnetic oxide of iron, or lodestone, since it is a natural magnet. [Pg 350]

Ferrous salts. These salts are obtained by dissolving iron in the appropriate acid, or, when insoluble, by precipitation. They are usually light green in color and crystallize well. In chemical reactions they are quite similar to the salts of magnesium and zinc, but differ from them in one important respect, namely, that they are easily changed into compounds in which the metal is trivalent. Thus ferrous chloride treated with chlorine or aqua regia is changed into ferric chloride:

$$FeCl_2 + Cl = FeCl_3.$$

Ferrous hydroxide exposed to moist air is rapidly changed into ferric hydroxide:

$$2Fe(OH)_2 + H_2O + O = 2Fe(OH)_3.$$

Ferrous sulphate *(copperas, green vitriol)* $(FeSO_4 \cdot 7H_2O)$. Ferrous sulphate is the most familiar ferrous compound. It is prepared commercially as a by-product in the steel-plate mills. Steel plates are cleaned by the action of dilute sulphuric acid upon them, and in the process some of the iron dissolves. The liquors are concentrated and the green vitriol separates from them.

Ferrous sulphide (FeS). Ferrous sulphide is sometimes found in nature as a golden-yellow crystalline mineral. It is formed as a black precipitate when a soluble sulphide and an iron salt are brought together in solution:

$$FeSO_4 + Na_2S = FeS + Na_2SO_4.$$

It can also be made as a heavy dark-brown solid by fusing together the requisite quantities of sulphur and iron. It is obtained as a by-product in the metallurgy of lead:

$$PbS + Fe = FeS + Pb.$$

[Pg 351]

It is used in the laboratory in the preparation of hydrosulphuric acid:

$$FeS + 2HCl = FeCl_2 + H_2S.$$

Iron disulphide *(pyrites)* (FeS_2). This substance bears the same relation to ferrous sulphide that hydrogen dioxide does to water. It occurs abundantly in nature in the form of brass-yellow cubical crystals and in compact masses. Sometimes the name "fool's gold" is applied to it from its superficial resemblance to the precious metal. It is used in very large quantities as a source of sulphur dioxide in the manufacture of sulphuric acid, since it burns readily in the air, forming ferric oxide and sulphur dioxide:

$$2FeS_2 + 11O = Fe_2O_3 + 4SO_2.$$

Ferrous carbonate ($FeCO_3$). This compound occurs in nature as siderite, and is a valuable ore. It will dissolve to some extent in water containing carbon dioxide, just as will calcium carbonate, and waters containing it are called chalybeate waters. These chalybeate waters are supposed to possess certain medicinal virtues and form an important class of mineral waters.

Ferric salts. Ferric salts are usually obtained by treating an acidified solution of a ferrous salt with an oxidizing agent:

$$2FeCl_2 + 2HCl + O = 2FeCl_3 + H_2O,$$

$$2FeSO_4 + H_2SO_4 + O = Fe_2(SO_4)_3 + H_2O.$$

They are usually yellow or violet in color, are quite soluble, and as a rule do not crystallize well. Heated with water in the absence of free acid, they hydrolyze even more readily than the salts of aluminium. The most familiar ferric salts are the chloride and the sulphate. [Pg 352]

Ferric chloride ($FeCl_3$). This salt can be obtained most conveniently by dissolving iron in hydrochloric acid and then passing chlorine into the solution:

$$Fe + 2HCl = FeCl_2 + 2H,$$

$$FeCl_2 + Cl = FeCl_3.$$

When the pure salt is heated with water it is partly hydrolyzed:

$$FeCl_3 + 3\,H_2O \longleftrightarrow Fe(OH)_3 + 3HCl.$$

This is a reversible reaction, however, and hydrolysis can therefore be prevented by first adding a considerable amount of the soluble product of the reaction, namely, hydrochloric acid.

Ferric sulphate ($Fe_2(SO_4)_3$). This compound can be made by treating an acid solution of green vitriol with an oxidizing agent. It is difficult to crystallize and hard to obtain in pure condition. When an alkali sulphate in proper quantity is added to ferric sulphate in solution an iron alum is formed, and is easily obtained in large crystals. The best known iron alums have the formulas $KFe(SO_4)_2 \cdot 12H_2O$ and $NH_4Fe(SO_4)_2 \cdot 12H_2O$. They are commonly used when a pure ferric salt is required.

Ferric hydroxide ($Fe(OH)_3$). When solutions of ferric salts are treated with ammonium hydroxide, ferric hydroxide is formed as a rusty-red precipitate, insoluble in water.

Iron cyanides. A large number of complex cyanides containing iron are known, the most important being potassium ferrocyanide, or yellow prussiate of potash ($K_4FeC_6N_6$), and potassium ferricyanide, or red prussiate of potash ($K_3FeC_6N_6$). These compounds are the potassium salts of the complex acids of the formulas $H_4FeC_6N_6$ [Pg 353] and $H_3FeC_6N_6$.

Oxidation of ferrous salts. It has just been seen that when a ferrous salt is treated with an oxidizing agent in the presence of a free acid a ferric salt is formed:

$$2FeSO_4 + H_2SO_4 + O = Fe_2(SO_4)_3 + H_2O.$$

In this reaction oxygen is used up, and the valence of the iron is changed from 2 to 3. The same equation may be written

$$2Fe^{++}, 2SO_4{}^- + 2H^+, SO_4{}^- + O = 2Fe^{+++}, 3SO_4{}^- + H_2O.$$

Hydrogen ions have been oxidized to water, while the charge of each iron ion has been increased from 2 to 3.

In a similar way the conversion of ferrous chloride into ferric chloride may be written

$$Fe^{++}, 2Cl^- + Cl = Fe^{+++}, + 3Cl^-.$$

Here again the valence of the iron and the charge on the iron ion has been increased from 2 to 3, though no oxygen has entered into the reaction. As a rule, however, changes of this kind are brought about by the use of an oxidizing agent, and are called oxidations.

The term "oxidation" is applied to all reactions in which the valence of the metal of a compound is increased, or, in other words, to all reactions in which the charge of a cation is increased.

Reduction of ferric salts. The changes which take place when a ferric salt is converted into a ferrous salt are the reverse of the ones just described. This is seen in the equation

$$FeCl_3 + H = FeCl_2 + HCl$$

In this reaction the valence of the iron has been changed from 3 to 2. The same equation may be written

$$Fe^{+++}, 3Cl^- + H = Fe^{++}, + H^+ + 3Cl$$

[Pg 354]

It will be seen that the charge of the iron ions has been diminished from 3 to 2. Since these changes are the reverse of the oxidation changes just considered, they are called reduction reactions. The term "reduction" is applied to all processes in which the valence of the metal of a compound is diminished, or, in other words, to all processes in which the charge on the cations is diminished.

NICKEL AND COBALT

These elements occur sparingly in nature, usually combined with arsenic or with arsenic and sulphur. Both elements have been found in the free state in meteorites. Like iron they form two series of compounds, but the salts corresponding to the ferrous salts are the most common, the ones corresponding to the ferric salts being difficult to obtain. Thus we have the chlorides $NiCl_2 \cdot 6H_2O$ and $CoCl_2 \cdot 6H_2O$; the sulphates $NiSO_4 \cdot 7H_2O$ and $CoSO_4 \cdot 7H_2O$; the nitrates $Ni(NO_3)_2 \cdot 6H_2O$ and $Co(NO_3)_2 \cdot 6H_2O$.

Nickel is largely used as an alloy with other metals. Alloyed with copper it forms coin metal from which five-cent pieces are made, with copper and zinc it forms German silver, and when added to steel in small quantities nickel steel is formed which is much superior to common steel for certain purposes. When deposited by electrolysis upon the surface of other metals such as iron, it forms a covering which will take a high polish and protects the metal from rust, nickel not being acted upon by moist air. Salts of nickel are usually green.

Compounds of cobalt fused with glass give it an intensely blue color. In powdered form such glass is sometimes used [Pg 355] as a pigment called smalt. Cobalt salts, which contain water of crystallization, are usually cherry red in color; when dehydrated they become blue.

EXERCISES

1. In the manufacture of cast iron, why is the air heated before being forced into the furnace?

2. Write the equations showing how each of the following compounds of iron could be obtained from the metal itself: ferrous chloride, ferrous hydroxide, ferrous sulphate, ferrous sulphide, ferrous carbonate, ferric chloride, ferric sulphate, ferric hydroxide.

3. Account for the fact that a solution of sodium carbonate, when added to a solution of a ferric salt, precipitates an hydroxide and not a carbonate.

4. Calculate the percentage of iron in each of the common iron ores.

5. One ton of steel prepared by the Bessemer process is found by analysis to contain 0.2% carbon. What is the minimum weight of carbon which must be added in order that the steel may be made to take a temper?

[Pg 356]

CHAPTER XXVIII

COPPER, MERCURY, AND SILVER

| | | | | | FORMULAS OF OXIDES | |
	SYMBOL	ATOMIC WEIGHT	DENSITY	MELTING POINT	"ous"	"ic"

Copper	Cu	63.6	8.89	1084°	Cu_2O	CuO
Mercury	Hg	200.00	13.596	-39.5°	Hg_2O	HgO
Silver	Ag	107.93	10.5	960°	Ag_2O	AgO

The family. By referring to the periodic arrangement of the elements (page 168), it will be seen that mercury is not included in the same family with copper and silver. Since the metallurgy of the three elements is so similar, however, and since they resemble each other so closely in chemical properties, it is convenient to class them together for study.

1. *Occurrence.* The three elements occur in nature to some extent in the free state, but are usually found as sulphides. Their ores are easy to reduce.

2. *Properties.* They are heavy metals of high luster and are especially good conductors of heat and electricity. They are not very active chemically. Neither hydrochloric nor dilute sulphuric acid has any appreciable action upon them. Concentrated sulphuric acid attacks all three, forming metallic sulphates and evolving sulphur dioxide, while nitric acid, both dilute and concentrated, converts them into nitrates with the evolution of oxides of nitrogen. [Pg 357]

3. *Two series of salts.* Copper and mercury form oxides of the types M_2O and MO, as well as two series of salts. In one series the metals are univalent and the salts have formulas like those of the sodium salts. They are called cuprous and mercurous salts. In the other series the metals are divalent and resemble magnesium salts in formulas. These are called cupric and mercuric salts. Silver forms only one series of salts, being always a univalent metal.

COPPER

Occurrence. The element copper has been used for various purposes since the earliest days of history. It is often found in the metallic state in nature, large masses of it occurring pure in the Lake Superior region and in other places to a smaller extent. The most valuable ores are the following:

Cuprite Cu_2O.

Chalcocite	Cu_2S.
Chalcopyrite	$CuFeS_2$.
Bornite	Cu_3FeS_3.
Malachite	$CuCO_3 \cdot Cu(OH)_2$.
Azurite	$2CuCO_3 \cdot Cu(OH)_2$.

Metallurgy of copper. Ores containing little or no sulphur are easy to reduce. They are first crushed and the earthy impurities washed away. The concentrated ore is then mixed with carbon and heated in a furnace, metallic copper resulting from the reduction of the copper oxide by the hot carbon.

Metallurgy of sulphide ores. Much of the copper of commerce is made from chalcopyrite and bornite, and these ores are more difficult to work. They are first roasted in the air, by which treatment much of the sulphur is burned to sulphur dioxide. The roasted ore is then [Pg 358] melted in a small blast furnace or in an open one like a puddling furnace. In melting, part of the iron combines with silica to form a slag of iron silicate. The product, called crude matte, contains about 50% copper together with sulphur and iron. Further purification is commonly carried on by a process very similar to the Bessemer process for steel. The converter is lined with silica, and a charge of matte from the melting furnace, together with sand, is introduced, and air is blown into the mass. By this means the sulphur is practically all burned out by the air, and the remaining iron combines with silica and goes off as slag. The copper is poured out of the converter and molded into anode plates for refining.

Refining of copper. Impure copper is purified by electrolysis. A large plate of it, serving as an anode, is suspended in a tank facing a thin plate of pure copper, which is the cathode. The tank is filled with a solution of copper sulphate and sulphuric acid to serve as the electrolyte. A current from a dynamo passes from the anode to the cathode, and the copper, dissolving from the anode, is deposited upon the cathode in pure form, while the impurities collect on the bottom of the tank. Electrolytic copper is one of the purest of commercial metals and is very nearly pure copper.

Recovery of gold and silver. Gold and silver are often present in small quantities in copper ores, and in electrolytic refining these metals collect in the muddy deposit on the bottom of the tank. The mud is carefully worked over from time to time and the precious metals extracted from it. A surprising amount of gold and silver is obtained in this way.

Properties of copper. Copper is a rather heavy metal of density 8.9, and has a characteristic reddish color. It is rather soft and is very malleable, ductile, and flexible, yet tough and strong; it melts at 1084°. As a conductor of heat and electrical energy it is second only to silver. [Pg 359]

Hydrochloric acid, dilute sulphuric acid, and fused alkalis are almost without action upon it; nitric acid and hot, concentrated sulphuric acid, however, readily dissolve it. In moist air it slowly becomes covered with a thin layer of green basic carbonate; heated in the air it is easily oxidized to black copper oxide (CuO).

Uses. Copper is extensively used for electrical purposes, for roofs and cornices, for sheathing the bottom of ships, and for making alloys. In the following table the composition of some of these alloys is indicated:

COMPOSITION OF ALLOYS OF COPPER IN PERCENTAGES

Aluminium bronze	copper (90 to 97%), aluminium (3 to 10%).
Brass	copper (63 to 73%), zinc (27 to 37%).
Bronze	copper (70 to 95%), zinc (1 to 25%), tin (1 to 18%).
German silver	copper (56 to 60%), zinc (20%), nickel (20 to 25%).
Gold coin	copper (10%), gold (90%).
Gun metal	copper (90%), tin (10%).
Nickel coin	copper (75%), nickel (25%)
Silver coin	copper (10%), silver (90%).

Electrotyping. Matter is often printed from electrotype plates which are prepared as follows. The matter is set up in type and wax is firmly pressed down upon the face of it until a clear impression is obtained. The impressed side of the wax is coated with graphite and the impression is made the cathode in an electrolytic cell containing a copper salt in solution. When connected with a current the copper is deposited as a thin sheet upon the letters in wax, and when detached is a perfect copy of the type, the under part of the letters being hollow. The sheet is strengthened by pouring on the under surface a suitable amount of molten metal (commercial lead is used). The sheet so strengthened is then used in printing.

Two series of copper compounds. Copper, like iron, forms two series of compounds: in the cuprous compounds it is univalent; in the cupric it is divalent. The cupric salts [Pg 360] are much the more common of the two, since the cuprous salts pass readily into cupric by oxidation.

Cuprous compounds. The most important cuprous compound is the oxide (Cu_2O), which occurs in nature as ruby copper or cuprite. It is a bright red substance and can easily be prepared by heating copper to a high temperature in a limited supply of air. It is used for imparting a ruby color to glass.

By treating cuprous oxide with different acids a number of cuprous salts can be made. Many of these are insoluble in water, the chloride (CuCl) being the best known. When suspended in dilute hydrochloric acid it is changed into cupric chloride, the oxygen taking part in the reaction being absorbed from the air:

$$2CuCl + 2HCl + O = 2CuCl_2 + H_2O.$$

Cupric compounds. Cupric salts are easily made by dissolving cupric oxide in acids, or, when insoluble, by precipitation. Most of them are blue or green in color, and the soluble ones crystallize well. Since they are so much more familiar than the cuprous salts, they are frequently called merely copper salts.

Cupric oxide (CuO). This is a black insoluble substance obtained by heating copper in excess of air, or by igniting the hydroxide or nitrate. It is used as an oxidizing agent.

Cupric hydroxide (Cu(OH)$_2$). The hydroxide prepared by treating a solution of a copper salt with sodium hydroxide is a light blue insoluble substance which easily loses water and changes into the oxide. Heat applied to the liquid containing the hydroxide suspended in it serves to bring about the reaction represented by the equation

$$Cu(OH)_2 = CuO + H_2O.$$

[Pg 361]

Cupric sulphate (*blue vitriol*) (CuSO$_4$·5H$_2$O). This substance, called blue vitriol or bluestone, is obtained as a by-product in a number of processes and is produced in very large quantities. It forms large blue crystals, which lose water when heated and crumble to a white powder. The salt finds many uses, especially in electrotyping and in making electrical batteries.

Cupric sulphide (CuS). The insoluble black sulphide (CuS) is easily prepared by the action of hydrosulphuric acid upon a solution of a copper salt:

$$CuSO_4 + H_2S = CuS + H_2SO_4.$$

It is insoluble in water and dilute acids.

MERCURY

Occurrence. Mercury occurs in nature chiefly as the sulphide (HgS) called cinnabar, and in globules of metal inclosed in the cinnabar. The mercury mines of Spain have long been famous, California being the next largest producer.

Metallurgy. Mercury is a volatile metal which has but little affinity for oxygen. Sulphur, on the other hand, readily combines with oxygen. These facts make the metallurgy of mercury very simple. The crushed ore, mixed with a small amount of carbon to reduce any oxide or sulphate that might be formed, is roasted in a current of air. The sulphur burns to sulphur dioxide, while the mercury is converted into vapor and is condensed in a series of condensing vessels. The metal is purified by distillation.

Properties. Mercury is a heavy silvery liquid with a density of 13.596. It boils at 357° and solidifies at -39.5°. [Pg 362] Small quantities of many metals dissolve in it, forming liquid alloys, while with larger quantities it forms solid alloys. The alloys of mercury are called amalgams.

Toward acids mercury conducts itself very much like copper; it is easily attacked by nitric and hot, concentrated sulphuric acids, while cold sulphuric and hydrochloric acids have no effect on it.

Uses. Mercury is extensively used in the construction of scientific instruments, such as the thermometer and barometer, and as a liquid over which to collect gases which are soluble in water. The readiness with which it alloys with silver and gold makes it very useful in the extraction of these elements.

Compounds of mercury. Like copper, mercury forms two series of compounds: the mercurous, of which mercurous chloride ($HgCl$) is an example; and the mercuric, represented by mercuric chloride ($HgCl_2$).

Mercuric oxide (HgO). Mercuric oxide can be obtained either as a brick-red or as a yellow substance. When mercuric nitrate is heated carefully the red modification is formed in accordance with the equation

$$Hg(NO_3)_2 = HgO + 2NO_2 + O.$$

The yellow modification is prepared by adding a solution of a mercuric salt to a solution of sodium or potassium hydroxide:

$$Hg(NO_3)_2 + 2NaOH = 2NaNO_3 + Hg(OH)_2,$$

$$Hg(OH)_2 = HgO + H_2O.$$

When heated the oxide darkens until it becomes almost black; at a higher temperature it decomposes into mercury and oxygen. It was by this reaction that oxygen was discovered. [Pg 363]

Mercurous chloride (*calomel*) (HgCl). Being insoluble, mercurous chloride is precipitated as a white solid when a soluble chloride is added to a solution of mercurous nitrate:

$$HgNO_3 + NaCl = HgCl + NaNO_3.$$

Commercially it is manufactured by heating a mixture of mercuric chloride and mercury. When exposed to the light it slowly changes into mercuric chloride and mercury:

$$2HgCl = HgCl_2 + Hg.$$

It is therefore protected from the light by the use of colored bottles. It is used in medicine.

Most mercurous salts are insoluble in water, the principal soluble one being the nitrate, which is made by the action of cold, dilute nitric acid on mercury.

Mercuric chloride (*corrosive sublimate*) (HgCl_2). This substance can be made by dissolving mercuric oxide in hydrochloric acid. On a commercial scale it is made by subliming a mixture of common salt and mercuric sulphate:

$$2NaCl + HgSO_4 = HgCl_2 + Na_2SO_4.$$

The mercuric chloride, being readily volatile, vaporizes and is condensed again in cool vessels. Like mercurous chloride it is a white solid, but differs from it in that it is soluble in water. It is extremely poisonous and in dilute solutions is used as an antiseptic in dressing wounds.

Mercuric sulphide (HgS). As cinnabar this substance forms the chief native compound of mercury, occurring in red crystalline masses. By passing hydrosulphuric acid into a solution of a mercuric salt it is precipitated as a black powder, insoluble in water and acids. By other means it can be prepared as a brilliant red powder known as vermilion, which is used as a pigment in fine paints. [Pg 364]

The iodides of mercury. If a solution of potassium iodide is added to solutions of a mercurous and a mercuric salt respectively, the corresponding iodides are precipitated. Mercuric iodide is the more important of the two, and as prepared above is a red powder which changes to yellow on heating to 150°. The yellow form on cooling changes back again to the red form, or may be made to do so by rubbing it with a knife blade or some other hard object.

SILVER

Occurrence. Silver is found in small quantities in the uncombined state; usually, however, it occurs in combination with sulphur, either as the sulphide (Ag_2S) or as a small constituent of other sulphides, especially those of lead and copper. It is also found alloyed with gold.

Metallurgy. *Parkes's process.* Silver is usually smelted in connection with lead. The ores are worked over together, as described under lead, and the lead and silver obtained as an alloy, the silver being present in small quantity. The alloy is melted and metallic zinc is stirred in. Zinc will alloy with silver but not with lead, and it is found that the silver leaves the lead and, in the form of an alloy with zinc, forms as a crust upon the lead and is skimmed off. This crust, which, of course, contains lead adhering to it, is partially melted and the most of the lead drained off. The zinc is removed by distillation, and the residue is melted on an open hearth in a current of air; by this means the zinc and lead remaining with the silver are changed into oxides and the silver remains behind unaltered.

Amalgamation process. In some localities the old amalgamation process is used. The silver ore is treated with common salt and ferrous compounds, which process converts the silver first into chloride and then into metallic silver. Mercury is then added and thoroughly mixed with the mass, forming an amalgam with the silver. After [Pg 365] some days the earthy materials are washed away and the heavier amalgam is recovered. The mercury is distilled off and the silver left in impure form.

Refining silver. The silver obtained by either of the above processes may still contain copper, gold, and iron, and is refined by "parting" with sulphuric acid. The metal is heated with strong sulphuric acid which dissolves the silver, copper, and iron present, but not the gold. In the solution of silver sulphate so obtained copper plates are suspended, upon which the pure silver precipitates, the copper going into solution as sulphate, as shown in the equation

$$Ag_2SO_4 + Cu = 2Ag + CuSO_4.$$

The solution obtained as a by-product in this process furnishes most of the blue vitriol of commerce. Silver is also refined by electrolytic methods similar to those used in refining copper.

Properties of silver. Silver is a heavy, rather soft, white metal, very ductile and malleable and capable of taking a high polish. It surpasses all other metals as a conductor of heat and electricity, but is too costly to find extensive use for such purposes. It melts at a little lower temperature than copper (961°). It alloys readily with other heavy metals, and when it is to be used for coinage a small amount of copper — from 8 to 10% — is nearly always melted with it to give it hardness.

It is not acted upon by water or air, but is quickly tarnished when in contact with sulphur compounds, turning quite black in time. Hydrochloric acid and fused alkalis do not act upon it, but nitric acid and hot, concentrated sulphuric acid dissolve it with ease. [Pg 366]

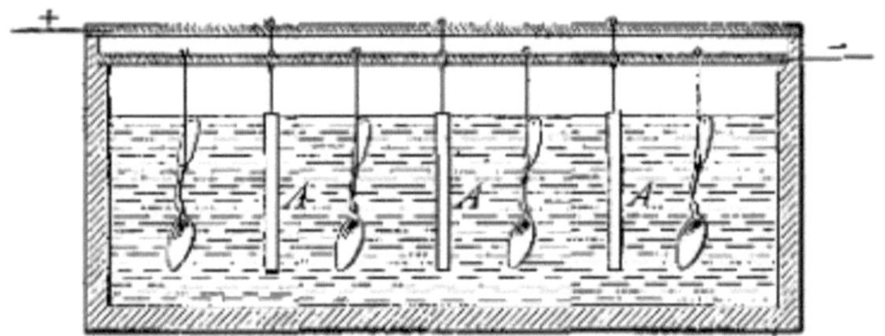

Fig. 88

Electroplating. Since silver is not acted upon by water or air, and has a pleasing appearance, it is used to coat various articles made of cheaper metals. Such articles are said to be silver plated. The process by which this is done is called electroplating. It is carried on as follows: The object to be plated (such as a spoon) is attached to a wire and dipped into a solution of a silver salt. Electrical connection is made in such a way that the article to be plated serves as the cathode, while the anode is made up of one or more plates of silver (Fig. 88, *A*). When a current is passed through the electrolyte silver dissolves from the anode plate and deposits on the cathode in the form of a closely adhering layer. By making the proper change in the electrolyte and anode plate objects may be plated with gold and other metals.

Compounds of silver. Silver forms two oxides but only one series of salts, namely, the one which corresponds to the mercurous and cuprous series.

Silver nitrate (*lunar caustic*) ($AgNO_3$). This salt is easily prepared by dissolving silver in nitric acid and evaporating the resulting solution. It crystallizes in flat plates, and when heated carefully can be melted without decomposition. When cast into sticks it is called lunar caustic, for it has a very corrosive action on flesh, and is sometimes used in surgery to burn away abnormal growths.

The alchemists designated the metals by the names of the heavenly bodies. The moon (luna) was the symbol for silver; hence the name "lunar caustic."

Silver sulphide (Ag_2S). This occurs in nature and constitutes one of the principal ores of silver. It can be [Pg 367] obtained in the form of a black solid by passing hydrosulphuric acid through a solution of silver nitrate.

Compounds of silver with the halogens. The chloride, bromide, and iodide of silver are insoluble in water and acids, and are therefore precipitated by bringing together a soluble halogen salt with silver nitrate:

$$AgNO_3 + KCl = AgCl + KNO_3.$$

They are remarkable for the fact that they are very sensitive to the action of light, undergoing a change of color and chemical composition when exposed to sunlight, especially if in contact with organic matter such as gelatin.

Photography. The art of photography is based on the fact that the halogen compounds of silver are affected by the light, particularly in the presence of organic matter. From a chemical standpoint the processes involved may be described under two heads: (1) the preparation of the negative; (2) the preparation of the print.

1. *Preparation of the negative.* The plate used in the preparation of the negative is made by spreading a thin layer of gelatin, in which silver bromide is suspended (silver iodide is sometimes added also), over a glass plate or celluloid film and allowing it to dry. When the plate so prepared is placed in a camera and the image of some object is focused upon it, the silver salt undergoes a change which is proportional at each point to the intensity of the light falling upon it. In this way an image of the object photographed is produced upon the plate, which is, however, invisible and is therefore called "latent." It can be made visible by the process of developing.

To develop the image the exposed plate is immersed in a solution of some reducing agent called the developer. The developer reduces that portion of the silver salt which has been affected by the light, depositing it in the form of black metallic silver which closely adheres to the plate.

The unaffected silver salt, upon which the developer has no action, must now be removed from the plate. This is done by immersing the plate in a solution of sodium thiosulphate (hypo). After the silver salt has been dissolved off, the plate is washed with water and [Pg 368] dried. The plate so prepared is called the negative because it is a picture of the object photographed, with the lights exactly reversed. This is called fixing the negative.

2. *Preparation of the print.* The print is made from paper which is prepared in the same way as the negative plate. The negative is placed upon this paper and exposed to the light in such a way that the light must pass through the negative before striking the paper. If the paper is coated with silver chloride, a visible image is produced, in which case a developer is not needed. The proofs are made in this way. In order to make them permanent the unchanged silver chloride must be dissolved off with sodium thiosulphate. The print is then toned by dipping it into a solution of gold or platinum salts. The silver on the print passes into solution, while the gold or platinum takes its place. These metals give a characteristic color or tone to the print, the gold making it reddish brown, while the platinum gives it a steel-gray tone. If a silver bromide paper is used in making the print, a latent image is produced which must be developed as in the case of the negative itself. The silver bromide is much more sensitive than the chloride, so that the printing can be done in artificial light. Since the darkest places on the negative cut off the most light, it is evident that the lights of the print will be the reverse of those of the negative, and will therefore correspond to those of the object photographed. The print is therefore called the positive.

EXERCISES

1. Account for the fact that copper has been used for so long a time.

2. Write equations for the action of concentrated sulphuric and nitric acids upon the metals of this family.

3. How would you account for the fact that normal copper sulphate is slightly acid to litmus?

4. Contrast the action of heat on cupric nitrate and mercuric nitrate.

5. State reasons why mercury is adapted for use in thermometers and barometers.

6. How could you distinguish between mercurous chloride and mercuric chloride?

7. Write equations for the preparation of mercuric and mercurous iodides. [Pg 369]

8. How would you account for the fact that solutions of the different salts of a metal usually have the same color?

9. Crude silver usually contains iron and lead. What would become of these metals in refining by parting with sulphuric acid?

10. In the amalgamation process for extracting silver, how does ferrous chloride convert silver chloride into silver? Write equation. Why is the silver sulphide first changed into silver chloride?

11. What impurities would you expect to find in the copper sulphate prepared from the refining of silver?

12. How could you prepare pure silver chloride from a silver coin?

13. Mercuric nitrate and silver nitrate are both white solids soluble in water. How could you distinguish between them?

14. Account for the fact that sulphur waters turn a silver coin black; also for the fact that a silver spoon is blackened by foods (eggs, for example) containing sulphur.

15. When a solution of silver nitrate is added to a solution of potassium chlorate no precipitate forms. How do you account for the fact that a precipitate of silver chloride is not formed?

[Pg 370]

CHAPTER XXIX

TIN AND LEAD

	SYMBOL	ATOMIC WEIGHT	DENSITY	MELTING POINT	COMMON OXIDES

Tin	Sn	119.0	7.35	235°	SnO SnO$_2$
Lead	Pb	206.9	11.38	327°	PbO Pb$_3$O$_4$ PbO$_2$

The family. Tin and lead, together with silicon and germanium, form a family in Group IV of the periodic table. Silicon has been discussed along with the non-metals, while germanium, on account of its rarity, needs only to be mentioned.

The other family of Group IV includes carbon, already described, and a number of rare elements.

TIN

Occurrence. Tin is found in nature chiefly as the oxide (SnO_2), called cassiterite or tinstone. The most famous mines are those of Cornwall in England, and of the Malay Peninsula and East India Islands; in small amounts tinstone is found in many other localities.

Metallurgy. The metallurgy of tin is very simple. The ore, separated as far as possible from earthy materials, is mixed with carbon and heated in a furnace, the reduction taking place readily. The equation is

$$SnO_2 + C = Sn + CO_2.$$

[Pg 371]

The metal is often purified by carefully heating it until it is partly melted; the pure tin melts first and can be drained away from the impurities.

Properties. Pure tin, called block tin, is a soft white metal with a silver-like appearance and luster; it melts readily (235°) and is somewhat lighter than copper, having a density of 7.3. It is quite malleable and can be rolled out into very thin sheets, forming tin foil; most tin foil, however, contains a good deal of lead.

Under ordinary conditions it is quite unchanged by air or moisture, but at a high temperature it burns in air, forming the oxide

406

SnO_2. Dilute acids have no effect upon it, but concentrated acids attack it readily. Concentrated hydrochloric acid changes it into the chloride

$$Sn + 2HCl = SnCl_2 + 2H.$$

With sulphuric acid tin sulphate and sulphur dioxide are formed:

$$Sn + 2H_2SO_4 = SnSO_4 + SO_2 + 2H_2O$$

Concentrated nitric acid oxidizes it, forming a white insoluble compound of the formula H_2SnO_3, called metastannic acid:

$$3Sn + 4HNO_3 + H_2O = 3H_2SnO_3 + 4NO.$$

Uses of tin. A great deal of tin is made into tin plate by dipping thin steel sheets into the melted metal. Owing to the way in which tin resists the action of air and dilute acids, tin plate is used in many ways, such as in roofing, and in the manufacture of tin cans, cooking vessels, and similar articles.

Many useful alloys contain tin, some of which have been mentioned in connection with copper. When tin is alloyed with other metals of low melting point, soft, easily [Pg 372] melted alloys are formed which are used for friction bearings in machinery; tin, antimony, lead, and bismuth are the chief constituents of these alloys. Pewter and soft solder are alloys of tin and lead.

Compounds of tin. Tin forms two series of compounds: the stannous, in which the tin is divalent, illustrated in the compounds SnO, SnS, $SnCl_2$; the stannic, in which it is tetravalent as shown in the compounds SnO_2, SnS_2. There is also an acid, H_2SnO_3, called stannic acid, which forms a series of salts called stannates. While this acid has the same composition as metastannic acid, the two are quite different in their chemical properties. This difference is probably due to the different arrangement of the atoms in the molecules of the two substances. Only a few compounds of tin need be mentioned.

Stannic oxide (SnO_2). Stannic oxide is of interest, since it is the chief compound of tin found in nature. It is sometimes found in good-sized crystals, but as prepared in the laboratory is a white powder. When fused with potassium hydroxide it forms potassium stannate, acting very much like silicon dioxide:

$$SnO_2 + 2KOH = K_2SnO_3 + H_2O.$$

Chlorides of tin. Stannous chloride is prepared by dissolving tin in concentrated hydrochloric acid and evaporating the solution to crystallization. The crystals which are obtained have the composition $SnCl_2 \cdot 2H_2O$, and are known as tin crystals. By treating a solution of stannous chloride with aqua regia, stannic chloride is formed:

$$SnCl_2 + 2Cl = SnCl_4.$$

The salt which crystallizes from such a solution has the composition [Pg 373] $SnCl_4 \cdot 5H_2O$, and is known commercially as oxymuriate of tin. If metallic tin is heated in a current of dry chlorine, the anhydrous chloride ($SnCl_4$) is obtained as a heavy colorless liquid which fumes strongly on exposure to air.

The ease with which stannous chloride takes up chlorine to form stannic chloride makes it a good reducing agent in many reactions, changing the higher chlorides of metals to lower ones. Thus mercuric chloride is changed into mercurous chloride:

$$SnCl_2 + 2HgCl_2 = SnCl_4 + 2HgCl.$$

If the stannous chloride is in excess, the reaction may go further, producing metallic mercury:

$$SnCl_2 + 2HgCl = SnCl_4 + 2Hg.$$

Ferric chloride is in like manner reduced to ferrous chloride:

$$SnCl_3 + 2FeCl_3 = SnCl_4 + 2FeCl_2.$$

The chlorides of tin, as well as the alkali stannates, are much used as mordants in dyeing processes. The hydroxides of tin and free stannic acid, which are easily liberated from these compounds, possess in very marked degree the power of fixing dyes upon fibers, as explained under aluminium.

LEAD

Occurrence. Lead is found in nature chiefly as the sulphide (PbS), called galena; to a much smaller extent it occurs as carbonate, sulphate, chromate, and in a few other forms. Practically all the lead of commerce is made from galena, two general methods of metallurgy being in use.

Metallurgy. 1. The sulphide is melted with scrap iron, when iron sulphide and metallic lead are formed; the [Pg 374] liquid lead, being the heavier, sinks to the bottom of the vessel and can be drawn off:

$$PbS + Fe = Pb + FeS.$$

2. The sulphide is roasted in the air until a part of it has been changed into oxide and sulphate. The air is then shut off and the heating continued, the reactions indicated in the following equations taking place:

$$2PbO + PbS = 3Pb + SO_2,$$

$$PbSO_4 + PbS = 2Pb + 2SO_2.$$

The lead so prepared usually contains small amounts of silver, arsenic, antimony, copper, and other metals. The silver is removed by Parkes's method, as described under silver, and the other metals in various ways. The lead of commerce is one of the purest commercial metals, containing as a rule only a few tenths per cent of impurities.

Properties. Lead is a heavy metal (den. = 11.33) which has a brilliant silvery luster on a freshly cut surface, but which soon tarnishes to a dull blue-gray color. It is soft, easily fused (melting at 327°), and quite malleable, but has little toughness or strength.

It is not acted upon to any great extent by the oxygen of the air under ordinary conditions, but is changed into oxide at a high temperature. With the exception of hydrochloric and sulphuric acids, most acids, even very weak ones, act upon it, forming soluble lead salts. Hot, concentrated hydrochloric and sulphuric acids also attack it to a slight extent.

Uses. Lead is employed in the manufacture of lead pipes and in large storage batteries. In the form of sheet lead it is used in lining the chambers of sulphuric acid [Pg 375] works and in the preparation of paint pigments. Some alloys of lead, such as solder and pewter (lead and tin), shot (lead and arsenic), and soft bearing metals, are widely used. Type metal consists of lead, antimony, and sometimes tin. Compounds of lead form several important pigments.

Compounds of lead. In nearly all its compounds lead has a valence of 2, but a few corresponding to stannic compounds have a valence of 4.

Lead oxides. Lead forms a number of oxides, the most important of which are litharge, red lead or minium, and lead peroxide.

1. *Litharge* (PbO). This oxide forms when lead is oxidized at a rather low temperature, and is obtained as a by-product in silver refining. It is a pale yellow powder, and has a number of commercial uses. It is easily soluble in nitric acid:

$$PbO + 2HNO_3 = Pb(NO_3)_2 + H_2O.$$

2. *Red lead, or minium* (Pb_3O_4). Minium is prepared by heating lead (or litharge) to a high temperature in the air. It is a heavy powder of a beautiful red color, and is much used as a pigment.

3. *Lead peroxide* (PbO_2). This is left as a residue when minium is heated with nitric acid:

$$Pb_3O_4 + 4HNO_3 = 2Pb(NO_3)_2 + PbO_2 + 2H_2O.$$

It is a brown powder which easily gives up a part of its oxygen and, like manganese dioxide and barium dioxide, is a good oxidizing agent.

Soluble salts of lead. The soluble salts of lead can be made by dissolving litharge in acids. Lead acetate ($Pb(C_2H_3O_2)_2 \cdot 3H_2O$), called sugar of lead, and lead [Pg 376] nitrate ($Pb(NO_3)_2$) are the most familiar examples. They are while crystalline solids and are poisonous in character.

Insoluble salts of lead; lead carbonate. While the normal carbonate of lead ($PbCO_3$) is found to some extent, in nature and can be prepared in the laboratory, basic carbonates of varying composition are much more easy to obtain. One of the simplest of these has the composition $2PbCO_3 \cdot Pb(OH)_2$. A mixture of such carbonates is called white lead. This is prepared on a large scale as a paint pigment and as a body for paints which are to be colored with other substances.

White lead. White lead is an amorphous white substance which, when mixed with oil, has great covering power, that is, it spreads out in an even waxy film, free from streaks and lumps, and covers the entire surface upon which it is spread. Its disadvantage as a pigment lies in the fact that it gradually blackens when exposed to sulphur compounds, which are often present in the air, forming black lead sulphide (PbS).

Technical preparation of white lead. Different methods are used in the preparation of white lead, but the old one known as the Dutch process is still the principal one employed. In this process, earthenware pots about ten inches high and of the shape shown in Fig. 89 are used. In the bottom *A* is placed a 3% solution of acetic acid (vinegar answers the purpose very well). The space above this is filled with thin, perforated, circular pieces of lead, supported by the flange *B* of the pot. These pots are placed close together on a bed of tan bark on the floor of a room known as the corroding room. They are covered over with boards, upon which tan bark is placed, and another row of pots is placed on this. In this way the room is filled. The white lead is formed by the fumes of the acetic acid, together with the carbon dioxide set free in the fermentation of the tan bark acting on the lead. About three months are required to complete the process.

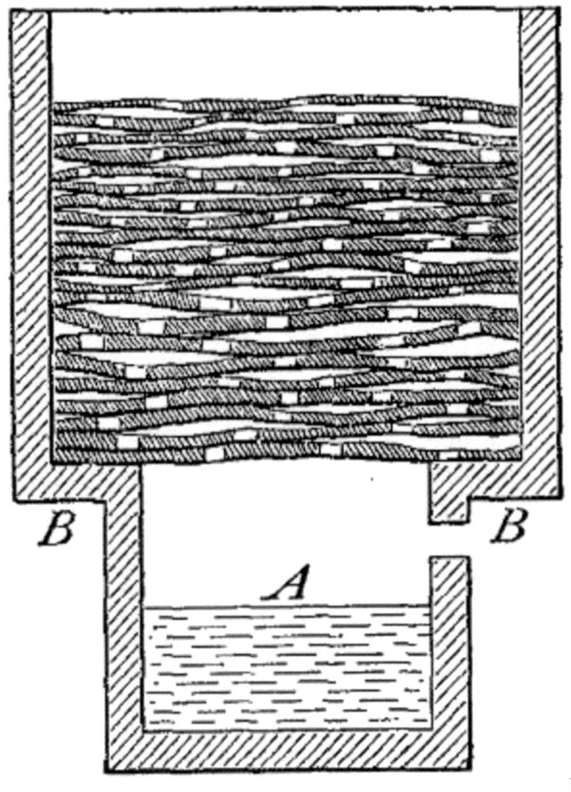

Fig. 89

[Pg 377]

Lead sulphide (PbS). In nature this compound occurs in highly crystalline condition, the crystals having much the same luster as

412

pure lead. It is readily prepared in the laboratory as a black precipitate, by the action of hydrosulphuric acid upon soluble lead salts:

$$Pb(NO_3)_2 + H_2S = PbS + 2HNO_3.$$

It is insoluble both in water and in dilute acids.

Other insoluble salts. Lead chromate ($PbCrO_4$) is a yellow substance produced by the action of a soluble lead salt upon a soluble chromate, thus:

$$K_2CrO_4 + Pb(NO_3)_2 = PbCrO_4 + 2\,KNO_3.$$

It is used as a yellow pigment. Lead sulphate ($PbSO_4$) is a white substance sometimes found in nature and easily prepared by precipitation. Lead chloride ($PbCl_2$) is likewise a white substance nearly insoluble in cold water, but readily soluble in boiling water.

Thorium and cerium. These elements are found in a few rare minerals, especially in the monazite sand of the Carolinas and Brazil. The oxides of these elements are used in the preparation of the Welsbach mantles for gas lights, because of the intense light given out when a mixture of the oxides is heated. These mantles contain the oxides of cerium and thorium in the ratio of about 1% of the former to 99% of the latter. Compounds of thorium, like those of radium, are found to possess radio-activity, but in a less degree.

EXERCISES

1. How could you detect lead if present in tin foil?

2. Stannous chloride reduces gold chloride ($AuCl_3$) to gold. Give equation.

3. What are the products of hydrolysis when stannic chloride is used as a mordant?

4. How could you detect arsenic, antimony, or copper in lead? [Pg 378]

5. Why is lead so extensively used for making water pipes?

6. What sulphates other than lead are insoluble?

7. Could lead nitrate be used in place of barium chloride in testing for sulphates?

8. How much lead peroxide could be obtained from 1 kg. of minium?

9. The purity of white lead is usually determined by observing the volume of carbon dioxide given off when it is treated with an acid. What acid should be used? On the supposition that it has the formula $2PbCO_3 \cdot Pb(OH)_2$, how nearly pure was a sample if 1 g. gave 30 cc. of carbon dioxide at 20° and 750 mm.?

10. Silicon belongs in the same family with tin and lead. In what respects are these elements similar?

11. What weight of tin could be obtained by the reduction of 1 ton of cassiterite?

12. What reaction would you expect to take place when lead peroxide is treated with hydrochloric acid?

13. White lead is often adulterated with barytes. Suggest a method for detecting it, if present, in a given example of white lead.

[Pg 379]

CHAPTER XXX

MANGANESE AND CHROMIUM

	SYMBOL	ATOMIC WEIGHT	DENSITY	MELTING POINT	FORMULAS OF ACIDS
Manganese	Mn	55.0	8.01	1900°	H_2MnO_4 and $HMnO_4$
Chromium	Cr	52.1	7.3	3000°	H_2CrO_4 and $H2Cr_2O_7$

General. Manganese and chromium, while belonging to different families, have so many features in common in their chemical conduct that they may be studied together with advantage. They differ from most of the elements so far studied in that they can act either as acid-forming or base-forming elements. As base-forming elements each of the metals forms two series of salts. In the one series, designated by the suffix "ous," the metal is divalent; in the other series, designated by the suffix "ic," the metal is trivalent. Only the manganous and the chromic salts, however, are of importance. The acids in which these elements play the part of a non-metal are unstable, but their salts are usually stable, and some of them are important compounds.

MANGANESE

Occurrence. Manganese is found in nature chiefly as the dioxide MnO_2, called pyrolusite. In smaller amounts it occurs as the oxides Mn_2O_3 and Mn_3O_4, and as the carbonate $MnCO_3$. Some iron ores also contain manganese. [Pg 380]

Preparation and properties. The element is difficult to prepare in pure condition and has no commercial applications. It can be prepared, however, by reducing the oxide with aluminium powder or by the use of the electric furnace, with carbon as the reducing agent. The metal somewhat resembles iron in appearance, but is harder, less fusible, and more readily acted upon by air and moisture. Acids readily dissolve it, forming manganous salts.

Oxides of manganese. The following oxides of manganese are known: MnO, Mn_2O_3, Mn_3O_4, MnO_2, and Mn_2O_7. Only one of these, the dioxide, needs special mention.

Manganese dioxide (*pyrolusite*) (MnO_2). This substance is the most abundant manganese compound found in nature, and is the ore from which all other compounds of manganese are made. It is a hard, brittle, black substance which is valuable as an oxidizing agent. It will be recalled that it is used in the preparation of chlorine and oxygen, in decolorizing glass which contains iron, and in the manufacture of ferromanganese.

Compounds containing manganese as a base-forming element. As has been stated previously, manganese forms two series of salts. The most important of these salts, all of which belong to the manganous series, are the following:

Manganous chloride	$MnCl_2\ 4H_2O.$
Manganous sulphide	$MnS.$
Manganous sulphate	$MnSO_4\ 4H_2O.$
Manganous carbonate	$MnCO_3.$
Manganous hydroxide	$Mn(OH)_2.$

The chloride and sulphate may be prepared by heating the dioxide with hydrochloric and sulphuric acids respectively:

$$MnO_2 + 4HCl = MnCl_2 + 2H_2O + 2Cl,$$

$$MnO_2 + H_2SO_4 = MnSO_4 + H_2O + O.$$

[Pg 381]

The sulphide, carbonate, and hydroxide, being insoluble, may be prepared from a solution of the chloride or sulphate by precipitation with the appropriate reagents. Most of the manganous salts are rose colored. They not only have formulas similar to the ferrous salts, but resemble them in many of their chemical properties.

Compounds containing manganese as an acid-forming element. Manganese forms two unstable acids, namely, manganic acid and permanganic acid. While these acids are of little interest, some of their salts, especially the permanganates, are important compounds.

Manganic acid and manganates. When manganese dioxide is fused with an alkali and an oxidizing agent a green compound is formed. The equation, when caustic potash is used, is as follows:

$$MnO_2 + 2KOH + O = K_2MnO_4 + H_2O.$$

The green compound (K_2MnO_4) is called potassium manganate, and is a salt of the unstable manganic acid (H_2MnO_4). The manganates are all very unstable.

Permanganic acid and the permanganates. When carbon dioxide is passed through a solution of a manganate a part of the manganese is changed into manganese dioxide, while the remainder forms a salt of the unstable acid $HMnO_4$, called permanganic acid. The equation is

$$3K_2MnO_4 + 2CO_2 = MnO_2 + 2KMnO_4 + 2K_2CO_3.$$

Potassium permanganate ($KMnO_4$) crystallizes in purple-black needles and is very soluble in water, forming an intensely purple solution. All other permanganates, as well as permanganic acid itself, give solutions of the same color. [Pg 382]

Oxidizing properties of the permanganates. The permanganates are remarkable for their strong oxidizing properties. When used as an oxidizing agent the permanganate is itself reduced, the exact character of the products formed from it depending upon whether the oxidation takes place (1) in an alkaline or neutral solution, or (2) in an acid solution.

1. *Oxidation in alkaline or neutral solution.* When the solution is either alkaline or neutral the potassium and the manganese of the permanganate are both converted into hydroxides, as shown in the equation

$$2KMnO_4 + 5H_2O = 2Mn(OH)_4 + 2KOH + 3O.$$

2. *Oxidation in acid solution.* When free acid such as sulphuric is present, the potassium and the manganese are both changed into salts of the acid:

$$2KMnO_4 + 3H_2SO_4 = K_2SO_4 + 2MnSO_4 + 3H_2O + 5O.$$

Under ordinary conditions, however, neither one of these reactions takes place except in the presence of a third substance which is capable of oxidation. The oxygen is not given off in the free state, as the equations show, but is used up in effecting oxidation.

Potassium permanganate is particularly valuable as an oxidizing agent not only because it acts readily either in acid or in alkaline solution, but also because the reaction takes place so easily that often it is not even necessary to heat the solution to secure action. The substance finds many uses in the laboratory, especially in analytical work. It is also used as an antiseptic as well as a disinfectant. [Pg 383]

CHROMIUM

Occurrence. The ore from which all chromium compounds are made is chromite, or chrome iron ore ($FeCr_2O_4$). This is found most abundantly in New Caledonia and Turkey. The element also occurs in small quantities in many other minerals, especially in crocoisite ($PbCrO_4$), in which mineral it was first discovered.

Preparation. Chromium, like manganese, is very hard to reduce from its ores, owing to its great affinity for oxygen. It can, however, be made by the same methods which have proved successful with manganese. Considerable quantities of an alloy of chromium with iron, called ferrochromium, are now produced for the steel industry.

Properties. Chromium is a very hard metal of about the same density as iron. It is one of the most infusible of the metals, requiring a temperature little short of 3000° for fusion. At ordinary temperatures air has little action on it; at higher temperatures, however, it burns brilliantly. Nitric acid has no action on it, but hydrochloric and dilute sulphuric acids dissolve it, liberating hydrogen.

Compounds containing chromium as a base-forming element. While chromium forms two series of salts, chromous salts are difficult to prepare and are of little importance. The most important of the chromic series are the following:

Chromic hydroxide $Cr(OH)_3$.

Chromic chloride $CrCl_3\ 6H_2O$.

Chromic sulphate $Cr_2(SO_4)_3$.

418

Chrome alums

Chromic hydroxide (Cr(OH)$_3$). This substance, being insoluble, can be obtained by precipitating a solution of the chloride or sulphate with a soluble hydroxide. It is a [Pg 384] greenish substance which, like aluminium hydroxide, dissolves in alkalis, forming soluble salts.

Dehydration of chromium hydroxide. When heated gently chromic hydroxide loses a part of its oxygen and hydrogen, forming the substance CrO·OH, which, like the corresponding aluminium compound, has more pronounced acid properties than the hydroxide. It forms a series of salts very similar to the spinels; chromite is the ferrous salt of this acid, having the formula Fe(CrO$_2$)$_2$. When heated to a higher temperature chromic hydroxide is completely dehydrated, forming the trioxide Cr$_2$O$_3$. This resembles the corresponding oxides of aluminium and iron in many respects. It is a bright green powder, and when ignited strongly becomes almost insoluble in acids, as is also the case with aluminium oxide.

Chromic sulphate (Cr$_2$(SO$_4$)$_3$). This compound is a violet-colored solid which dissolves in water, forming a solution of the same color. This solution, however, turns green on heating, owing to the formation of basic salts. Chromic sulphate, like ferric and aluminium sulphates, unites with the sulphates of the alkali metals to form alums, of which the best known are potassium chrome alum (KCr(SO$_4$)$_2$·12H$_2$O) and ammonium chrome alum (NH$_4$Cr(SO$_4$)$_2$·12H$_2$O).

These form beautiful dark purple crystals and have some practical uses in the tanning industry and in photography. A number of the salts of chromium are also used in the dyeing industry, for they hydrolyze like aluminium salts and the hydroxide forms a good mordant.

Hydrolysis of chromium salts. When ammonium sulphide is added to a solution of a chromium salt, such as the sulphate, chromium hydroxide precipitates instead of the sulphide. This is due to the fact that chromic sulphide, like aluminium sulphide, hydrolyzes in the presence of water, forming chromic hydroxide and hydrosulphuric acid. Similarly, a soluble carbonate precipitates a basic carbonate of chromium.

[Pg 385]

Compounds containing chromium as an acid-forming element. Like manganese, chromium forms two unstable acids, namely, chromic acid and dichromic acid. Their salts, the chromates and dichromates, are important compounds.

Chromates. When a chromium compound is fused with an alkali and an oxidizing agent a chromate is produced. When potassium hydroxide is used as the alkali the equation is

$$2Cr(OH)_3 + 4KOH + 3O = 2K_2CrO_4 + 5H_2O.$$

This reaction recalls the formation of a manganate under similar conditions.

Properties of chromates. The chromates are salts of the unstable chromic acid (H_2CrO_4), and as a rule are yellow in color. Lead chromate ($PbCrO_4$) is the well-known pigment chrome yellow. Most of the chromates are insoluble and can therefore be prepared by precipitation. Thus, when a solution of potassium chromate is added to solutions of lead nitrate and barium nitrate respectively, the reactions expressed by the following equations occur:

$$Pb(NO_3)_2 + K_2CrO_4 = PbCrO_4 + 2KNO_3,$$

$$Ba(NO_3)_2 + K_2CrO_4 = BaCrO_4 + 2KNO_3.$$

The chromates of lead and barium separate as yellow precipitates. The presence of either of these two metals can be detected by taking advantage of these reactions.

Dichromates. When potassium chromate is treated with an acid the potassium salt of the unstable dichromic acid ($H_2Cr_2O_7$) is formed:

$$2K_2CrO_4 + H_2SO_4 = K_2Cr_2O_7 + K_2SO_4 + H_2O.$$

[Pg 386]

The relation between the chromates and dichromates is the same as that between the phosphates and the pyrophosphates. Potassium dichromate might therefore be called potassium pyrochromate.

Potassium dichromate ($K_2Cr_2O_7$). This is the best known dichromate, and is the most familiar chromium compound. It forms large crystals of a brilliant red color, and is rather sparingly soluble in water. When treated with potassium hydroxide it is converted into the chromate

$$K_2Cr_2O_7 + 2KOH = 2K_2CrO_4 + H_2O.$$

When added to a solution of lead or barium salt the corresponding chromates (not dichromates) are precipitated. With barium nitrate the equation is

$$2Ba(NO_3)_2 + K_2Cr_2O_7 + H_2O = 2BaCrO_4 + 2KNO_3 + 2HNO_3.$$

Potassium dichromate finds use in many industries as an oxidizing agent, especially in the preparation of organic substances, such as the dye alizarin, and in the construction of several varieties of electric batteries.

Sodium chromates. The reason why the potassium salt rather than the sodium compound is used is that sodium chromate and dichromate are so soluble that it is hard to prepare them pure. This difficulty is being overcome now, and the sodium compounds are replacing the corresponding potassium salts. This is of advantage, since a sodium salt is cheaper than a potassium salt, so far as raw materials go.

Oxidizing action of chromates and dichromates. When a dilute solution of a chromate or dichromate is acidified with an acid, such as sulphuric acid, no reaction apparently takes place. However, if there is present a third substance capable of oxidation, the chromium compound gives up a [Pg 387] portion of its oxygen to this substance. Since the chromate changes into a dichromate in the presence of an acid, it will be sufficient to study the action of the dichromates alone. The reaction takes place in two steps. Thus, when a solution of ferrous sulphate is added to a solution of potassium dichromate acidified with sulphuric acid, the reaction is expressed by the following equations:

$$(1)\ K_2Cr_2O_7 + 4H_2SO_4 = K_2SO_4 + Cr_2(SO_4)_3 + 4H_2O + 3O,$$

$$(2)\ 6FeSO_4 + 3H_2SO_4 + 3O = 3Fe_2(SO_4)_3 + 3H_2O.$$

The dichromate decomposes in very much the same way as a permanganate does, the potassium and chromium being both changed into salts in which they play the part of metals, while part of the oxygen of the dichromate is liberated.

By combining equations (1) and (2), the following is obtained:

$$K_2Cr_2O_7 + 7H_2SO_4 + 6FeSO_4 = K_2SO_4 + Cr_2(SO_4)_3 + 3Fe_2(SO_4)_3 + 7H_2O.$$

This reaction is often employed in the estimation of iron in iron ores.

Potassium chrome alum. It will be noticed that the oxidizing action of potassium dichromate leaves potassium sulphate and chromium sulphate as the products of the reaction. On evaporating the solution these substances crystallize out as potassium chrome alum, which substance is produced as a by-product in the industries using potassium dichromate for oxidizing purposes.

Chromic anhydride (CrO_3). When concentrated sulphuric acid is added to a strong solution of potassium dichromate, and the liquid allowed to stand, deep red needle-shaped crystals appear which have the formula [Pg 388] CrO_3. This oxide of chromium is called chromic anhydride, since it combines readily with water to form chromic acid:

$$CrO_3 + H_2O = H_2CrO_4.$$

It is therefore analogous to sulphur trioxide which forms sulphuric acid in a similar way:

$$SO_3 + H_2O = H_2SO_4.$$

Chromic anhydride is a very strong oxidizing agent, giving up oxygen and forming chromic oxide:

$$2CrO_3 = Cr_2O_3 + 3O.$$

Rare elements of the family. Molybdenum, tungsten, and uranium are three rather rare elements belonging in the same family with chromium, and form many compounds which are similar in formulas to the corresponding compounds of chromium. They can play the part of metals and also form acids resembling chromic acid in formula. Thus we have molybdic acid (H_2MoO_4), the ammonium salt of which is $(NH_4)_2MoO_4$. This salt has the property of combining with phosphoric acid to form a very complex substance which is insoluble in nitric acid. On this account molybdic acid is often used in the estimation of the phosphoric acid present in a substance. Like chromium, the metals are difficult to prepare in pure condition. Alloys with iron can be prepared by reducing the mixed oxides with carbon in an electric furnace; these alloys are used to some extent in preparing special kinds of steel.

EXERCISES

1. How does pyrolusite effect the decolorizing of glass containing iron?

2. Write the equations for the preparation of manganous chloride, carbonate, and hydroxide.

3. Write the equations representing the reactions which take place when ferrous sulphate is oxidized to ferric sulphate by potassium permanganate in the presence of sulphuric acid. [Pg 389]

4. In the presence of sulphuric acid, oxalic acid is oxidized by potassium permanganate according to the equation

$$C_2H_2O_4 + O = 2CO_2 + H_2O.$$

Write the complete equation.

5. 10 g. of iron were dissolved in sulphuric acid and oxidized to ferric sulphate by potassium permanganate. What weight of the permanganate was required?

6. What weight of ferrochromium containing 40% chromium must be added to a ton of steel to produce an alloy containing 1% of chromium?

7. Write the equation representing the action of ammonium sulphide upon chromium sulphate.

8. Potassium chromate oxidizes hydrochloric acid, forming chlorine. Write the complete equation.

9. Give the action of sulphuric acid on potassium dichromate (*a*) in the presence of a large amount of water; (*b*) in the presence of a small amount of water.

[Pg 390]

CHAPTER XXXI

GOLD AND THE PLATINUM FAMILY

	SYMBOL	ATOMIC WEIGHT	DENSITY	HIGHEST OXIDE	HIGHEST CHLORIDE	MELTING POINT
uthenium	Ru	101.7	12.26	RuO_4	$RuCl_4$	Electric arc
hodium	Rh	103.	12.1	RhO_2	$RhCl_2$	Electric arc
alladium	Pd	106.5	11.8	PdO_2	$PdCl_4$	1500°
idium	Ir	193.	22.42	IrO_2	$IrCl_4$	1950°
smium	Os	191.	22.47	OsO_4	$OsCl_4$	Electric arc
atinum	Pt	194.8	21.50	PtO_2	$PtCl_4$	1779°
old	Au	197.2	19.30	Au_2O_3	$AuCl_3$	1064°

The family. Following iron, nickel, and cobalt in the eighth column of the periodic table are two groups of three elements each. The metals of the first of these groups—ruthenium, rhodium, and palladium—have atomic weights near 100 and densities near 12. The metals of the other group—iridium, osmium, and platinum—have atomic weights near 200 and densities near 21. These six rare elements have very similar physical properties and resemble each other chemically not only in the type of compounds which they form but also in the great variety of them. They occur closely associated in nature, usually as alloys of platinum in the form of irregular metallic grains in sand and gravel. Platinum is by far the most abundant of the six.

Although the periodic classification assigns gold to the silver-copper group, its physical as well as many of its [Pg 391] chemical properties much more closely resemble those of the platinum metals, and it can he conveniently considered along with them. The four elements gold, platinum, osmium, and iridium are the heaviest substances known, being about twice as heavy as lead.

PLATINUM

Occurrence. About 90% of the platinum of commerce comes from Russia, small amounts being produced in California, Brazil, and Australia.

Preparation. Native platinum is usually alloyed with gold and the platinum metals. To separate the platinum the alloy is dissolved in aqua regia, which converts the platinum into chloroplatinic acid (H_2PtCl_6). Ammonium chloride is then added, which precipitates the platinum as insoluble ammonium chloroplatinate:

$$H_2PtCl_6 + 2NH_4Cl = (NH_4)_2PtCl_6 + 2HCl.$$

Some iridium is also precipitated as a similar compound. On ignition the double chloride is decomposed, leaving the platinum as a spongy metallic mass, which is melted in an electric furnace and rolled or hammered into the desired shape.

Physical properties. Platinum is a grayish-white metal of high luster, and is very malleable and ductile. It melts in the oxyhydrogen blowpipe and in the electric furnace; it is harder than gold and is a good conductor of electricity. In finely divided form it has the ability to absorb or occlude gases, especially oxygen and hydrogen. These gases, when occluded, are in a very active condition resembling the nascent state, and can combine with each other at ordinary [Pg 392] temperatures. A jet of hydrogen or coal gas directed upon spongy platinum is at once ignited.

Platinum as a catalytic agent. Platinum is remarkable for its property of acting as a catalytic agent in a large number of chemical reactions, and mention has been made of this use of the metal in connection with the manufacture of sulphuric acid. When desired for this purpose some porous or fibrous substance, such as asbestos, is soaked in a solution of platinic chloride and then ignited. The platinum compound is decomposed and the platinum deposited in very finely divided form. Asbestos prepared in this way is called platinized asbestos. The catalytic action seems to be in part connected with the property of absorbing gases and rendering them nascent. Some other metals possess this same power, notably palladium, which is remarkable for its ability to absorb hydrogen.

Chemical properties. Platinum is a very inactive element chemically, and is not attacked by any of the common acids. Aqua regia slowly dissolves it, forming platinic chloride ($PtCl_4$), which in turn unites with the hydrochloric acid present in the aqua regia, forming the compound chloroplatinic acid (H_2PtCl_6). Platinum is attacked by fused alkalis. It combines at higher temperatures with carbon and phosphorus and alloys with many metals. It is readily attacked by chlorine but not by oxidizing agents.

Applications. Platinum is very valuable as a material for the manufacture of chemical utensils which are required to stand a high temperature or the action of strong reagents. Platinum crucibles, dishes, forceps, electrodes, and similar articles are indispensable in the chemical laboratory. In the industries it is used for such purposes as the manufacture of pans for evaporating sulphuric acid, wires for sealing through incandescent light bulbs, and for making a great variety of instruments. Unfortunately the supply [Pg 393] of the metal is very limited, and the cost is steadily advancing, so that it is now more valuable than gold.

Compounds. Platinum forms two series of salts of which platinous chloride ($PtCl_2$) and platinic chloride ($PtCl_4$) are examples. Platinates are also known. While a great variety of compounds of platinum have been made, the substance is chiefly employed in the metallic state.

Platinic chloride ($PtCl_4$). Platinic chloride is an orange-colored, soluble compound made by heating chloroplatinic acid in a current of chlorine. If hydrochloric acid is added to a solution of the substance, the two combine, forming chloroplatinic acid (H_2PtCl_6):

$$2HCl + PtCl_4 = H_2PtCl_6.$$

The potassium and ammonium salts of this acid are nearly insoluble in water and alcohol. The acid is therefore used as a reagent to precipitate potassium in analytical work. With potassium chloride the equation is

$$2KCl + H_2PtCl_6 = K_2PtCl_6 + 2HCl.$$

Other metals of the family. The other members of the family have few applications. Iridium is used in the form of a platinum alloy, since the alloy is much harder than pure platinum and is even less fusible. This alloy is sometimes used to point gold pens. Osmium tetroxide (OsO_4) is a very volatile liquid and is used under the name of osmic acid as a stain for sections in microscopy.

GOLD

Occurrence. Gold has been found in many localities, the most famous being South Africa, Australia, Russia, and the United States. In this country it is found in Alaska and in nearly half of the states of the union, notably [Pg 394] in California, Colorado, and Nevada. It is usually found in the native condition, frequently alloyed with silver; in combination it is sometimes found as telluride ($AuTe_2$), and in a few other compounds.

Mining. Native gold occurs in the form of small grains or larger nuggets in the sands of old rivers, or imbedded in quartz veins in rocks. In the first case it is obtained in crude form by placer mining. The sand containing the gold is shaken or stirred in troughs of running waters called sluices. This sweeps away the sand but allows the heavier gold to sink to the bottom of the sluice. Sometimes the sand containing the gold is washed away from its natural location into the sluices by powerful streams of water delivered under pressure from pipes. This is called hydraulic mining. In vein mining the gold-bearing quartz is mined from the veins, stamped into fine powder in stamping mills, and the gold extracted by one of the processes to be described.

Extraction. 1. *Amalgamation process.* In the amalgamation process the powder containing the gold is washed over a series of copper plates whose surfaces have been amalgamated with mercury. The gold sticks to the mercury or alloys with it, and after a time the gold and mercury are scraped off and the mixture is distilled. The mercury distills off and the gold is left in the retort ready for refining.

2. *Chlorination process.* When gold occurs along with metallic sulphides it is often extracted by chlorination. The ore is first roasted, and is then moistened and treated with chlorine. This dissolves the gold but not the metallic oxides:

$$Au + 3Cl = AuCl_3.$$

[Pg 395]

The gold chloride, being soluble, is extracted from the mixture with water, and the gold is precipitated from the solution, usually by adding ferrous sulphate:

$$AuCl_3 + 3FeSO_4 = Au + FeCl_3 + Fe_2(SO_4)_3.$$

3. *Cyanide process.* This process depends upon the fact that gold is soluble in a solution of potassium cyanide in the presence of the oxygen of the air. The powder from the stamping mills is treated with a very dilute potassium cyanide solution which extracts the gold:

$$2Au + 4KCN + H_2O + O = 2KOH + 2KAu(CN)_2.$$

From this solution the gold can be obtained by electrolysis or by precipitation with metallic zinc:

$$2KAu(CN)_2 + Zn = K_2Zn(CN)_4 + 2Au.$$

Refining of gold. Gold is refined by three general methods:

1. *Electrolysis.* When gold is dissolved in a solution of potassium cyanide, and the solution electrolyzed, the gold is deposited in very pure condition on the cathode.

2. *Cupellation.* When the gold is alloyed with easily oxidizable metals, such as copper or lead, it may be refined by cupellation. The alloy is fused with an oxidizing flame on a shallow hearth made of bone ash, which substance has the property of absorbing metallic oxides but not the gold. Any silver which may be present remains alloyed with the gold.

3. *Parting with sulphuric acid.* Gold may be separated from silver, as well as from many other metals, by heating the alloy with concentrated sulphuric acid. This dissolves the silver, while the gold is not attacked. [Pg 396]

Physical properties. Gold is a very heavy bright yellow metal, exceedingly malleable and ductile, and a good conductor of electricity. It is quite soft and is usually alloyed with copper or silver to give it the hardness required for most practical uses. The degree of fineness is expressed in terms of carats, pure gold being twenty-four carats; the gold used for jewelry is usually eighteen carats, eighteen parts being gold and six parts copper or silver. Gold coinage is 90% gold and 10% copper.

Chemical properties. Gold is not attacked by any one of the common acids; aqua regia easily dissolves it, forming gold chloride ($AuCl_3$), which in turn combines with hydrochloric acid to form chlorauric acid ($HAuCl_4$). Fused alkalis also attack it. Most oxidizing agents are without action upon it, and in general it is not an active element.

Compounds. The compounds of gold, though numerous and varied in character, are of comparatively little importance and need not be described in detail. The element forms two series of salts in which it acts as a metal: in the aurous series the gold is univalent, the chloride having the formula $AuCl$; in the auric series it is trivalent, auric chloride having the formula $AuCl_3$. Gold also acts as an acid-forming element, forming such compounds as potassium aurate ($KAuO_2$). Its compounds are very easily decomposed, however, metallic gold separating from them.

EXERCISES

1. From the method of preparation of platinum, what metal is likely to be alloyed with it?

2. The "platinum chloride" of the laboratory is made by dissolving platinum in aqua regia. What is the compound?

3. How would you expect potassium aurate and platinate to be formed? What precautions would this suggest in the use of platinum vessels?

4. Why must gold ores be roasted in the chlorination process?

[Pg 397]

CHAPTER XXXII

SOME SIMPLE ORGANIC COMPOUNDS

Division of chemistry into organic and inorganic. Chemistry is usually divided into two great divisions,—organic and inorganic. The original significance of these terms was entirely different from the meaning which they have at the present time.

1. *Original significance.* The division into organic and inorganic was originally made because it was believed that those substances which constitute the essential parts of living organisms were built up under the influence of the life force of the organism. Such substances, therefore, should be regarded as different from those compounds prepared in the laboratory or formed from the inorganic or mineral constituents of the earth. In accordance with this view organic chemistry included those substances formed by living organisms. Inorganic chemistry, on the other hand, included all substances formed from the mineral portions of the earth.

In 1828 the German chemist Wöhler prepared urea, a typical organic compound, from inorganic materials. The synthesis of other so-called organic compounds followed, and at present it is known that the same chemical laws apply to all substances whether formed in the living organism or prepared in the laboratory from inorganic constituents. The terms "organic" and "inorganic" have therefore lost their original significance. [Pg 398]

2. *Present significance.* The great majority of the compounds found in living organisms contain carbon, and the term "organic chemistry," as used at present, includes not only these compounds but all compounds of carbon. *Organic chemistry* has become, therefore, *the*

chemistry of the compounds of carbon, all other substances being treated under the head of inorganic chemistry. This separation of the compounds of carbon into a group by themselves is made almost necessary by their great number, over one hundred thousand having been recorded. For convenience some of the simpler carbon compounds, such as the oxides and the carbonates, are usually discussed in inorganic chemistry.

The grouping of compounds in classes. The study of organic chemistry is much simplified by the fact that the large number of bodies included in this field may be grouped in classes of similar compounds. It thus becomes possible to study the properties of each class as a whole, in much the same way as we study a group of elements. The most important of these classes are the *hydrocarbons,* the *alcohols,* the *aldehydes,* the *acids,* the *ethereal salts,* the *ethers,* the *ketones,* the *organic bases,* and the *carbohydrates.* A few members of each of these classes will now be discussed briefly.

THE HYDROCARBONS

Carbon and hydrogen combine to form a large number of compounds. These compounds are known collectively as the *hydrocarbons.* They may be divided into a number of groups or series, each being named from its first member. Some of the groups are as follows: [Pg 399]

METHANE SERIES

CH_4	methane
C_2H_6	ethane
C_3H_8	propane
C_4H_{10}	butane
C_5H_{12}	pentane
C_6H_{14}	hexane
C_7H_{16}	heptane
C_8H_{18}	octane

ETHYLENE SERIES

C_2H_4	ethylene
C_3H_6	propylene
C_4H_8	butylene

BENZENE SERIES

C_6H_6	benzene
C_7H_8	toluene
C_8H_{10}	xylene

ACETYLENE SERIES

C_2H_2	acetylene
C_3H_4	allylene

Only the lower members (that is, those which contain a small number of carbon atoms) of the above groups are given. The methane series is the most extensive, all of the compounds up to $C_{24}H_{50}$ being known.

It will be noticed that the successive members of each of the above series differ by the group of atoms (CH_2). Such a series is called an *homologous series*. In general, it may be stated that the members of an homologous series show a regular gradation in most physical properties and are similar in chemical properties. Thus in the methane group the first four members are gases at ordinary temperatures; those containing from five to sixteen carbon atoms are liquids, the boiling points of which increase with the number of carbon atoms present. Those containing more than sixteen carbon atoms are solids.

Sources of the hydrocarbons. There are two chief sources of the hydrocarbons, namely, (1) crude petroleum and (2) coal tar.

1. *Crude petroleum.* This is a liquid pumped from wells driven into the earth in certain localities. Pennsylvania, Ohio, Kansas, California, and Texas are the chief [Pg 400] oil-producing regions in the United States. The crude petroleum consists largely of liquid hydro-

carbons in which are dissolved both gaseous and solid hydrocarbons. Before being used it must be refined. In this process the petroleum is run into large iron stills and subjected to fractional distillation. The various hydrocarbons distill over in the general order of their boiling points. The distillates which collect between certain limits of temperature are kept separate and serve for different uses; they are further purified, generally by washing with sulphuric acid, then with an alkali, and finally with water. Among the products obtained from crude petroleum in this way are the naphthas, including benzine and gasoline, kerosene or coal oil, lubricating oils, vaseline, and paraffin. None of these products are definite chemical compounds, but each consists of a mixture of hydrocarbons, the boiling points of which lie within certain limits.

2. *Coal tar.* This product is obtained in the manufacture of coal gas, as already explained. It is a complex mixture and is refined by the same general method used in refining crude petroleum. The principal hydrocarbons obtained from the coal tar are benzene, toluene, naphthalene, and anthracene. In addition to the hydrocarbons, coal tar contains many other compounds, such as carbolic acid and aniline.

Properties of the hydrocarbons. The lower members of the first two series of hydrocarbons mentioned are all gases; the succeeding members are liquids. In some series, as the methane series, the higher members are solids. The preparation and properties of methane and acetylene have been discussed in a previous chapter. Ethylene is present in small quantities in coal gas and may be [Pg 401] obtained in the laboratory by treating alcohol (C_2H_6O) with sulphuric acid:

$$C_2H_6O = C_2H_4 + H_2O.$$

Benzene, the first member of the benzene series, is a liquid boiling at 80°.

The hydrocarbons serve as the materials from which a large number of compounds can be prepared; indeed, it has been proposed to call organic chemistry *the chemistry of the hydrocarbon derivatives.*

Substitution products of the hydrocarbons. As a rule, at least a part of the hydrogen in any hydrocarbon can be displaced by an equivalent amount of certain elements or groups of elements. Thus the compounds CH_3Cl, CH_2Cl_2, $CHCl_3$, CCl_4 can be obtained from methane by treatment with chlorine. Such compounds are called *substitution products.*

Chloroform ($CHCl_3$). This can be made by treating methane with chlorine, as just indicated, although a much easier method consists in treating alcohol or acetone (which see) with bleaching powder. Chloroform is a heavy liquid having a pleasant odor and a sweetish taste. It is largely used as a solvent and as an anæsthetic in surgery.

Iodoform (CHI_3). This is a yellow crystalline solid obtained by treating alcohol with iodine and an alkali. It has a characteristic odor and is used as an antiseptic.

ALCOHOLS

When such a compound as CH_3Cl is treated with silver hydroxide the reaction expressed by the following equation takes place:

$$CH_3Cl + AgOH = CH_3OH + AgCl.$$

[Pg 402]

Similarly C_2H_5Cl will give C_2H_5OH and AgCl. The compounds CH_3OH and C_2H_5OH so obtained belong to the class of substances known as *alcohols.* From their formulas it will be seen that they may be regarded as derived from hydrocarbons by substituting the hydroxyl group (OH) for hydrogen. Thus the alcohol CH_3OH may be regarded as derived from methane (CH_4) by substituting the group OH for one atom of hydrogen. A great many alcohols are known, and, like the hydrocarbons, they may be grouped into series. The relation between the first three members of the methane series and the corresponding alcohols is shown in the following table:

CH$_4$	(methane)	CH$_3$OH	(methyl alcohol).
C$_2$H$_6$	(ethane)	C$_2$H$_5$OH	(ethyl alcohol).
C$_3$H$_8$	(propane)	C$_3$H$_7$OH	(propyl alcohol).

Methyl alcohol (*wood alcohol*) (CH$_3$OH). When wood is placed in an air-tight retort and heated, a number of compounds are evolved, the most important of which are the three liquids, methyl alcohol, acetic acid, and acetone. Methyl alcohol is obtained entirely from this source, and on this account is commonly called *wood alcohol*. It is a colorless liquid which has a density of 0.79 and boils at 67°. It burns with an almost colorless flame and is sometimes used for heating purposes, in place of the more expensive ethyl alcohol. It is a good solvent for organic substances and is used especially as a solvent in the manufacture of varnishes. It is very poisonous.

Ethyl alcohol (*common alcohol*) (C$_2$H$_5$OH). 1. *Preparation.* This compound may be prepared from glucose (C$_6$H$_{12}$O$_6$), a sugar easily obtained from starch. If some baker's yeast is added to a solution of glucose and the temperature is maintained at about 30°, bubbles of gas are [Pg 403] soon evolved, showing that a change is taking place. The yeast contains a large number of minute organized bodies, which are really forms of plant life. The plant grows in the glucose solution, and in so doing secretes a substance known as *zymase*, which breaks down the glucose in accordance with the following equation:

$$C_6H_{12}O_6 = 2C_2H_5OH + 2CO_2.$$

Laboratory preparation of alcohol. The formation of alcohol and carbon dioxide from glucose may be shown as follows: About 100 g. of glucose are dissolved in a liter of water in flask *A* (Fig. 90). This flask is connected with the bottle *B*, which is partially filled with limewater. The tube *C* contains solid sodium hydroxide. A little baker's yeast is now added to the solution in flask *A*, and the apparatus is connected, as shown in the figure. If the temperature is maintained at about 30°, the reaction soon begins. The bubbles of gas escape through the limewater in *B*. A precipitate of calcium carbonate soon forms in the limewater, showing the presence of carbon dioxide. The sodium hydroxide in tube C prevents the carbon dioxide in the air from acting on the limewater. The alcohol remains in the flask A and may be separated by fractional distillation.

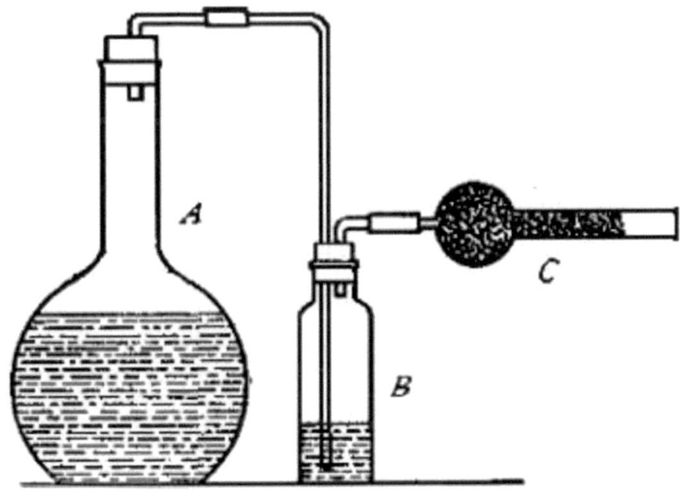

Fig. 90

2. *Properties.* Ethyl alcohol is a colorless liquid with a pleasant odor. It has a density of 0.78 and boils at 78°. It resembles methyl alcohol in its general properties. It is sometimes used as a source of heat, since its flame is very hot and does not deposit carbon, as the flame from oil does. When taken into the system in small quantities [Pg 404] it causes intoxication; in larger quantities it acts as a poison. The intoxicating properties of such liquors as beer, wine, and whisky are due to the alcohol present. Beer contains from 2 to 5% of alcohol, wine from 5 to 20%, and whisky about 50%. The ordinary alcohol of the druggist contains 94% of alcohol and 6% of water. When this is boiled with lime and then distilled nearly all the water is removed, the distillate being called *absolute alcohol.*

Commercial preparation of alcohol. Alcohol is prepared commercially from starch obtained from corn or potatoes. The starch is first converted into a sugar known as maltose, by the action of *malt,* a substance prepared by moistening barley with water, allowing it to germinate, and then drying it. There is present in the malt a substance known as diastase, which has the property of changing starch into maltose. This sugar, like glucose, breaks down into alcohol and carbon dioxide in the presence of yeast. The resulting alcohol is separated by fractional distillation.

Denatured alcohol. The 94% alcohol is prepared at present at a cost of about 35 cents per gallon, which is about half the cost of the preparation of methyl alcohol. The government, however, imposes a tax on all ethyl alcohol which amounts to $2.08 per gallon on the 94% product. This increases its cost to such an extent that it is not economical to use it for many purposes for which it is adapted, such as a solvent in the preparation of paints and varnishes and as a material for the preparation of many important organic compounds. By an act of Congress in 1906, the tax was removed from *denatured* alcohol, that is alcohol mixed with some substance which renders it unfit for the purposes of a beverage but will not impair its use for manufacturing purposes. Some of the European countries have similar laws. The substances ordinarily used to denature alcohol are wood alcohol and pyridine, the latter compound having a very offensive odor.

Fermentation. The reaction which takes place in the preparation of ethyl alcohol belongs to the class of changes known under the general name of fermentation. Thus we say that the yeast causes the glucose to ferment, and the process is known as alcoholic fermentation. There are many kinds of fermentations, and each is thought to be due to the presence of a definite substance known [Pg 405] as an *enzyme*, which acts by catalysis. In many cases, as in alcoholic fermentation, the change is brought about by the action of minute forms of life. These probably secrete the enzymes which cause the fermentation to take place. Thus the yeast plant is supposed to bring about alcoholic fermentation by secreting the enzyme known as zymase.

Glycerin ($C_3H_5(OH)_3$). This compound may be regarded as derived from propane (C_3H_8) by displacing three atoms of hydrogen by three hydroxyl groups, and must therefore be regarded as an alcohol. It is formed in the manufacture of soaps, as will be explained later. It is an oily, colorless liquid having a sweetish taste. It is used in medicine and in the manufacture of the explosives nitroglycerin and dynamite.

ALDEHYDES

When alcohols are treated with certain oxidizing agents two hydrogen atoms are removed from each molecule of the alcohol. The resulting compounds are known as aldehydes. The relation of the aldehydes derived from methyl and ethyl alcohol to the alcohols themselves may be shown as follows:

Alcohols {CH_3OH {C_2H_5OH Corresponding aldehydes   {CH_2O {C_2H_4O

The first of these (CH_2O) is a gas known as formaldehyde. Its aqueous solution is largely used as an antiseptic and disinfectant under the name of *formalin*. Acetaldehyde (C_2H_4O) is a liquid boiling at 21°.

ACIDS

Like the other classes of organic compounds, the organic acids may be arranged in homologous series. One of the most important of these series is the *fatty-acid series*, the [Pg 406] name having been given to it because the derivatives of certain of its members are constituents of the fats. Some of the most important members of the series are given in the following table. They are all monobasic, and this fact is expressed in the formulas by separating the replaceable hydrogen atom from the rest of the molecule:

$H \cdot CHO_2$ formic acid, a liquid boiling at 100°.

$H \cdot C_2H_3O$ acetic acid, a liquid boiling at 118°.

$H \cdot C_3H_5O_2$ propionic acid, a liquid boiling at 140°.

$H \cdot C_4H_7O_2$ butyric acid, a liquid boiling at 163°.

$H \cdot C_{16}H_{31}O_2$ palmitic acid, a solid melting at 62°.

$H \cdot C_{18}H_{35}O_2$ stearic acid, a solid melting at 69°.

Formic acid (H·CHO$_2$). The name "formic" is derived from the Latin *formica*, signifying ant. This name was given to the acid because it was formerly obtained from a certain kind of ants. It is a colorless liquid and occurs in many plants such as the stinging nettles. The inflammation caused by the sting of the bee is due to formic acid.

Acetic acid (H·C$_2$H$_3$O$_2$). Acetic acid is the acid present in vinegar, the sour taste being due to it. It can be prepared by either of the following methods.

1. *Acetic fermentation.* This consists in the change of alcohol into acetic acid through the agency of a minute organism commonly called mother of vinegar. The change is represented by the following equation:

$$C_2H_5OH + 2O = HC_2H_3O_2 + H_2O.$$

The various kinds of vinegars are all made by this process. In the manufacture of cider vinegar the sugar present in the cider first undergoes alcoholic fermentation; the resulting alcohol then undergoes acetic fermentation. The amount of acetic acid present in vinegars varies from 3 to 6%. [Pg 407]

2. *From the distillation of wood.* The liquid obtained by heating wood in the absence of air contains a large amount of acetic acid, and this can be separated readily in a pure state. This is the most economical method for the preparation of the concentrated acid.

Acetic acid is a colorless liquid and has a strong pungent odor. Many of its salts are well-known compounds. Lead acetate (Pb(C$_2$H$_3$O$_2$)$_2$) is the ordinary *sugar of lead*. Sodium acetate (NaC$_2$H$_3$O$_2$) is a white solid largely used in making chemical analyses. Copper acetate (Cu(C$_2$H$_3$O$_2$)$_2$) is a blue solid. When copper is acted upon by acetic acid in the presence of air a green basic acetate of copper is formed. This is commonly known as verdigris. All acetates are soluble in water.

Butyric acid (H·C$_4$H$_7$O$_2$). Derivatives of butyric acid are present in butter and impart to it its characteristic flavor.

Palmitic and stearic acids. Ordinary fats consist principally of derivatives of palmitic and stearic acids. When the fats are heated with sodium hydroxide the sodium salts of these acids are formed. If hydrochloric acid is added to a solution of the sodium salts, the free palmitic and stearic acids are precipitated. They are white solids, insoluble in water. Stearic acid is often used in making candles.

Acids belonging to other series. In addition to members of the fatty-acid series, mention may be made of the following well-known acids.

Oxalic acid ($H_2C_2O_4$). This is a white solid which occurs in nature in many plants, such as the sorrels. Its ammonium salt ((NH_4)$_2C_2O_4$) is used as a reagent for the detection of calcium. When added to a solution of a calcium [Pg 408] compound the white, insoluble calcium oxalate (CaC_2O_4) precipitates.

Tartaric acid ($H_2 \cdot C_4H_4O_6$). This compound occurs either in a free state or in the form of its salts in many fruits. The potassium acid salt ($KHC_4H_4O_6$) occurs in the juice of grapes. When the juice ferments in the manufacture of wine, this salt, being insoluble in alcohol, separates out on the sides of the cask and in this form is known as argol. This is more or less colored by the coloring matter of the grape. When purified it forms a white solid and is sold under the name of cream of tartar. The following are also well-known salts of tartaric acid: potassium sodium tartrate (Rochelle salt) ($KNaC_4H_4O_6$), potassium antimonyl tartrate (tartar emetic) ($KSbOC_4H_4O_6$).

Cream of tartar baking powders. The so-called cream of tartar baking powders consist of a mixture of cream of tartar, bicarbonate of soda, and some starch or flour. When water is added to this mixture the cream of tartar slowly acts upon the soda present liberating carbon dioxide in accordance with the following equation:

$$KHC_4H_4O_6 + NaHCO_3 = KNaC_4H_4O_6 + H_2O + CO_2.$$

The carbon dioxide evolved escapes through the dough, thus making it light and porous.

Citric acid ($H_3 \cdot C_6H_5O_7$). This acid occurs in many fruits, especially in lemons. It is a white solid, soluble in water, and is often used as a substitute for lemons in making lemonade.

Lactic acid (H·C$_3$H$_5$O$_3$). This is a liquid which is formed in the souring of milk.

Oleic acid (H·C$_{18}$H$_{33}$O$_2$). The derivatives of this acid constitute the principal part of many oils and liquid fats. The acid itself is an oily liquid. [Pg 409]

ETHEREAL SALTS

When acids are brought in contact with alcohols under certain conditions a reaction takes place similar to that which takes place between acids and bases. The following equations will serve as illustrations:

$$KOH + HNO_3 = KNO_3 + H_2O,$$

$$CH_3OH + HNO_3 = CH_3NO_3 + H_2O.$$

The resulting compounds of which methyl nitrate (CH$_3$NO$_3$) may be taken as the type belong to the class known as *ethereal salts*, the name having been given them because some of them possess pleasant ethereal odors. It will be seen that the ethereal salts differ from ordinary salts in that they contain a hydrocarbon radical, such as CH$_3$, C$_2$H$_5$, C$_3$H$_5$, in place of a metal.

The nitrates of glycerin (*nitroglycerin*). Nitric acid reacts with glycerin in the same way that it reacts with a base containing three hydroxyl groups such as Fe(OH)$_3$:

$$Fe(OH)_3 + 3HNO_3 = Fe(NO_3)_3 + 3H_2O,$$

$$C_3H_5(OH)_3 + 3HNO_3 = C_3H_5(NO_3)_3 + 3H_2O.$$

The resulting nitrate (C$_3$H$_5$(NO$_3$)$_3$) is the main constituent of *nitroglycerin*, a slightly yellowish oil characterized by its explosive properties. Dynamite consists of porous earth which has absorbed nitroglycerin, and its strength depends on the amount present. It is used much more largely than nitroglycerin itself, since it does not explode so readily by concussion and hence can be transported with safety.

The fats. These are largely mixtures of the ethereal salts known respectively as olein, palmitin, and stearin. [Pg 410] These salts may be regarded as derived from oleic, palmitic, and stearic acids respectively, by replacing the hydrogen of the acid with the glycerin radical C_3H_5. Since this radical is trivalent and oleic, palmitic, and stearic acids contain only one replaceable hydrogen atom to the molecule, it is evident that three molecules of each acid must enter into each molecule of the ethereal salt. The formulas for the acids and the ethereal salts derived from each are as follows:

$HC_{18}H_{33}O_2$	(oleic acid)
$C_8H_6(C_{18}H_{33}O_2)_3,$	(olein)
$HC_{16}H_{31}O_2$	(palmitic acid)
$C_3H_5(C_{16}H_{31}O_2)_3$	(palmitin)
$HC_{18}H_{35}O_2$	(stearic acid)
$C_3H_5(C_{18}H_{35}O_2)_3$	(stearin)

Olein is a liquid and is the main constituent of liquid fats. Palmitin and stearin are solids.

Butter fat and oleomargarine. Butter fat consists principally of olein, palmitin, and stearin. The flavor of the fat is due to the presence of a small amount of butyrin, which is an ethereal salt of butyric acid. Oleomargarine differs from butter mainly in the fact that a smaller amount of butyrin is present. It is made from the fats obtained from cattle and hogs. This fat is churned up with milk, or a small amount of butter is added, in order to furnish sufficient butyrin to impart the butter flavor.

Saponification. When an ethereal salt is heated with an alkali a reaction expressed by the following equation takes place:

$$C_2H_5NO_3 + KOH = C_2H_5OH + KNO_3.$$

This process is known as *saponification*, since it is the one which takes place in the manufacture of soaps. The ordinary soaps are made by heating fats with a solution of [Pg 411] sodium hydroxide. The reactions involved may be illustrated by the following equation representing the reaction between palmitin and sodium hydroxide:

$$C_3H_5(C_{16}H_{31}O_2)_3 + 3\ NaOH = 3\ NaC_{16}H_{31}O_2 + C_3H_5(OH)_3.$$

In accordance with this equation the ethereal salts in the fats are converted into glycerin and the sodium salts of the corresponding acids. The sodium salts are separated and constitute the soaps. These salts are soluble in water. When added to water containing calcium salts the insoluble calcium palmitate and stearate are precipitated. Magnesium salts act in a similar way. It is because of these facts that soap is used up by hard waters.

ETHERS

When ethyl alcohol is heated to 140° with sulphuric acid the reaction expressed by the following equation takes place:

$$2C_2H_5OH = (C_2H_5)_2O + H_2O.$$

The resulting compound, $(C_2H_5)_2O$, is ordinary ether and is the most important member of the class of compounds called *ethers*. Ordinarily ether is a light, very inflammable liquid boiling at 35°. It is used as a solvent for organic substances and as an anæsthetic in surgical operations.

KETONES

The most common member of this group is acetone (C_3H_6O), a colorless liquid obtained when wood is heated in the absence of air. It is used in the preparation of other organic compounds, especially chloroform. [Pg 412]

ORGANIC BASES

This group includes a number of compounds, all of which contain nitrogen as well as carbon. They are characterized by combining directly with acids to form salts, and in this respect they resemble ammonia. They may, indeed, be regarded as derived from ammonia by displacing a part or all of the hydrogen present in ammonia by hydrocarbon radicals. Among the simplest of these compounds may be mentioned methylamine (CH_3NH_2) and ethylamine ($C_2H_5NH_2$). These two compounds are gases and are formed in the distillation of wood and bones. Pyridine (C_5H_6N) and quinoline (C_9H_7N) are liquids present in small amounts in coal tar, and also in the liquid obtained by the distillation of bones. Most of the compounds now classified under the general name of *alkaloids* (which see) also belong to this group.

CARBOHYDRATES

The term "carbohydrate" is applied to a class of compounds which includes the sugars, starch, and allied bodies These compounds contain carbon, hydrogen, and oxygen the last two elements generally being present in the proportion in which they combine to form water. The most important members of this class are the following:

Cane sugar	$C_{12}H_{22}O_{11}.$
Milk sugar	$C_{12}H_{22}O_{11}.$
Dextrose	$C_6H_{12}O_6.$
Levulose	$C_6H_{12}O_6.$
Cellulose	$C_6H_{10}O_5.$
Starch	$C_6H_{10}O_5.$

Cane sugar ($C_{12}H_{22}O_{11}$). This is the well-known substance commonly called sugar. It occurs in many plants [Pg 413] especially in the sugar cane and sugar beet. It was formerly obtained almost entirely from the sugar cane, but at present the greatest amount of it comes from the sugar beet. The juice from the cane or beet contains the sugar in solution along with many impurities. These impurities are removed, and the resulting solution is then evaporated until the sugar crystallizes out. The evaporation is conducted in closed vessels from which the air is partially exhausted. In this way the boiling point of the solution is lowered and the charring of the sugar is prevented. It is impossible to remove all the sugar from the solution. In preparing sugar from sugar cane the liquors left after separating as much of it as possible from the juice of the cane constitute ordinary molasses. Maple sugar is made by the evaporation of the sap obtained from a species of the maple tree. Its sweetness is due to the presence of cane sugar, other products present in the maple sap imparting the distinctive flavor.

When a solution of cane sugar is heated with hydrochloric or other dilute mineral acid, two compounds, dextrose and levulose, are formed in accordance with the following equation:

$$C_{12}H_{22}O_{11} + H_2O = C_6H_{12}O_6 + C_6H_{12}O_6.$$

This same change is brought about by the action of an enzyme present in the yeast plant. When yeast is added to a solution of cane sugar fermentation is set up. The cane sugar, however, does not ferment directly: the enzyme in the yeast first transforms the sugar into dextrose and levulose, and these sugars then undergo alcoholic fermentation.

When heated to 160° cane sugar melts; if the temperature is increased to about 215°, a partial decomposition [Pg 414] takes place and a brown substance known as caramel forms. This is used largely as a coloring matter.

Milk sugar ($C_{12}H_{22}O_{11}$). This sugar is present in the milk of all mammals. The average composition of cow's milk is as follows:

Water	87.17%
Casein (nitrogenous matter)	3.56
Butter fat	3.64
Milk sugar	4.88

When *rennin*, an enzyme obtained from the stomach of calves, is added to milk, the casein separates and is used in the manufacture of cheese. The remaining liquid contains the milk sugar which separates on evaporation; it resembles cane sugar in appearance but is not so sweet or soluble. The souring of milk is due to the fact that the milk sugar present undergoes *lactic fermentation* in accordance with the equation

$$C_{12}H_{22}O_{11} + H_2O = 4C_3H_6O_3.$$

The lactic acid formed causes the separation of the casein, thus giving the well-known appearance of sour milk.

Isomeric compounds. It will be observed that cane sugar and milk sugar have the same formulas. Their difference in properties is due to the different arrangement of the atoms in the molecule. Such compounds are said to be isomeric. Dextrose and levulose are also isomeric.

Dextrose (*grape sugar, glucose*) ($C_6H_{12}O_6$). This sugar is present in many fruits and is commonly called grape sugar because of its presence in grape juice. It can be obtained by heating cane sugar with dilute acids, as [Pg 415] explained above; also by heating starch with dilute acids, the change being as follows:

$$C_6H_{10}O_5 + H_2O = C_6H_{12}O_6.$$

Pure dextrose is a white crystalline solid, readily soluble in water, and is not so sweet as cane sugar. In the presence of yeast it undergoes alcoholic fermentation. It is prepared from starch in large quantities, and being less expensive than cane sugar, is used as a substitute for it in the manufacture of jellies, jams, molasses, candy, and other sweets. The product commonly sold under the name of *glucose* contains about 45% of dextrose.

Levulose (*fruit sugar*)($C_6H_{12}O_6$). This sugar is a white solid which occurs along with dextrose in fruits and honey. It undergoes alcoholic fermentation in the presence of yeast.

Cellulose ($C_6H_{10}O_5$). This forms the basis of all woody fibers. Cotton and linen are nearly pure cellulose. It is insoluble in water, alcohol, and dilute acids. Sulphuric acid slowly converts it into dextrose. Nitric acid forms nitrates similar to nitroglycerin in composition and explosive properties. These nitrates are variously known as nitrocellulose, pyroxylin, and gun cotton. When exploded they yield only colorless gases; hence they are used especially in the manufacture of smokeless gunpowder. *Collodion* is a solution of nitrocellulose in a mixture of alcohol and ether. *Celluloid* is a mixture of nitrocellulose and camphor. *Paper* consists mainly of cellulose, the finer grades being made from linen and cotton rags, and the cheaper grades from straw and wood.

Starch ($C_6H_{10}O_5$). This is by far the most abundant carbohydrate found in nature, being present especially in [Pg 416] seeds and tubers. In the United States it is obtained chiefly from corn, nearly 80% of which is starch. In Europe it is obtained principally from the potato. It consists of minute granules and is practically insoluble in cold water. These granules differ somewhat in appearance, according to the source of the starch, so that it is often possible to determine from what plant the starch was obtained. When heated with water the granules burst and the starch partially dissolves. Dilute acids, as well as certain enzymes, convert it into dextrose or similar sugars. When seeds germinate the starch present is converted into soluble sugars, which are used as food for the growing plant.

Chemical changes in bread making. The average composition of wheat flour is as follows:

Water.	13.8%
Protein (nitrogenous matter)	7.9
Fats	1.4
Starch	76.4
Mineral matter	0.5

In making bread the flour is mixed with water and yeast, and the resulting dough set aside in a warm place for a few hours. The yeast first converts a portion of the starch into dextrose or a similar sugar, which then undergoes alcoholic fermentation. The carbon dioxide formed escapes through the dough, making it light and porous. The yeast plant thrives best at about 30°; hence the necessity for having the dough in a warm place. If the temperature rises above 50°, the vitality of the yeast is destroyed and fermentation ceases. In baking the bread, the heat expels the alcohol and also expands the bubbles of carbon dioxide caught in the dough, thus increasing its lightness. [Pg 417]

SOME DERIVATIVES OF BENZENE

Attention has been called to the complex nature of coal tar. Among the compounds present are the hydrocarbons, benzene, toluene, naphthalene, and anthracene. These compounds are not only useful in themselves but serve for the preparation of many other important compounds known under the general name of coal-tar products.

Nitrobenzene (*oil of myrbane*) ($C_6H_5NO_2$). When benzene is treated with nitric acid a reaction takes place which is expressed by the following equation:

$$C_6H_6 + HNO_3 = C_6H_5NO_2 + H_2O.$$

The product $C_6H_5NO_2$ is called nitrobenzene. It is a slightly yellowish poisonous liquid, with a characteristic odor. Its main use is in the manufacture of aniline.

Aniline ($C_6H_5NH_2$). When nitrobenzene is heated with iron and hydrochloric acid the hydrogen evolved by the action of the iron upon the acid reduces the nitrobenzene in accordance with the following equation:

$$C_6H_5NO_2 + 6H = C_6H_5NH_2 + 2H_2O.$$

The resulting compound is known as aniline, a liquid boiling at 182°. When first prepared it is colorless, but darkens on standing. Large quantities of it are used in the manufacture of the *aniline or coal-tar dyes*, which include many important compounds.

Carbolic acid (C_6H_5OH). This compound, sometimes known as *phenol*, occurs in coal tar, and is also prepared from benzene. It forms colorless crystals which are very soluble in water. It is strongly corrosive and very poisonous. [Pg 418]

Naphthalene and anthracene. These are hydrocarbons occurring along with benzene in coal tar. They are white solids, insoluble in water. The well-known *moth balls* are made of naphthalene. Large quantities of naphthalene are used in the preparation of *indigo*, a dye formerly obtained from the indigo plant, but now largely prepared by laboratory methods. Similarly anthracene is used in the preparation of the dye *alizarin*, which was formerly obtained from the madder root.

THE ALKALOIDS

This term is applied to a group of compounds found in many plants and trees. They all contain nitrogen, and most of them are characterized by their power to combine with acids to form salts. This property is indicated by the name alkaloids, which signifies alkali-like. The salts are soluble in water, and on this account are more largely used than the free alkaloids, which are insoluble in water. Many of the alkaloids are used in medicine, some of the more important ones being given below.

Quinine. This alkaloid occurs along with a number of others in the bark of certain trees which grow in districts in South America and also in Java and other tropical islands. It is a white solid, and its sulphate is used in medicine in the treatment of fevers.

Morphine. When incisions are made in the unripe capsules of one of the varieties of the poppy plant, a milky juice exudes which soon thickens. This is removed and partially dried. The resulting substance is the ordinary *opium* which contains a number of alkaloids, the principal one being morphine. This alkaloid is a white solid and is of great service in medicine. [Pg 419]

Among the other alkaloids may be mentioned the following: *Nicotine*, a very poisonous liquid, the salts of which occur in the leaves of the tobacco plant; *cocaine*, a crystalline solid present in coca leaves and used in medicine as a local anæsthetic; *atropine*, a solid present in the berry of the deadly nightshade, and used in the treatment of diseases of the eye; *strychnine*, a white, intensely poisonous solid present in the seeds of the members of the *Strychnos* family. [Pg 421]

ANNOUNCEMENTS

AN ELEMENTARY STUDY OF CHEMISTRY

By WILLIAM McPHERSON, Professor of Chemistry in Ohio State University, and WILLIAM E. HENDERSON, Associate Professor of Chemistry in Ohio State University.

12mo. Cloth. 434 pages. Illustrated. List price, $1.25; mailing price, $1.40

This book is the outgrowth of many years of experience in the teaching of elementary chemistry. In its preparation the authors have steadfastly kept in mind the limitations of the student to whom chemistry is a new science. They have endeavored to present the subject in a clear, well-graded way, passing in a natural and logical manner from principles which are readily understood to those which are more difficult to grasp. The language is simple and as free as possible from unusual and technical phrases. Those which are unavoidable are carefully defined. The outline is made very plain, and the paragraphing is designed to be of real assistance to the student in his reading.

The book is in no way radical, either in the subject-matter selected or in the method of treatment. At the same time it is in thorough harmony with the most recent developments in chemistry, both in respect to theory and discovery. Great care has been taken in the theoretical portions to make the treatment simple and well within the reach of the ability of an elementary student. The most recent discoveries have been touched upon where they come within the scope of an elementary text. Especial attention has been given to the practical applications of chemistry, and to the description of the manufacturing processes in use at the present time.

EXERCISES IN CHEMISTRY. By WILLIAM McPHERSON and WILLIAM E. HENDERSON.

(In press.)

GINN & COMPANY PUBLISHERS

A FIRST COURSE IN PHYSICS

By ROBERT A. MILLIKAN, Associate Professor of Physics, and HENRY G. GALE, Assistant Professor of Physics in The University of Chicago

12mo, cloth, 488 pages, illustrated, $1.25

A LABORATORY COURSE IN PHYSICS

FOR SECONDARY SCHOOLS

By ROBERT A. MILLIKAN and HENRY G. GALE 12mo, flexible cloth, 134 pages, illustrated, 40 cents

This one-year course in physics has grown out of the experience of the authors in developing the work in physics at the School of Education of The University of Chicago, and in dealing with the physics instruction in affiliated high schools and academies.

The book is a simple, objective presentation of the subject as opposed to a formal and mathematical one. It is intended for the third-year high-school pupils and is therefore adapted in style and method of treatment to the needs of students between the ages of fifteen and eighteen. It especially emphasizes the historical and practical aspects of the subject and connects the study very intimately with facts of daily observation and experience.

The authors have made a careful distinction between the class of experiments which are essentially laboratory problems and those which belong more properly to the classroom and the lecture table.

The former are grouped into a Laboratory Manual which is designed for use in connection with the text. The two books are not, however, organically connected, each being complete in itself.

All the experiments included in the work have been carefully chosen with reference to their usefulness as effective classroom demonstrations.

GINN AND COMPANY PUBLISHERS

APPENDIX A

LIST OF THE ELEMENTS, THEIR SYMBOLS, AND ATOMIC WEIGHTS

The more important elements are marked with an asterisk

$O = 16$

*Antimony	Sb	120.2
*Argon	A	39.9
*Arsenic	As	75.0
*Barium	Ba	137.4
Beryllium	Be	9.1
*Bismuth	Bi	208.5
*Boron	B	11.0
*Bromine	Br	79.96
*Cadmium	Cd	112.4
Cæsium	Cs	132.9
*Calcium	Ca	40.1
*Carbon	C	12.00
Cerium	Ce	140.25
*Chlorine	Cl	35.45
*Chromium	Cr	52.1
*Cobalt	Co	59.0
Columbium	Cb	94.0
*Copper	Cu	63.6
Erbium	Er	166.0

*Fluorine	F	19.0
Gadolinium	Gd	156.0
Gallium	Ga	70.0
Germanium	Ge	72.5
*Gold	Au	197.2
Helium	He	4.0
*Hydrogen	H	1.008
Indium	In	115.0
*Iodine	I	126.97
Iridium	Ir	193.0
*Iron	Fe	55.9
Krypton	Kr	81.8
Lanthanum	La	138.9
*Lead	Pb	206.9
Lithium	Li	7.03
*Magnesium	Mg	24.36
*Manganese	Mn	55.0
*Mercury	Hg	200.0
Molybdenum	Mo	96.0
Neodymium	Nd	143.6
Neon	Ne	20.0
*Nickel	Ni	58.7
*Nitrogen	N	14.04
Osmium	Os	191.0
*Oxygen	O	16.00

Palladium	Pd	106.5
*Phosphorus	P	31.0
*Platinum	Pt	194.8
*Potassium	K	39.15
Praseodymium	Pr	140.5
Radium	Ra	225.0
Rhodium	Rh	103.0
Rubidium	Rb	85.5
Ruthenium	Ru	101.7
Samarium	Sm	150.3
Scandium	Sc	44.1
Selenium	Se	79.2
*Silicon	Si	28.4
*Silver	Ag	107.93
*Sodium	Na	23.05
*Strontium	Sr	87.6
*Sulphur	S	32.06
Tantalum	Ta	183.0
Tellurium	Te	127.6
Terbium	Tb	160.0
Thallium	Tl	204.1
Thorium	Th	232.5
Thulium	Tm	171.0
*Tin	Sn	119.0
Titanium	Ti	48.1

Tungsten	W	184.0
Uranium	U	238.5
Vanadium	V	51.2
Xenon	Xe	128.0
Ytterbium	Yb	173.0
Yttrium	Yt	89.0
*Zinc	Zn	65.4
Zirconium	Zr	90.6

APPENDIX B

Tension of Aqueous Vapor expressed in Millimeters of Mercury

TEMPERATURE	PRESSURE
16	13.5
17	14.4
18	15.3
19	16.3
20	17.4
21	18.5
22	19.6
23	20.9
24	22.2
25	23.5

Weight of 1 Liter of Various Gases measured under Standard Conditions

Acetylene	1.1614

Air	1.2923
Ammonia	0.7617
Carbon dioxide	1.9641
Carbon monoxide	1.2499
Chlorine	3.1650
Hydrocyanic acid	1.2036
Hydrochloric acid	1.6275
Hydrogen	0.08984
Hydrosulphuric acid	1.5211
Methane	0.7157
Nitric oxide	1.3410
Nitrogen	1.2501
Nitrous oxide	1.9677
Oxygen	1.4285
Sulphur dioxide	2.8596

Densities and Melting Points of Some Common Elements

	DENSITY	MELTING POINT
Aluminium	2.68	640
Antimony	6.70	432
Arsenic	5.73	—
Barium	3.75	—
Bismuth	9.80	270
Boron	2.45	—
Cadmium	8.67	320
Cæsium	1.88	26.5

Calcium	1.54	—
Carbon, Diamond	3.50	—
" Graphite	2.15	—
" Charcoal	1.80	—
Chromium	7.30	3000
Cobalt	8.55	1800
Copper	8.89	1084
Gold	19.30	1064
Iridium	22.42	1950
Iron	7.93	1800
Lead	11.38	327
Lithium	0.59	186
Magnesium	1.75	750
Manganese	8.01	1900
Mercury	13.596	-39.5
Nickel	8.9	1600
Osmium	22.47	—
Palladium	11.80	1500
Phosphorus	1.80	45
Platinum	21.50	1779
Potassium	0.87	62.5
Rhodium	12.10	—
Rubidium	1.52	38.5
Ruthenium	12.26	—
Silicon	2.35	—

Silver	10.5	960
Sodium	0.97	97.6
Strontium	2.50	—
Sulphur	2.00	114.8
Tin	7.35	235
Titanium	3.50	—
Zinc	7.00	420